New Challenges and Industrial Applications for Corrosion Prevention and Control

Younes El Kacimi
Ibn Tofail University, Morocco

Savas Kaya
Cumhuriyet University, Turkey

Rachid Touir
Ibn Tofail University, Morocco

IGI Global
PUBLISHER of TIMELY KNOWLEDGE

A volume in the Advances in Chemical and Materials Engineering (ACME) Book Series

Published in the United States of America by
IGI Global
Engineering Science Reference (an imprint of IGI Global)
701 E. Chocolate Avenue
Hershey PA, USA 17033
Tel: 717-533-8845
Fax: 717-533-8661
E-mail: cust@igi-global.com
Web site: http://www.igi-global.com

Library of Congress Cataloging-in-Publication Data

Names: El Kacimi, Younes, 1980- editor. | Kaya, Savas, 1989- editor. | Touir, Rachid, 1977- editor.
Title: New challenges and industrial applications for corrosion prevention and control / Younes El Kacimi, Savas Kaya, and Rachid Touir, editors.
Description: Hershey, PA : Engineering Science Reference, an imprint of IGI Global, [2020] | Includes bibliographical references and index. | Summary: "This book examines new challenges and industrial applications for corrosion prevention and control"-- Provided by publisher.
Identifiers: LCCN 2019043920 (print) | LCCN 2019043921 (ebook) | ISBN 9781799827757 (hardcover) | ISBN 9781799827764 (paperback) | ISBN 9781799827771 (ebook)
Subjects: LCSH: Steel--Corrosion. | Corrosion and anti-corrosives.
Classification: LCC TA467 .N473 2020 (print) | LCC TA467 (ebook) | DDC 660/.28304--dc23
LC record available at https://lccn.loc.gov/2019043920
LC ebook record available at https://lccn.loc.gov/2019043921

This book is published in the IGI Global book series Advances in Chemical and Materials Engineering (ACME) (ISSN: 2327-5448; eISSN: 2327-5456)

British Cataloguing in Publication Data
A Cataloguing in Publication record for this book is available from the British Library.

All work contributed to this book is new, previously-unpublished material. The views expressed in this book are those of the authors, but not necessarily of the publisher.

For electronic access to this publication, please contact: eresources@igi-global.com.

Advances in Chemical and Materials Engineering (ACME) Book Series

J. Paulo Davim
University of Aveiro, Portugal

ISSN:2327-5448
EISSN:2327-5456

Mission

The cross disciplinary approach of chemical and materials engineering is rapidly growing as it applies to the study of educational, scientific and industrial research activities by solving complex chemical problems using computational techniques and statistical methods.

The **Advances in Chemical and Materials Engineering (ACME) Book Series** provides research on the recent advances throughout computational and statistical methods of analysis and modeling. This series brings together collaboration between chemists, engineers, statisticians, and computer scientists and offers a wealth of knowledge and useful tools to academics, practitioners, and professionals through high quality publications.

Coverage

- Multifuncional and Smart Materials
- Industrial Chemistry
- Computational methods
- Electrochemical and Corrosion
- Metallic Alloys
- Ductility and Crack-Resistance
- Ceramics and glasses
- Polymers Engineering
- Wear of Materials
- Artificial Intelligence Methods

IGI Global is currently accepting manuscripts for publication within this series. To submit a proposal for a volume in this series, please contact our Acquisition Editors at Acquisitions@igi-global.com or visit: http://www.igi-global.com/publish/.

The Advances in Chemical and Materials Engineering (ACME) Book Series (ISSN 2327-5448) is published by IGI Global, 701 E. Chocolate Avenue, Hershey, PA 17033-1240, USA, www.igi-global.com. This series is composed of titles available for purchase individually; each title is edited to be contextually exclusive from any other title within the series. For pricing and ordering information please visit http://www.igi-global.com/book-series/advances-chemical-materials-engineering/73687. Postmaster: Send all address changes to above address.

Titles in this Series

For a list of additional titles in this series, please visit:
http://www.igi-global.com/book-series/advances-chemical-materials-engineering/73687

Inspiration and Design for Bio-Inspired Surfaces in Tribology Emerging Research and Opportunities
Hisham Abdel-Aal (Drexel University, USA)
Engineering Science Reference • © 2021 • 367pp • H/C (ISBN: 9781799816478) • US $195.00

Technology Development for Adsorptive Heat Energy Converters Emerging Research and Opportunities
Kostyantyn M. Sukhyy (Ukrainian State University of Chemical Engineering, Ukraine) Elena A. Belyanovskaya (Ukrainian State University of Chemical Engineering, Ukraine) and Mikhailo P. Sukhyy (Ukrainian State University of Chemical Engineering, Ukraine)
Engineering Science Reference • © 2021 • 200pp • H/C (ISBN: 9781799844327) • US $195.00

Advanced Applications of 2D Nanostructures
Subhash Singh (National Institute of Technology Jamshedpur, India) and Kartikey Verma (Indian Institute of Technology, Kanpur, India)
Engineering Science Reference • © 2020 • 300pp • H/C (ISBN: 9781799829911) • US $225.00

Innovative Applications of Layered Materials and Structures
Sarmila Sahoo (Heritage Institute of Technology, Kolkata, India)
Engineering Science Reference • © 2020 • 300pp • H/C (ISBN: 9781799835875) • US $225.00

Using Lasers as Safe Alternatives for Adhesive Bonding Emerging Research and Opportunities
Barbara Ewa Ciecińska (Rzeszow University of Technology, Poland)
Engineering Science Reference • © 2020 • 248pp • H/C (ISBN: 9781799846345) • US $175.00

Natural Fiber-Reinforced Geopolymer Composites Synthesis, Properties, and Applications
It-Meng (Jim) Low (Curtin University, Australia) Thamer Alomayri (Umm Al-Qura Univeristy, Saudi Arabia) and Hasan Assaedi (Umm Al-Qura Univeristy, Saudi Arabia)
Engineering Science Reference • © 2020 • 230pp • H/C (ISBN: 9781799834595) • US $185.00

Nanotechnologies and Clusters in the Spaces of Higher Dimension Emerging Research and Opportunities
Gennadiy Vladimirovich Zhizhin (Independent Researcher, Russia)
Engineering Science Reference • © 2020 • 240pp • H/C (ISBN: 9781799837848) • US $195.00

Processing, Structure, and Applications of High-Temperature Materials
Brahma Raju Golla (National Institute of Technology Warangal, India)
Engineering Science Reference • © 2020 • 300pp • H/C (ISBN: 9781799831792) • US $195.00

IGI Global
PUBLISHER of TIMELY KNOWLEDGE

701 East Chocolate Avenue, Hershey, PA 17033, USA
Tel: 717-533-8845 x100 • Fax: 717-533-8661
E-Mail: cust@igi-global.com • www.igi-global.com

Table of Contents

Detailed Table of Contents

This chapter studies the operational parameters effects, such as immersion time, corrosion products, and pH, on the inhibition efficiency against corrosion for low carbon steel in simulated cooling water system of the mixture 1 (10-3 M 3-MPOX + 20 ppm CTAB), which corresponds to the best inhibition efficiency. The results show that the corrosion inhibition performance of the studied mixture reinforces with immersion time and has a maximum in the pH range 6.5–7.5. In addition, the analysis of the results shows that there is a dependence of the pH value of the solution and the corrosion current density (icorr). Indeed, a correlation can be established between log (icorr) and the pH using the polynomial equation. Finally, the mixture 1 provides good protection for the attack materials. These studies were evaluated from the polarization curves (PP), electrochemical impedance spectroscopy (EIS).

In this study, scale inhibition performance Trisodium citrate (TSC) was studied using static scale inhibition method, and its corrosion inhibition performance was also investigated by weight loss method and electrochemical measurement. Result showed that TSC could form a layer of corrosion and scale produTSC on mild steel surface in synthetic cooling water. Its inhibition increases with concentration and reaches a maximum of 89% and 95% at 10-3 M for corrosion and scale, respectively. It is noted that TSC takes its inhibition at high temperature and act by physic-sorption process. It is found that the TSC obey to the Langmuir isotherm. Scanning electron microscopy (SEM) showed that TSC inhibits the scale and corrosion by the formation of a protective layer on the mild steel surface.

Section 2
Biomaterials Corrosion and Protection

The goal of this chapter is to define concepts and methods currently used to study the corrosion behavior of biomaterials in physiological conditions. One of the interesting points of corrosion of biomaterials is that they must be designed to fulfill different physical and chemical requirements within the human body. For instance, ceramic biomaterials are designed, sometimes, to accomplish a bone growth task. Alternatively, bioceramics are also employed to provide high wear resistance to implant surfaces with the lowest corrosion activity in biological environment. Depending on the part of the human body where the biomaterial is implanted, the corrosion process and chemical interaction with body fluids can be accelerated, and consequently, the product of these reactions can have a negative effect in the health of the patients. For this reason, the chapter is also focused on explaining how to study the interaction between biomaterial surface and body fluids as well as the existing methods to prevent corrosion phenomena that could lead to affect human health.

Section 3
Application of Eco-Friendly Corrosion Inhibitors

In this chapter, the authors present a synthesis of some works carried out within the framework of the study of the corrosion inhibiting activity of a set of monosaccharide derivatives as new biodegradable, nontoxic corrosion inhibitors, and respectful of the environment. The comparative study that they present reveals the existence of more than 20 compounds derivatives of monosaccharide used as corrosion inhibitors in neutral and acidic mediums, and which have given rise to very good corrosion inhibiting efficiencies. These studies of the corrosion inhibiting effect were studied using potentiodynamic polarization curves and electrochemical impedance spectroscopy methods.

Section 4
Corrosion Inhibitors for Industrial Cleaning, Oil Refinery Equipment, and Pickling Process

Imidazole is part of the theophylline molecule, found in tea leaves and coffee beans, which stimulates the central nervous system. Imidazole and its derivatives are an important class of heterocyclic compounds that exhibit a wide range of biological and pharmacological activities. Furthermore, synthetic Imidazole are present in many fungicides, analgesic, anti-inflammatory, antibacterial, antitumor activities, antiprotozoal, and antihypertensive medications. Apart of its use for biological and pharmaceutical purpose, it also has varying applications in industries: the imidazole has been used extensively as a corrosion inhibitor on metals such as iron, copper, zinc, carbon steel, and their alloys. Imidazole and its derivatives have been reported as effective corrosion inhibitors due to their excellent excited state intra-molecular proton transfer, which elevates the performances of their adsorption into the metallic surface. In this chapter, the authors have reported a review of some works that investigated the effects of some Imidazole molecule on corrosion inhibition properties.

The inhibition performance of two imidazole derivatives, IM-Cl and IM-CH3, on the corrosion behavior of MS in 1M HCl acid solution was studied through weight loss method and electrochemical tests. The results obtained from the electrochemical methods show that inhibition efficiency increases with the increase of inhibitors concentration. The adsorption of inhibitors on the mild steel surface obeys Langmuir adsorption isotherm. The corrosion protection was once also investigated with the aid of UV–Vis spectrophotometry. SEM-EDX was performed and discussed for surface study of uninhibited and inhibited mild steel samples.

This chapter describes some recent good works in the field of metallic corrosion monitoring using Triazepine carboxylate inhibitors in hydrochloric acidic media. The different sections of this chapter cover electrochemical measurements and theoretical investigations. This review reveals Triazepine carboxylate compounds as very good inhibitors for mild steel in hydrochloric medium. The inhibition action of all Triazepine carboxylates compound studied was performed via adsorption on mild steel surface. Comparison between several adsorption isotherms reveal that the adsorption was spontaneous and followed Langmuir isotherm in HCl for all inhibitors and all studied temperatures. Kinetic and thermodynamic parameters for all inhibitors led to suggest the occurrence of chemical mechanism and also the spontaneity of the adsorption process on mild steel surface. The corrosion inhibition mechanism was also compared and discussed with the light of some Triazepine carboxyale compounds constituents.

The corrosion inhibition of mild steel in hydrochloric, sulfuric, and phosphoric acids solutions containing a mineral compound-based phosphate (apatite) was evaluated by potentiodynamic polarization and

Foreword

My first practical contact with corrosion inhibition and its evaluation was during my scientific career in several University laboratories, namely: Mohemmed First University in Oujda, Ibn Tofail University in Kenitra, Mohammed V University in Rabat and university Ain Chock in Casablanca Rabat, Morocco, since 2010. My main research interests are in the field of Physical Chemistry, especially electrochemistry (Corrosion and corrosion inhibition in all media for metals and alloys using various techniques). He is currently interested in using some quantum chemical methods like density functional theory (DFT), Molecular Dynamics Simulation and other semi-empirical methods to study corrosion inhibition efficiencies of compounds. Recently, he is interested to organic solar cells, the elaboration and characterization of thin films via electrodeposition and their applications. At that time I was using several methods like as gravimetric, potentiodyanamic polarization and surface analysis. Since then, the technique and equipment have evolved significantly. I was continually using several methods in subsequent years in the study of corrosion in several medium and its inhibition by organic and inorganic to determine the kinetics of hydrogen evolution, adsorption and absorption into metals, impedance of porous electrodes, and electrocatalytic reactions. After a series of seminars on this domain in several laboratories, I thought about to applying these investigations in industrial domain. For this, I thought about writing some chapters regroup several aspects about some fundamental aspect of corrosion and its inhibition in simple and some others industrial media. This book brings together according to this table of contents and according to my previous reviews and conferences in various universities, some information about corrosion and its inhibitions by non-toxic and biodegradable compounds in acidic and industrial environments.

The purpose of this book is to present the concept of corrosion, its inhibition in acidic medium and in cooling water system, biomaterials and certain applications. The available books were written either by chemists or engineers and professors, and I wanted to present it from the chemist's point of view. So, the book contains five sections, which contains ten chapters. In fact, the authors presented corrosion control in cooling water systems in first section. In this part, the main problems of cooling water systems are cited and studied by chemical and electrochemical measurements, and surface analysis. It also presented its prevention by inhibitors and formulations. To understand another concept of corrosion, the authors presented biomaterials corrosion and protection in second section. So, in this section the authors define concepts and methods currently used to study the corrosion behavior of biomaterials in physiological conditions. Additionally, this book presented several examples for protection like as application of eco-friendly corrosion inhibitorsin section three. Thus, the protection against corrosion is based on the use of organic or inorganic chemical compounds, but unfortunately these later sometimes show signs of toxicity during or after use. Also, the application of synthetic eco-friendly compounds as monosaccharide and their derivatives are being used as green corrosion inhibitors. This section followed by corrosion inhibitors for industrial cleaning, oil refinery equipment and pickling process were presented in section four. So, the authors presented the use of acidic median industrial applications, particularly during cleaning processes including industrial acid cleaning, acid pickling and acid-descaling. They cited several types of inhibitors are being employed for inhibition of metallic dissolution in these media in which use of organic and/or inorganic inhibitors is one of the most frequent and economic methods. This section describes some recent major works in the field of metallic corrosion monitoring using novel organic and mineral inhibitors in acidic media. Finally, the corrosion methods and control in civil engineering and others specific industrial applications of stainless steel industrial were presented in section five. In this section of book an explanation of the fundamentals of the corrosion of steel in concrete. Thus, it gives an overview of various corrosion monitoring techniques for steel reinforced concrete, its different corrosion control methods, and different methods for the anti-corrosion of steel in concrete.

I think that it will be useful to students and researchers who are just starting to study the materials corrosion and their protections and to others already using it in their research.

Abdelkader Zarrouk
Mohammed V University, Rabat, Morocco

Preface

Metals and their alloys offer excellent mechanical properties, such as high specific strength and good ductility. Metals is applied to approximately the entire industrial field for the reason of its excellent properties. However, its corrosion usually results in enormous financial losses and potential security problems where the protection technology has attracted the interest of several researchers and industrials.

Corrosion studies have attracted considerable interest in the areas of materials chemistry and industrial chemistry. Corrosion is the undesirable process that is the progressive destruction of materials by chemical reaction with their environment. This chemical process can also be expressed as electrochemical oxidation of metals in reaction with an oxidant such as oxygen.

In recent years, scientists and industrial studying corrosion and corrosion protection of metals using many theoretical and experimental tools. In this forthcoming comprehensive book on corrosion, the authors will present new and up-to-date studies and techniques used to prevent the corrosion of metals.

THE CHALLENGES

This book is written with a new approach and new philosophy and it is hoped that it will fulfill their aspirations. This book will aim to provide relevant theoretical frameworks and the latest empirical research findings in the corrosion engineering area. It will be written for provides a medium for the communication of ideas, best practices, developments in all aspects of this field and includes both fundamental and industrial aspects of corrosion sciences engineering. Occurrence of corrosion and its practical control is an area of study covering a wide range of scientific knowledge.

This book devoted to critical reviews and original articles that are key to advancing the understanding and application of corrosion science and engineering in the service of society. Chapters may be of a theoretical, experimental or practical nature, provided that they make a significant contribution to knowledge in the corrosion sciences fields.

The target audience of this book will be composed of professionals and researchers working in the field of corrosion and corrosion protection of materials in various disciplines.

ORGANIZATION OF THE BOOK

The book is organized into 10 chapters. Furthermore, in order to cover all aspects of corrosion, we have chosen to divide this book into five sections. Each of these sections contains several chapters, which detail feedback from expert and specialist authors in their fields. The sections are:

Section 1: Industrial Corrosion Control in Cooling Water Systems
Section 2: Biomaterials Corrosion and Protection
Section 3: Application of Eco-Friendly Corrosion Inhibitors
Section 4: Corrosion Inhibitors for Industrial Cleaning, Oil Refinery Equipment, and Pickling Process
Section 5: Corrosion Methods and Control in Civil Engineering and Other Specific Industrial Applications of Stainless Steel

The purpose of the sections making up this book, is presented and explained below:

Section 1, "Industrial Corrosion Control in Cooling Water Systems," presents two of the main problems of cooling water systems are corrosion and scale phenomena. These problems have a great economic impact, since the first involves deterioration of a metallic surface, whereas the second causes loss of capacity for thermal exchange. These problems occur simultaneously and hence a study of their development and inhibition under real conditions becomes an issue of great interest.

Many inhibitors and formulations have been used for water treatment in cooling water systems in order to solve the above cited problems. So, the use of inorganic for corrosion and scale inhibition is increasingly being curtailed by recent environmental restrictions. In many part of the world, high levels of inorganic product are also restricted by law. Thus, inhibitors are diminishing, because of the concern over their toxic effects on aquatic and possibly animal life. For this, the use of eco-friendly compounds was recommended.

The purpose of the present section is to define the main problems of cooling water systems and methods currently used to their study. It contains two chapters indicating the available information about scale and corrosion and their inhibitions by using eco-friendly inhibitors or formulations under similar conditions to typical operating conditions. These studies were carried out by using chemical, electrochemical measurements and Scanning electron microscopy analysis. Alternatively, the influence of some parameters influencing the working of the cooling water system was also studied and discussed.

Section 1 includes two conceptual chapters which presents good practices related to the thematic. Biomaterials is a field within materials science that has been established to attend the development of materials to treat the diseases of the body. The goal of the Section 2 is to define concepts and methods currently used to study the corrosion behavior of biomaterials in physiological conditions. One of the interesting points of corrosion of biomaterials is that they must be designed to fulfill different physical and chemical requirements within the human body. For instance, ceramic biomaterials are designed, sometimes, to accomplish a bone growth task. Alternatively, bio-ceramics are also employed to provide high wear resistance to implant surfaces with the lowest corrosion activity in biological environment. Depending on the part of the human body where the biomaterial is implanted, the corrosion process and chemical interaction with body fluids can be accelerated and, consequently, the product of these reactions can have a negative effect in the health of the patients. For this reason, the present book chapter is also focused on the interaction of biomaterial with body fluids as well as the methods to prevent corrosion phenomena that could lead to affect human health.

Section 3, "Application of Eco-Friendly Corrosion Inhibitors," includes the application of novel ecological inhibitors. Among the proposed solutions to combat the corrosion phenomenon, there is the protection of coatings, protection against oxidation and protection against corrosion. Currently, researchers are focusing more on inhibiting corrosion because it has the advantage of being the simplest and most economical technique. Protection against corrosion is based on the use of organic or inorganic chemical compounds, but unfortunately these later sometimes show signs of toxicity during or after use. Also, the application of synthetic compounds as corrosion inhibitors is limited because of the increasing ecological awareness and strict environmental regulations as most of the traditional corrosion inhibitors are synthesized using expensive and toxic chemicals and solvents. In view of this, recently, Monosaccharide and their derivatives are being used as corrosion inhibitors as they are available at the economic and commercial level. Because of the natural and biological origin they can be regarded as one of the greenest sources of the chemical compounds used as corrosion inhibitors.

Section 4 begins where Section 3 left off, though this section pronounces that Acidic media have been employed extensively in industry particularly during cleaning processes including industrial acid cleaning, acid pickling and acid-descaling. Hydrochloric, sulphuric, nitric, phosphoric, and acetic acids are a few of the common aggressive solutions that are used for cleaning purposes. Section 4, "Corrosion Inhibitors for Industrial Cleaning, Oil Refinery Equipment, and Pickling Process," presents extensive coverage of the various types of inhibitors are being employed for inhibition of metallic dissolution in acidic media in which use of organic and/or inorganic inhibitors is one of the most frequent and economic methods. This section describes some recent major works in the field of metallic corrosion monitoring using novel organic and mineral inhibitors in acidic media.

The first chapter, "Review of Application of Imidazole and Imidazole Derivatives as Corrosion Inhibitors of Metals," focuses on review of some works that investigated the effects of some Imidazole molecules on corrosion inhibition properties using weight loss, Tafel polarization and electrochemical impedance spectroscopy (EIS). They reported that the adsorption of some one of this derivatives into the metals surface follows the Langmuir isotherm and making a strong bond sometimes and linked with physical adsorption another time. Other Imidazole derivatives follow the Freundlich and Flory–Huggins isotherm. The Electrochemical data's and thermodynamics parameters for some Triphenylimidazole compounds have also been presented and discussed. This is belongs to that Imidazole and its derivatives are an important class of heterocyclic compounds that exhibit a wide range of biological and pharmacological activities. Furthermore, synthetic Imidazole are present in many fungicides, analgesic, anti-inflammatory, antibacterial, antitumor activities, antiprotozoal, and antihypertensive medications.

Another chapter of interest is "Novel Compounds of Imidazole Derivatives as Effective Corrosion Inhibitors for Mild Steel in Hydrochloric Acid Solution: Electrochemical and Spectroscopic Characterization," which highlights the inhibition performance of two newly imidazole derivatives on the corrosion behavior of mild steel in hydrochloric acidic media. The corrosion protection was investigated through weight loss method, electrochemical tests, UV–Vis spectrophotometry and SEM-EDX images.

The third chapter in the section is titled "Triazepines Compounds as Novel Synthesized Corrosion Inhibitors: State of Play and Perspectives." It investigates the importance of some recent good works in the field of metallic corrosion monitoring using Triazepine carboxylate inhibitors in hydrochloric acidic media. The potentiodynamic polarization curves and electrochemical impedance spectroscopy investigations used to evaluate the corrosion inhibition performance of Triazepine carboxylate compounds as very good inhibitors for mild steel in Hydrochloric medium. The inhibition action of all triazepine carboxylates compound studied was performed via adsorption on mild steel surface. Comparison between several adsorption isotherms, reveal that the adsorption was spontaneous and followed Langmuir isotherm in HCl for all inhibitors and at all studied temperatures. Kinetic and thermodynamic parameters for all inhibitors led to suggest the occurrence of chemical mechanism and also the spontaneity of the adsorption process on mild steel surface. The corrosion inhibition mechanism was also compared and discussed with the light of some Triazepine carboxyale compounds constituents. The effect of molecular structure on the inhibition efficiency has been explored by quantum chemical computations and obvious correlations were observed. The binding energies of tested Triazepine carboxyale compounds on Fe (110) surfaces were calculated using molecular dynamics simulation.

The fourth chapter is followed by "A New Class of Corrosion Inhibitors of Mild Steel in Industrial Sulphuric, Phosphoric, and Hydrochloric Acids Media." which present the comparison of corrosion inhibition of mild steel in hydrochloric, sulfuric and phosphoric acids solutions containing a mineral compound-based phosphate. The performance was evaluated using potentiodynamic polarization and electrochemical impedance spectroscopy methods. Potentiodynamic polarization measurements reveal that the inhibition efficiency increased with the concentration of the apatite which appears to be a cathodic type inhibitor in the three mediums. Electrochemical impedance spectroscopy confirms this result; indeed, the transfer resistance increases with apatite concentration. We note that the double layer capacitance decreases simultaneously suggesting the formation of an adsorbed layer on the mild steel surface. The inhibition mechanism has been elucidated by a thermodynamic study which showed that the film was formed by physisorption. The adsorption model obeys to the Langmuir adsorption isotherm. The parameters of activation energy was evaluated and discussed.

Finally, Section 5 presents corrosion methods used for civil engineering and others specific industrial application who use stainless steel. This section includes two chapters, the first one is dedicated to review of corrosion methods for stainless steel. Stainless steel is known to be important in most structural applications due to its high mechanical strength, hardness and machinability. It is considered a versatile steels that has good formability, can be welded as it is and is non-magnetic austenitic structure. Its corrosion resistance decreases at higher temperatures when exposed to water for prolonged periods of time. It is sensitive to pitting, crevice and stress corrosion cracking at elevated temperatures. In some cases, the resistance is improved by addition of corrosion inhibitor that negatively affects its formability and welding advantages. Therefore, other methods of corrosion protection are desired. This chapter provides in-depth review of corrosion protections materials and methods that have been used for protecting 304 Stainless Steel in different specific applications. Also provides systemic analysis of the potentiodynamic polarization method to compare the corrosion potential of different materials.

The last chapter is devoted to corrosion assessment and control techniques for reinforced concrete structures.

Poor durability of concrete is a major cause of problems in modern building and civil engineering structures in all countries: the annual cost of investigating and repairing deteriorating reinforced concrete structures runs into many millions of pounds. This book explains the fundamentals of the corrosion of steel in concrete. Thus, it gives an overview of various corrosion monitoring techniques for steel reinforced concrete, and its different corrosion control methods. In addition, this section presents also different methods for the anti-corrosion of steel in concrete by using polymers and electroless deposition. It is comprehensive and provides a basis for the practicing engineer to design concrete structures which avoid the problem using modern concepts and specifications. A limited discussion of corrosion measurement and repairs is also provided.

The editors are grateful to the many contributors who have aided the production of this book. We are confident that the readers, researchers, industrials and professor will find the *New Challenges and Industrial Applications for Corrosion Prevention and Control* book of great interest. Therefore, it was very meaningful to edit the articles for this book.

With cooperation of editor's friends, I am grateful to faculty members of Ibn Tofail university in Morocco, Cumhuriyet University, Sivas, Turkey for their participations and to be strong supporters with discussion to the contents of this book. We believe that this will make significant contribution to improving knowledge and skill for new generation corrosion engineering students and also industrials teams.

I would like to thank IGI Global team for their generous cooperation at every stage of the book production.

Younes El Kacimi
University Ibn Tofail, Morocco

Savas Kaya
Cumhuriyet University, Turkey

Rachid Touir
University Ibn Tofail, Morocco

Acknowledgment

The editors are grateful to the many contributors who have aided the production of this Book. We are confident that the readers, researchers, industrials, and professor will find the *New Challenges and Industrial Applications for Corrosion Prevention and Control* book of great interest. Therefore, it was very meaningful to edit the articles for this book.

With cooperation of editor's friends, I am grateful to faculty members of Ibn Tofail University in Morocco, Cumhuriyet University, Sivas, Turkey for their participations and to be strong supporters with discussion to the contents of this book. We believe that this will make significant contribution to improving knowledge and skill for new generation corrosion engineering students and also industrials teams.

I would like to thank IGI Global team for their generous cooperation at every stage of the book production.

Section 1
Industrial Corrosion Control in Cooling Water Systems

Chapter 1
Protection of Low Carbon Steel in Industrial Cooling Water System by New Formulation

Abdelhadi Rochdi
Ibn Tofail University, Morocco

Rachid Touir
https://orcid.org/0000-0003-2796-2157
CRMEF, Morocco & Ibn Tofail University, Kenitra, Morocco

Mohamed Ebn Touhami
Laboratory of Materials Engineering and Environment: Modeling and Application, Faculty of Science, Ibn Tofail University, Morocco

ABSTRACT

This chapter studies the operational parameters effects, such as immersion time, corrosion products, and pH, on the inhibition efficiency against corrosion for low carbon steel in simulated cooling water system of the mixture 1 (10^{-3} M 3-MPOX + 20 ppm CTAB), which corresponds to the best inhibition efficiency. The results show that the corrosion inhibition performance of the studied mixture reinforces with immersion time and has a maximum in the pH range 6.5–7.5. In addition, the analysis of the results shows that there is a dependence of the pH value of the solution and the corrosion current density (i_{corr}). Indeed, a correlation can be established between log (i_{corr}) and the pH using the polynomial equation. Finally, the mixture 1 provides good protection for the attack materials. These studies were evaluated from the polarization curves (PP), electrochemical impedance spectroscopy (EIS).

DOI: 10.4018/978-1-7998-2775-7.ch001

INTRODUCTION

The purpose of the present book chapter entitled: "Protection of low carbon steel in industrial cooling water system by new formulation" is to prevent the main problems of cooling water systems. This book chapter presents the main problems reencountered in cooling water systems like as corrosion, scale and bio-corrosion phenomena (Scheme 1). In addition, the book chapter aims to propose a formulation in order to protect low carbon steel in cooling water system against the above problems using electrochemical measurements and SEM/EDAX analysis. So, the effect of some parameters was also studied and discussed.

BACKGROUND

Steel is a material widely used in various industrial fields [Benarioua et al., 2019; Lafont et al., 1993; Abouchane et al., 2015]. It is found in cooling water systems because of their interesting properties including electrical and thermal conductivities high, good chemical weld ability and low cost. Despite these properties, this material may suffer a significant degradation whose forms and their characteristics differ according to several parameters: composition of the middle, temperature, pH, flow velocity...

The quality of the heat exchange in a cooling circuit is a crucial element. Scale, corrosion, and bio-coorrosion affect the performance of the cooling process That is why researchers are currently working on the development of formulations of water treatment composed mainly of scale inhibitor, corrosion inhibitor, oxidizing and non-oxidizing biocide, Legionella, biodispersant and dispersants for mineral and organic matter [Abousalem et al., 2019; Ramesh et al., 2004].

For instance, in our recently papers [Rochdi et al., 2014, 2015], the oxadiazoles derivatives effect on corrosion and scale inhibition of brass and low carbon steel in simulated cooling water was compared. It was found that these compounds acte as mixed inhibitors-type and their inhibition efficiency depended on their molecular structure of the oxadiazoles compounds and more precisely on the position of the methyl group in the cycle. Indeed, the sequence of the undertaken oxadiazole inhibitors is then, 3-MPOX ≥4-MPOX >2-MPOX. In addition, their inhibitions were enhanced by the non-oxidizing biocide CTAB addition. So, a large plate of passivation (about 500 mV/SCE of wide) was registered in the case of mixture 1 (10^{-3} M 3-MPOX + 20ppm CTAB) which reinforced the low carbon steel or brass resistance against pitting corrosion.

The originality of this work lies in the fact that the proposed formulation protects both the low carbon steel as brass [Rochdi et al., 2014, 2015] and therefore can be used in a cooling system constitutes, as is the case in general, the two materials. In such a case, the industrial use of formulations containing two inhibitors; one for the brass and the other for the steel.

MAIN FOCUS OF THE CHAPTER

In continuation of our work, we evaluate the operational parameters effect, such as immersion time, corrosion products and pH on the inhibition efficiency against corrosion for low carbon steel in simulated cooling water system of the mixture 1 (10^{-3} M 3-MPOX + 20 ppm CTAB) which corresponds to the best inhibition efficiency. These studies were evaluated from the polarization curves (PP), electrochemical impedance spectroscopy (EIS).

Scheme 1. Cooling water Triangle

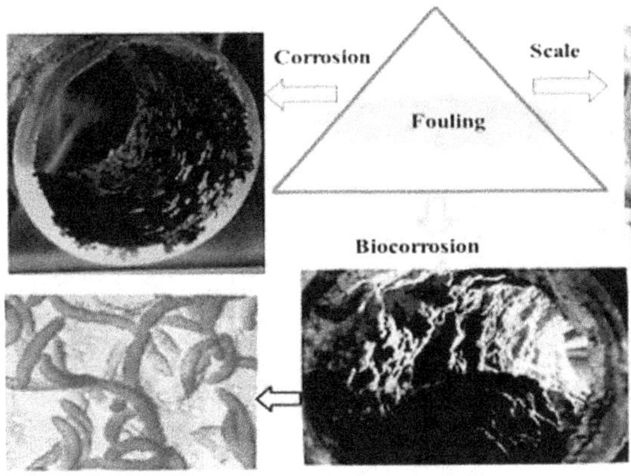

1. EXPERIMENTAL PROCEDURES:

1.1. Electrochemical Cell and Materials

The electrolysis cell was a borrosilcate glass (Pyrex®) cylinder closed by cap with five apertures. Three of them were used for the electrode insertions. The working electrode was low carbon steel previously used [Rochdi et al., 2015; Touir et all,.2008, Dkhireche et al., 2013] and its composition is summarized in Table 1. The investigated area of the surface was 0.8 cm^2. Prior to immersion test, the electrode was abraded using emery paper up to 1200 grade, cleaned with acetone, washed with distilled water, and dried finally. The rotating electrode (INF-EL-EDI 101) was used for electrochemical measurements. The rotation speed was fixed at 1000 rpm with CTV 101 (Radiometer analytical).

Pt plate as the counter while a saturated calomel electrode (SCE) was used as the reference electrode. All potentials are referred with respect to this electrode. For long exposure experiments, between two measurements, the reference electrode was removed from the test solution to minimize its contamination by chloride ions and the working electrode was set in rotation for one hour before each test.

The simulated cooling water solution is used previously [Rochdi et al., 2015; Touir et all,.2008, Dkhireche et al, 2013, El Bakri et al., 2016] and its composition is displayed in Table 1. This composition represents the average amount of various ions present in the waters used in Moroccan cooling waters system. According to the literature [Ramesh et al.,2004; Saremi et al., 2006] this solution was corrosive and scaling. The temperature was adjusted at 32 ± 2 °C. The electrolyte was in contact with air without any purging of dissolved oxygen.

2,5-bis(2-methylphenyl)-1,3,4-oxadiazole (2-MPOX), 2,5-bis(3-methylphenyl)-1,3,4-oxadiazole (3-MPOX), and 2,5-bis(4-methylphenyl)-1,3,4-oxadiazole (4-MPOX) were used as corrosion and scale inhibitors. The cetyltrimethylammonium bromide (CTAB) was added as corrosion and biofilm inhibitor. There structure formulas were presented in Table 1. All assessments were carried out at temperature and pH of 32±1 °C and 7.35±0.02, respectively

Scheme 2. SEM micrographs of copper electrode surface after immersion in cooling water solution: (a) the blank solution and (b) the presence of mixture VATP and biocide CTAB

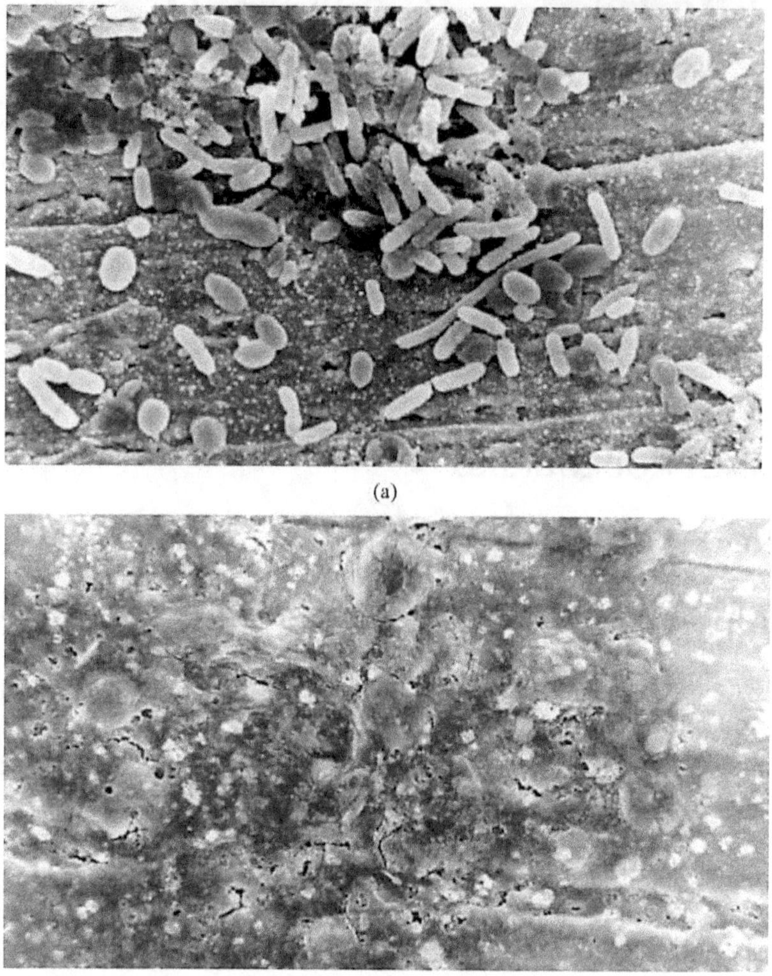

(a)

(b)

[Ramesh et all., 2004].

1.2. Polarization Measurements

The working electrode was immersed in test solution under rotation at 1000 rpm during one hour until the steady state corrosion potential (E_{corr}) was reached. The cathodic polarization curve was recorded

Table 1. Composition of simulated cooling water

Salts	$MgCl_2$, $6H_2O$	$CaCl_2$,$2H_2O$	$Ca(NO_3)_2$,$6H_2O$	$MgSO_4$,$7H_2O$	$NaHCO_3$
Concentration (mM)	2.85	3.64	0.566	1.56	3.77
Conductivity (µS cm^{-1}) at 299 K	1678				

Table 2. Molecular structures, names and abbreviations of the studied oxadiazole derivatives and biocide.

Inhibitor Formula	Name	Abbreviation
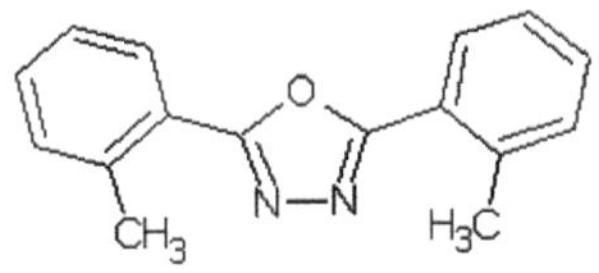	2,5-bis(2-methylphenyl)-1,3,4-oxadiazole	2-MPOX
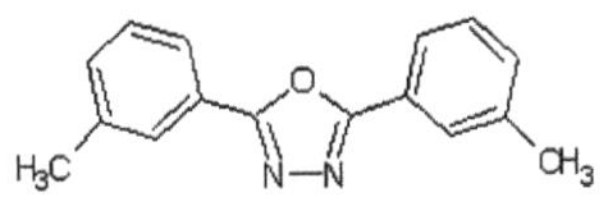	2,5-bis(3-methylphenyl)-1,3,4-oxadiazole	3-MPOX
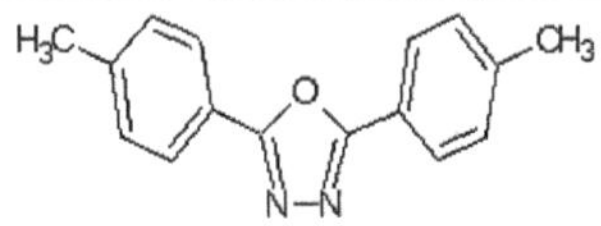	2,5-bis(4-methylphenyl)-1,3,4-oxadiazole	4-MPOX
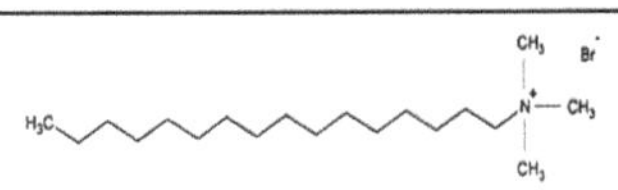	Cetyltrimethylammonium bromide	CTAB

by polarization from E_{corr} towards more negative direction with a sweep rate of 1 mV/s. After this scan, the same electrode was remained in solution until the obtaining of the steady state corrosion potential (E_{corr} ± 0.002 V), and then the anodic polarization curve was recorded from E_{corr} to positive direction with the same sweep rate. The obtained polarization curves were corrected for ohmic drop with the electrolyte resistance determined by electrochemical impedance spectroscopy. These measurements were carried out using Potentiostat/Galvanostat/Voltalab PGZ 100 monitored by a personal computer. For each concentration three independent experiments were performed. The mean values and the standard deviations are reported as well.

The overall current density values, *i*, were considered as the sum of two contributions, anodic and cathodic current i_a and i_c, respectively. For the potential domain not too far from the open circuit, it may be considering that both processes follow the Tafel law [Stern et al.,1957]. Thus, it can be derive:

$$i = i_a + i_b = i_{corr} \times \left\{ \exp\left[b_a \times \left(E - E_{corr} \right) \right] - \exp\left[b_c \times \left(E - E_{corr} \right) \right] \right\} \quad (1)$$

where i_{corr} is the corrosion current density (A cm^{-2}), b_a and b_c are the Tafel constant of anodic and cathodic reactions (V^{-1}), respectively. These constant are linked to the Tafel slope β (V/dec) in usual logarithmic scale by:

$$\beta = \frac{\ln(10)}{b} = \frac{2.303}{b} \quad (2)$$

The corrosion parameters were then evaluated by means of nonlinear least square method by applying this equation using Origin software. However, for this calculation, the potential range applied was limited to ±0.100V around the E_{corr} else a significant systematic divergence was sometimes observed for both anodic and cathodic branches.

The inhibition efficiency (η_I) was calculated using the following equation:

$$\eta_I = \frac{i^0_{corr} - i_{corr}}{i^0_{corr}} \times 100 \qquad (3)$$

where i^0_{corr} and i_{corr} are the corrosion current density values without and with the inhibitor, respectively.

1.3. EIS Measurements

The electrochemical impedance spectroscopy measurements were carried out using a transfer function analyzer (Voltalab PGZ 100, Radiometer Analytical), over the frequency range from 100 kHz to 0.01 Hz with 10 points per decade. The applied amplitude of AC signal was 10 mV_{rms}. All experiments were performed at the open circuit potential. The obtained impedance data were analyzed in term of equivalent electrical circuit using Bouckamp's program [Bouckamp,1993]. The inhibition efficiency was evaluated from R_p (which is obtained from the diameter of the semicircle in the Nyquist representation) with the relationship:

$$\eta_Z = \frac{R_p - R^0_p}{R_p} \times 100 \qquad (4)$$

where R^0_p and R_p are the resistance polarization values in the absence and in the presence of inhibitor, respectively.

2. RESULTS AND DISCUSSION

2.1. Parameters Effect on the Formulation Performance

a. Influence of Immersion Time

Fig. 1 shows the Nyquist diagrams of low carbon steel in cooling water solution containing mixture 1 (10^{-3} M 3-MPOX + 20ppm CTAB) after the low carbon steel electrode was exposed to the solution for different immersion time. We noted from EIS results the total impedance (the diameter of impedance from high frequency to low frequency) of the system gradually increased with increasing of immersion time, this evolution suggests a strong inhibition of the dissolution processes occurring on the surface [Gao et al., 2007]. Whatever it during the immersion time, the impedance diagrams are characterized by three time constants. at high frequencies, the first loop might be attributed to the adsorption of the inhibitor molecules, and that at the medium frequencies can be related to the relaxation of the double layer in parallel with the charge transfer resistance. However, at the low frequencies, the observed loop was generally assigned to a slow process such as diffusion. It is known that, the anodic dissolution of steel and the cathodic oxygen reduction reactions take place on the metal surface simultaneously. The diffusion process may be due to either the transportation of corrosive ions and soluble corrosion products at the metal/solution or the diffusion of dissolved oxygen to the steel surface [Barcia et al, 2002]. However, the second one is more probable [Cubillos et al., 2010]. These observations provide an equivalent electrical circuit shown in Figure 2. Where, R_s represents the ohmic resistance between the working and the reference electrode; R_{ct} is the charge transfer resistance related to the corrosion reaction at OCP; R_f resistance associated with the layer of products formed, Z_w is Warburg impedance. Q_f and Q_{ct} are the constant phase elements used to replace the film capacitance and the double layer capacitance, respectively. Constant phase elements have widely been used to account for deviations brought about by surface roughness [Felhősi et al., 2002; Georges et al., 2008]. The impedance of CPE is given by equation:

$$Z_{CPE}(\omega) = Q^{-1}(j\omega)^{-n} \quad (5)$$

Where Q is the constant phase element (*CPE*), ω is the sine wave modulation angular frequency (in rad s^{-1}), $j^2 = -1$ is the imaginary number and n is the *CPE* exponent. According on n, can represent a CPE can be resistance [$Z_{CPE} = R$, $n = 0$], pure capacity [$Z_{CPE} = C$, $n = 1$], inductance [$Z_{CPE} = L$, $n = -1$] or Warburg impedance for ($n = 0.5$) [Tsuchiya et al., 2002].

The parameters evolutions with time are summarized in Table 2. The analysis of this parameters revealed that the R_{ct} value increased steadily to achieve a higher value (167016 Ohm.cm^2) after 24h immersion time. This increase illustrates formation of a stable protective film on low carbon steel surface, which is reinforced with the immersion time. In addition, it is also noted that the electrolyte resistance, R_s, values increase with immersion time. This increase in R_s can be explained by the protective effect of the mixture, which is reinforced with the immersion time. It is also noted that the C_{dl} decreases with the immersion time. The change in C_{dl} values was caused by to gradual replacement of water molecule by the chloride anions and by the adsorption of the organic molecules on the metal surface, decreasing the extent of dissolution reaction [Touir et al, 2008]. All of these results show that the corrosion inhibition performance of the studied mixture reinforce with immersion time.

b. Effect of Corrosion Products on the Formulation Performance

To examine the impact of the mixture 1 of the aging of our electrode has been investigated using electrochemical impedance spectroscopy. In other words, we try to look at the possible interaction between our molecules formulation (mixture 1) and products corrosion and scale formed in time. This study may identify the idea or not adding the mixture 1 in cooling water systems in operation without resorting to a complete shutdown of the installation. Such cases are often at the basis of many direct and indirect losses, mainly economic. This consists of tries to maintain the steel electrode in corrosive solution (no mixture 1) for a period of 6h immersion and thereafter add the formulation. Fig.3 shows the impedance Nyquist plots reflecting this manipulation after two immersion times 0.5 h and 24 h.

At high frequency, the first observed loop was attributed to the adsorbed film resistance due to adsorption of the molecules inhibitors and corrosion and/or scale products and the second was attributed to the influence of charge transfer resistance. The third was attributed to the Warburg impedance such as obtained in our previous work [Touir et al., 2014], indicates that the products prevented the diffusion of O_2 to the steel substrate and/or suppressed the cathodic reduction of corrosion products. The same equivalent circuit shown in Fig. 2 was used to simulate the impedance spectra. The relevant simulated impedance parameters are given in Table.3. It is noted that the used formulation takes its performance

Figure 1. Nyquist diagrams for low carbon steel in cooling water and presence of mixture 1 (10^{-3} M 3-MPOX + 20ppm CTAB) for different immersion time (T = 305 K; Ω = 1000 rpm).

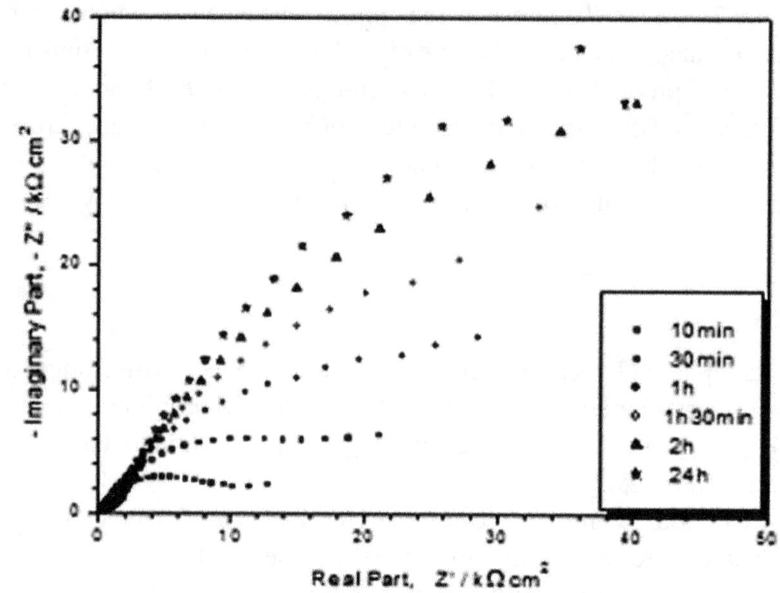

Table 3. Data obtained from the impedance diagrams of low carbon steel in 3% NaCl and presence of mixture 1 (10^{-3} M 3-MPOX + 20ppm CTAB) for different immersion time.

Immersion Time	R_s Ω.cm²	R_f Ω.cm²	C_f µF.cm^{-2}	n_f	R_{ct} Ω.cm²	C_{dl} µF.cm^{-2}	n_{tc}
10 min	215	-	-	0.90	11823	150	0.82
30 min	209	142	11	0.92	25791	125	0.85
1h	254	638	3.3	0.94	101700	25	0.89
1h 30 min	199	1014	2.47	0.95	115360	23	0.90
2h	190	1413	1.77	0.96	122080	24	0.90
24h	157	1320	0.85	0.96	167016	17	0.91

Figure 2. Equivalent circuit proposed for fitting the impedance spectra obtained on low carbon steel surface in presence of mixture 1 (10^{-3} M 3-MPOX + 20ppm CTAB).

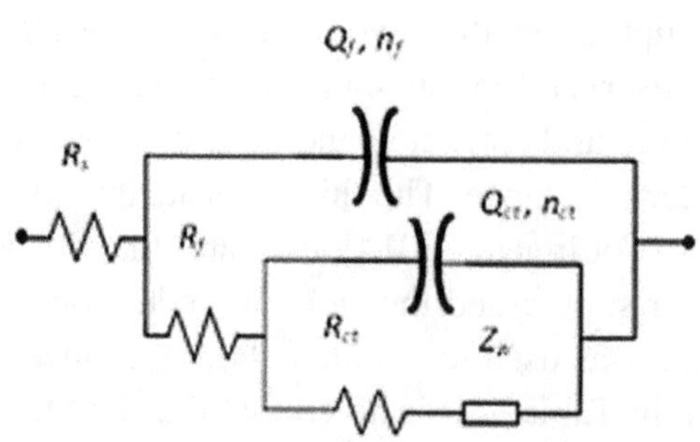

Figure 3. Impedance Nyquist plots of low carbon steel in simulated solution with mixture 1 after 6 h electrode attack and for different immersion time.

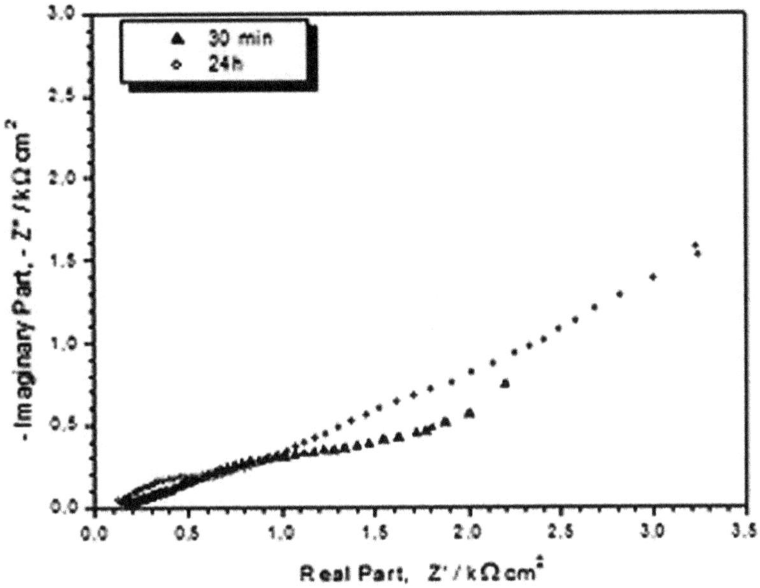

in the presence of the corrosion and scale products indicating its adsorption on covered surface by these products. It can still consider a probable interaction between formulation molecules and corrosion and scale products leading to the reinforcement of adsorbed film on metallic surface. About this result, we can conclude that the investigated formulation provides good protection for the attacks materials indicating its validation for cooling water already damaged.

c. Effect of pH Solution

The ability of an inhibitor to form a protective film on the surface was also evaluated by the polarization curves to pH values of 5.5, 6.5, 7.5, 8.5 and 9.5. This is to show how pH range of inhibitor molecules of mixture 1 is effective and if it are pH for which protection is optimal or minimal. The potentiodynamic polarization curves of low carbon steel after 1 h of immersion in the simulated cooling water solution with mixture 1 are presented in Fig. 4. The examination of this figure rises as the pH value clearly changed the type of the kinetics of electrochemical processes involved. The electrochemical parameters resulting from these curves are illustrated in Table 4.

Table 4. Impedance data of low carbon steel in simulated cooling water solution with mixture 1 after 6 h electrode attack.

Temps	R_s $\Omega.cm^2$	R_f $\Omega.cm^2$	C_f $\mu F.cm^{-2}$	R_{ct} $\Omega.cm^2$	C_{dl} $\mu F.cm^{-2}$
30min	194	594	20	980	156
24h	151	1058	0.6	1543	65.6

Figure 4. Potentiodynamic polarization curves of low carbon steel in simulated cooling water with mixture 1 at various pH values (T = 305 K; Ω = 1000 rpm).

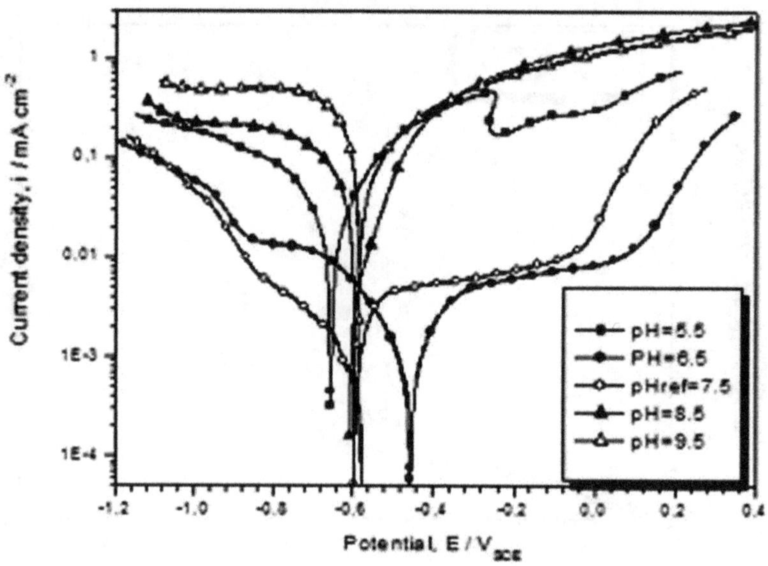

Table 4. Electrochemical parameters of low carbon steel in simulated cooling water with mixture 1 at various pH values (T = 305 K; Ω = 1000 rpm).

pH Solution	E_{corr} mV/ECS	I_{corr} µA.cm^{-2}	η_{PP} %
5.5	-657	14.5	87.5
6.5	-454	0.71	99.3
7.5	-400	0.64	99.37
8.5	-605	46.2	55.8
9.5	-588	72.2	29.2

The analysis of this table shows that there is a dependence of the pH value of the solution simulated and the corrosion current density (I_{corr}). Indeed, a correlation can be established between log (I_{corr}) and the pH (fig.5.) using the following polynomial equation [Garcés et al., 2011]:

$$\log\left(I_{corr}\right) = 124{,}67837 - 49{,}26169pH + 6{,}34854pH^2 - 0{,}26456pH^3 \quad (5)$$

The highest inhibition efficiency could be obtained in the pH range 6.5–7.5. But, when the pH is decreased from 6.5 to 5.5, the inhibition efficiency is reduced to 87.5% and on increasing pH range from 7.5 to 9.5, the inhibition efficiency is reduced to 29%. The reasons for decrease in inhibition efficiency in more alkaline and acidic environments are explained under the mechanistic aspects [Prabakaran et al., 2013].

Figure 5. Polynomial correlation between log i_{corr} and pH.

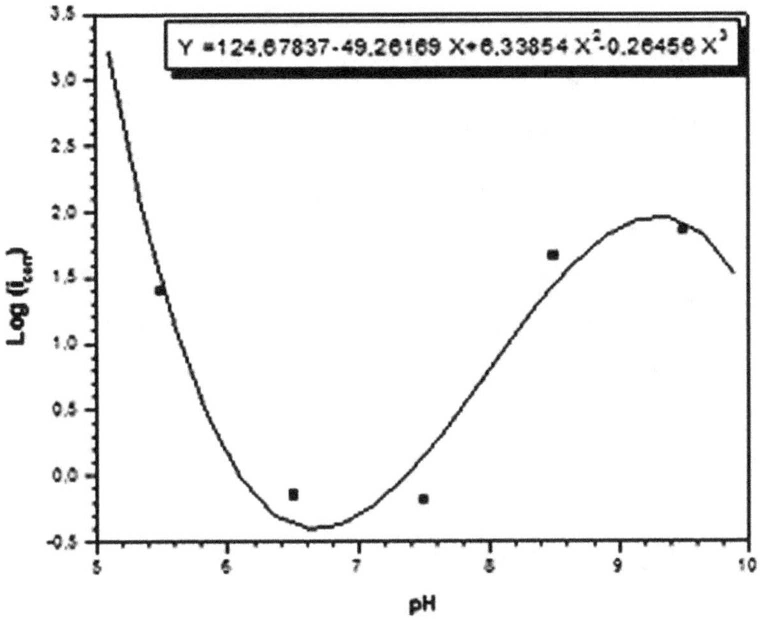

3. SEM AND EDAX RESULTS

SEM and EDX results (Scheme 3) shows the scanning electronic microscopy (SEM) images of low carbon steel surface that have been exposed to the simulated cooling solution (blank solution) during 2 days. The image of the surface exposed to the inhibitor-free aggressive solution, shows heterogeneous layer of corrosion products (Scheme 3.a and 3.c) Indeed, a qualitative EDX analysis (Scheme 4) reveals that characteristic scaling elements are identified (Ca, Mg, O and C) on this layer. In addition, at an enlargement of 24000x, we manage to detect bacteria in the form of strains (Scheme 3.b) which can be classified in the category of BSR through an EDX analysis which shows a level of carbon and sulfur (S) quite high (Scheme 4).

In addition, SEM micrographs in the presence of mixture 1 (Scheme 5) show that the surface is free from any corrosion and scale product and reveal the formation of a protective layer. This is confirmed by the increase in the iron-related peak and disappearance of the peaks of Calcium (Ca), magnesium (Mg) and sulfur (S) (Scheme 6). These results confirm the mixture 1 provides good protection against corrosion, scale and biocorrosion.

4. CONCLUSION

Inhibitive performance of the mixture 1 (10^{-3} M 3-MPOX + 20 ppm CTAB) on the corrosion of steel in industrial cooling water was studied at different immersion time and different pH solution using potentiodynamic polarization curves (PP) and electrochemical impedance spectroscopy techniques (EIS). Indeed, the inhibition efficiency of mixture 1 was found to enhance remarkably with immersion time and

Scheme 3. (a), (b), (c) Surface morphologies of ordinary steel immersed for 2 days in the simulated cooling water solution (blank solution)

[Rochdi, 2014].

Scheme 4. EDX analyses of low carbon steel electrode surface after 2 days of immersion in simulated cooling water solution (blank solution)

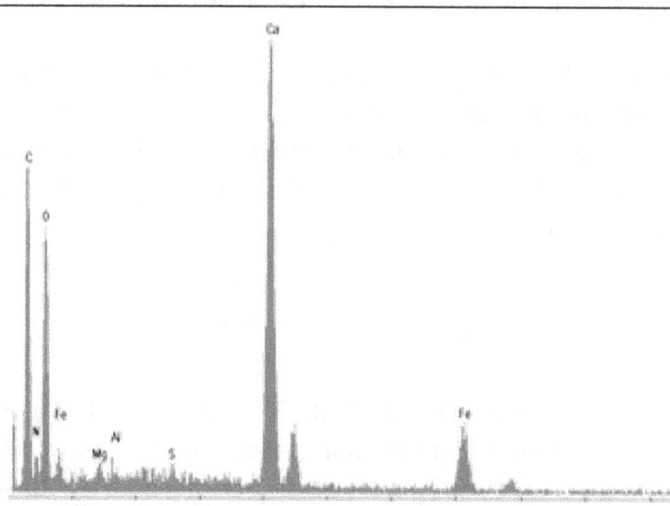

[Rochdi, 2014].

Scheme 5. (d) Surface morphologies of ordinary steel immersed for 2 days in the simulated cooling water solution with mixture 1 (10^{-3} M 3-MPOX + 20 ppm CTAB)

[Rochdi, 2014].

Scheme 6. EDX analyses of low carbon steel electrode surface after 2 days of immersion in simulated cooling water solution with mixture 1 (10^{-3} M 3-MPOX + 20 ppm CTAB)

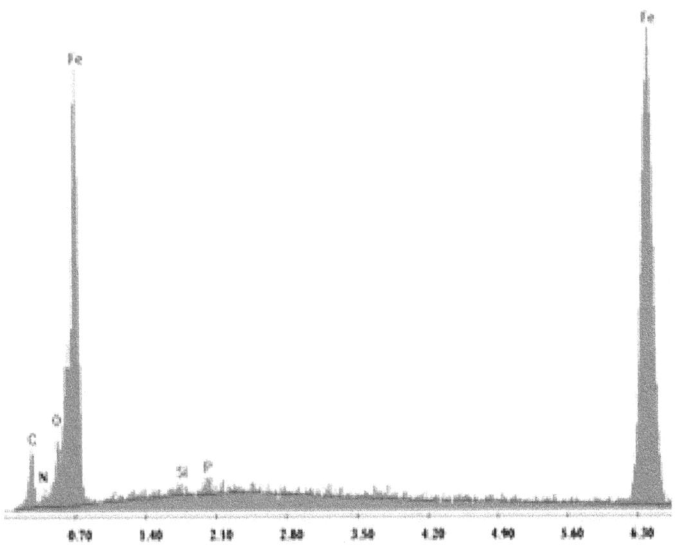

[Rochdi, 2014].

has a maximum around pH value 7. In addition, the corrosion current density (I_{corr}) varies polynomially with the pH value of the solution. Finally the mixture 1 provides good protection for the damaged materials indicating its validation for cooling water already damaged. All of these results show the industrial interest of our formulation.

REFERENCES

Abouchane, M., El Bakri, M., Touir, R., Rochdi, A., Elkhattabi, O., Touhami, M. E., ... Mernari, B. (2015). Corrosion inhibition and adsorption behavior of triazoles derivatives on mild steel in 1 MH-3PO4 and synergistic effect of iodide ions. *Research on Chemical Intermediates*, *41*(4), 1907–1923. doi:10.100711164-013-1319-5

Abousalem, A. S., Ismail, M. A., & Fouda, A. S. (2019). A complementary experimental and in silico studies on the action of fluorophenyl-2, 2′-bichalcophenes as ecofriendly corrosion inhibitors and biocide agents. *Journal of Molecular Liquids*, *276*, 255–274. doi:10.1016/j.molliq.2018.11.125

Barcia, O. E., D'Elia, E., Frateur, I., Mattos, O. R., Pébère, N., & Tribollet, B. (2002). Application of the impedance model of de Levie for the characterization of porous electrodes. *Electrochimica Acta*, *47*(13-14), 2109–2116. doi:10.1016/S0013-4686(02)00081-6

Benarioua, M., Mihi, A., Bouzeghaia, N., & Naoun, M. (2019). Mild steel corrosion inhibition by Parsley (Petroselium Sativum) extract in acidic media. *Egyptian Journal of Petroleum*, *28*(2), 155–159. doi:10.1016/j.ejpe.2019.01.001

Bouckamp, A. (1993). Users manual equivalent circuit. Faculty of Chemical Technology, 4.

Cubillos, M., Sancy, M., Pavez, J., Vargas, E., Urzua, R., Henríquez-Roman, J., Tribollet, B., Zagal, J. H., & Páez, M. A. (2010). Influence of 8-aminoquinoline on the corrosion behaviour of copper in 0.1 M NaCl. *Electrochimica Acta*, *55*(8), 2782–2792. doi:10.1016/j.electacta.2009.12.064

Dkhireche, N., Dahami, A., Rochdi, A. J. Hmimou, J., Touir, R. M., Ebn Touhami. M., El Bakri, M., El Hallaoui, A., Anouar, A., & Takenouti, H. (2013). Corrosion and scale inhibition of low carbon steel in cooling water system by 2-propargyl-5-o-hydroxyphenyltetrazole. *Journal of Industrial and Engineering Chemistry*, *19*, 1996–2003.

El Bakri, M., Touir, R., Tazouti, A., Dkhireche, N., Touhami, M. E., Rochdi, A., & Zarrouk, A. (2016). Corrosion inhibition study of brass in simulated cooling water by triazole derivatives, cetyltrimethyl-ammonium bromide and their mixture. *Arabian Journal for Science and Engineering*, *41*(1), 75–88. doi:10.100713369-014-1532-6

Felhősi, I., Telegdi, J., Pálinkás, G., & Kálmán, E. (2002). Kinetics of self-assembled layer formation on iron. *Electrochimica Acta*, *47*(13-14), 2335–2340. doi:10.1016/S0013-4686(02)00084-1

Gao, G., & Liang, C. H. (2007). 1, 3-Bis-diethylamino-propan-2-ol as volatile corrosion inhibitor for brass. *Corrosion Science*, *49*(9), 3479–3493. doi:10.1016/j.corsci.2007.03.030

Garcés, P., Saura, P., Zornoza, E., & Andrade, C. (2011). Influence of pH on the nitrite corrosion inhibition of reinforcing steel in simulated concrete pore solution. *Corrosion Science*, *53*(12), 3991–4000. doi:10.1016/j.corsci.2011.08.002

Georges, C., Rocca, E., & Steinmetz, P. (2008). Synergistic effect of tolutriazol and sodium carboxylates on zinc corrosion in atmospheric conditions. *Electrochimica Acta*, *53*(14), 4839–4845. doi:10.1016/j.electacta.2008.01.073

Lafont, M., Pebere, N., Moran, F., & Bleriot, P. (1993). Inhibition de la corrosion d'un acier au carbone par des produits dérivés de phosphonates en association avec des sels de zinc. *Revue des sciences de l'eau/Journal of Water Science*, *6*(1), 97-112.

Prabakaran, M., Venkatesh, M., Ramesh, S., & Periasamy, V. (2013). Corrosion inhibition behavior of propyl phosphonic acid–Zn^{2+} system for carbon steel in aqueous solution. *Applied Surface Science*, *276*, 592–603. doi:10.1016/j.apsusc.2013.03.138

Ramesh, S., & Rajeswari, S. (2004). Corrosion inhibition of mild steel in neutral aqueous solution by new triazole derivatives. *Electrochimica Acta*, *49*(5), 811–820. doi:10.1016/j.electacta.2003.09.035

Ramesh, S., Rajeswari, S., & Maruthamuthu, S. (2004). Corrosion inhibition of copper by new triazole phosphonate derivatives. *Applied Surface Science*, *229*(1-4), 214–225. doi:10.1016/j.apsusc.2004.01.063

Rochdi, A., Kassou, O., Dkhireche, N., Touir, R., El Bakri, M., Touhami, M. E., ... Hammouti, B. (2014). Inhibitive properties of 2, 5-bis (n-methylphenyl)-1, 3, 4-oxadiazole and biocide on corrosion, biocorrosion and scaling controls of brass in simulated cooling water. *Corrosion Science*, *80*, 442–452. doi:10.1016/j.corsci.2013.11.067

Rochdi, A., Touir, R., El Bakri, M., Touhami, M. E., Bakkali, S., & Mernari, B. (2015). Protection of low carbon steel by oxadiazole derivatives and biocide against corrosion in simulated cooling water system. *Journal of Environmental Chemical Engineering*, *3*(1), 233–242. doi:10.1016/j.jece.2014.11.020

Saremi, M., Dehghanian, C., & Sabet, M. M. (2006). The effect of molybdate concentration and hydrodynamic effect on mild steel corrosion inhibition in simulated cooling water. *Corrosion Science*, *48*(6), 1404–1412. doi:10.1016/j.corsci.2005.06.009

Stern, M., & Geary, A. L. (1957). Electrochemical polarization I. A theoretical analysis of the shape of polarization curves. *Journal of the Electrochemical Society*, *104*(1), 56–63. doi:10.1149/1.2428496

Touir, R., Cenoui, M., El Bakri, M., & Touhami, M. E. (2008). Sodium gluconate as corrosion and scale inhibitor of ordinary steel in simulated cooling water. *Corrosion Science*, *50*(6), 1530–1537. doi:10.1016/j.corsci.2008.02.011

Touir, R., Dkhireche, N., Ebn Touhami, M., El Bakri, M., Rochdi, A., & Belakhmima, R. A. (2014). Study of the mechanism action of sodium gluconate used for the protection of scale and corrosion in cooling water system. *Journal of Saudi Chemical Society*, *18*(6), 873–881. doi:10.1016/j.jscs.2011.10.020

Tsuchiya, H., Fujimoto, S., Chihara, O., & Shibata, T. (2002). Semiconductive behavior of passive films formed on pure Cr and Fe–Cr alloys in sulfuric acid solution. *Electrochimica Acta*, *47*(27), 4357–4366. doi:10.1016/S0013-4686(02)00508-X

Chapter 2
Tri-Sodium Citrate as Corrosion and Scale Inhibitor of Mild Steel in Synthetic Cooling Water System

Rachid Touir
https://orcid.org/0000-0003-2796-2157
CRMEF, Morocco & Ibn Tofail University, Kenitra, Morocco

Hayat Larhzil
CRMEF, Meknès, Morocco

ABSTRACT

In this study, scale inhibition performance Trisodium citrate (TSC) was studied using static scale inhibition method, and its corrosion inhibition performance was also investigated by weight loss method and electrochemical measurement. Result showed that TSC could form a layer of corrosion and scale produTSC on mild steel surface in synthetic cooling water. Its inhibition increases with concentration and reaches a maximum of 89% and 95% at 10-3 M for corrosion and scale, respectively. It is noted that TSC takes its inhibition at high temperature and act by physic-sorption process. It is found that the TSC obey to the Langmuir isotherm. Scanning electron microscopy (SEM) showed that TSC inhibits the scale and corrosion by the formation of a protective layer on the mild steel surface.

INTRODUCTION

The purpose of the present book chapter entitled: "Tri-sodium citrate as corrosion and scale inhibitor of mild steel in synthetic cooling water system" is to define the main problems of cooling water systems and methods currently used to study these problems. This book chapter will be build up using available information about scale and corrosion in the literature, but also counting case-studies carried out by our research team for this book chapter. One of the interesting points of corrosion and scale of mild steel

DOI: 10.4018/978-1-7998-2775-7.ch002

in cooling water is that they have a great economic impact. In addition, the book chapter aims to study chemically and electrochemically the influence of Tri-sodium citrate on corrosion and scale processes in synthetic cooling water solution under similar conditions to typical operating conditions.

BACKGROUND

Two of the main problems of cooling water systems are corrosion and scale phenomena. These problems have a great economic impact, since the first involves deterioration of a metallic surface, whereas the second causes loss of capacity for thermal exchange. These problems occur simultaneously and hence a study of their development and inhibition under real conditions becomes an issue of great interest (Touir et al., 2009; Choi et al., 2002; Konstantinos et al, 2007).

Many inhibitors have been used to solve the above cited problems. Particularly, fatty amine coupled with phosphonocarboxylic acid salts (Ochoa et al. 2004), gluconate sodium (Touir et al., 2008, 2010), monosaccharide derivatives (Touir et al., 2009), citrate (Farr & Saremi, 1982) and phosphonate (Touir et al., 2010; Dkhireche et al. 2012), oxadiazole derivatives (Rochdi et al. 2011) oxazoline (Labriti et al. 2012) were used as the standard corrosion inhibitors. Others organic compounds associated with molybdate were used for water treatment in cooling water systems (Touir et al., 2009; Cenoui et al., 2010). However, some Investigations were carried out to compare the effectiveness of sodium nitrite, tri-sodium citrate (TSC) and TSC–zinc acetate to inhibit the corrosion of steel rebar in simulated concrete interstitial solution contaminated with chloride using electrochemical techniques, FT-IR spectral and SEM/EDAX analysis in order to understanding the mechanism of corrosion inhibition (Maliekkal et al., 2018).

However, the use of inorganic for corrosion and scale inhibition is increasingly being curtailed by recent environmental restrictions. In many part of the world, high levels of inorganic product are also restricted by law. Thus inhibitors are diminishing, because of the concern over their toxic effects on aquatic and possibly animal life (Lake, 1988). However, the tri-sodium citrate, a natural non-toxic compound, easy handling, non-toxic, stable in air and inexpensive, is used as a catalyst for an efficient for synthesis of biological and pharmacological products (Zheng & Li, 2011) and in food industrial (Garimella Purna et al., 2006). In addition, some research was focused to use tri-sodium citrate as an effective anticoagulation for continuous renal replacement therapy (CRRT) for patients with contraindications to heparin (Ashita et al., 2001).

MAIN FOCUS OF THE CHAPTER

As a result, the current trend for inhibitors use is towards more environmental-friendly 'green' chemicals. In response to environmental guidelines, this book chapter presents the results of 'green' chemicals study that was undertaken to investigate the ability of new non-toxic inhibitors (Tri-sodium citrate) to inhibit the corrosion and scale problems found in cooling systems. The goal of the present book chapter is to study the effect of Tri-sodium citrate on the corrosion and scale processes in cooling media under conditions similar to typical operating conditions using chemical and electrochemical measurements and Scanning electron microscopy analysis. Alternatively, the influence of temperature and the mechanism of TSC on mild steel were also studied and discussed.

1. EXPERIMENTAL PROCEDURE

a. Electrochemical Cell

The electrolysis cell was Pyrex of cylinder closed by cap containing five openings. Three of them were used for the electrodes. The working electrode was a mild steel with surface area of 0.8 cm^2 and its chemical composition is shown in Table 1.

Before each experiment, the electrode was polished using emery paper until 1200 grade. After this, the electrode was cleaned ultrasonically with double-distillate water, degreased with ethanol and dried with hot air. The rotating disk electrode used was an INF-EL-EDI 101 type. The rotational speed adjustment was made with an independent electronic amplifier of the CTV 101 type.

A saturated calomel electrode (SCE) was used as a reference and all potentials were given with reference to it. The counter electrode was a platinum plate of surface area of 2 cm^2.

The synthetic cooling water solution used here was presented in Ref. (Touir et al., 2008), and shown in Table 2. This composition represents the medium of the chemical elements presented in the real solution using in Moroccan cooling water system. The organic compound tested is a tri-sodium citrate anion which its molecule formula is shown in Figure 1. The temperature and pH were adjusted to 305 K and 7.3±0.02, respectively. The choice of this temperature is justified by the fact that it is the temperature of Moroccan cooling water circulation.

Figure 1.

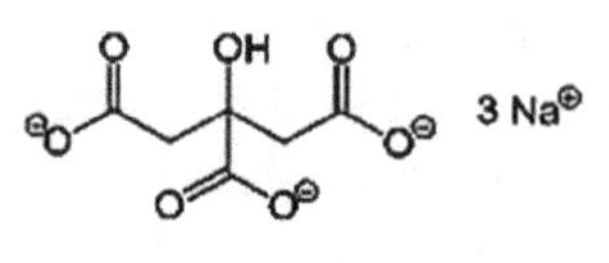

Table 1. Chemical composition of mild steel in wt.%.

C	Si	Mn	Cr	Mo	Ni	Al	Cu	Co	V	W	Fe
0.11	0.24	0.47	0.12	0.02	0.1	0.03	0.14	<0.0012	<0.003	0.06	Balance

Table 2. Composition of synthetic cooling water.

Salts	$MgCl_2$, $6H_2O$	$CaCl_2$,$2H_2O$	$Ca(NO_3)_2$,$6H_2O$	$MgSO_4$,$7H_2O$	$NaHCO_3$
Concentration (mM)	2.85	3.64	0.566	1.56	3.77
Conductivity (μS cm^{-1}) at 299 K	1678				

b. Scale Tests

The inhibitor was evaluated according to NACE Standard TM 0374-95 (NACE Standard TM0374-95, 1997) in flasks to which calcium and M-alkalinity were added. The protocol consists to mixer two equal volumes of 500 mg/L of solution calcium (like ion) with a total solution of alkalinity 1200 mg/L to prepare starting from a report/ratio 80/20 of bicarbonate/carbonate (NACE solution). The flasks were incubated for 24 h at 55 °C. After the incubation period, the flasks were removed from the heat source (an oven) and an aliquot was filtered through a 0.45 μm membrane. The filtrate was titrated for calcium (the calcium concentration is determined by a chemical proportioning with EDTA). The percent of calcium inhibition (η_{Ca} %) was determined according to NACE Standard (Amjad, 1995; Kumar et al., 2010):

$$\eta_{Ca}\% = \frac{Ca_i - Ca_b}{Ca_c - Ca_b} \times 100 \tag{1}$$

Where,

Ca_i: Concentration of Ca^{2+} in the presence of inhibitor after tests,
Ca_b: Concentration of Ca^{2+} in the absence of inhibitor after tests,
Ca_c: Concentration of Ca^{2+} before tests.

c. Weight Loss Measurements

Rectangular mild steel specimens of size (2.9 cm × 1.6 cm × 0.3 cm) were used as samples for corrosion tests. After having been polished with emery paper (400–1200 grade) successively and weighted, the samples were immersed for 24 h in 100 mL in uninhibited and inhibited solutions at 305 K, controlled thermostatically, and under air atmosphere. At the end of experiments, the specimens were cleaned according to ASTM G1-03 and reweighed up to 10^{-4} g for determining corrosion rate (ASTM G1-03, 2011).

The inhibition efficiency (η_w%) is defined as follows:

$$\eta_w\% = \frac{w_0 - w}{w_0} \times 100 \tag{2}$$

where w_0 and w are the weight loss values without and with TSC, respectively.

d. Potentiodynamic Polarization Measurements

The working electrode was immersed in the test solution during 1 h until a steady-state open circuit potential (E_{OCP}) was obtained. The disk rotation speed was 1000 rpm. The cathodic polarization curve was recorded by polarization from E_{OCP} to negative direction under potentiodynamic conditions corresponding to 0.5 mV s^{-1} (sweep rate) and under air atmosphere after 1 h of immersion time until reaching steady state. After this scan, the anodic polarization curve was recorded by polarization from the E_{ocp} to positive direction. The potentiodynamic measurements were carried out using a Potentiostat type,

VoltaLab PGZ 100, controlled by a personal computer. Since the conductivity of medium studied is low, the polarization curves i(E) were corrected from ohmic drop; that is, by $R_s \times i$. The solution resistance, R_s, was determined by the below electrochemical impedance spectroscopy. On the Nyquist diagrams, R_s is the abscissa of high-frequency intersection of the diagram with the real axis.

The evaluation of corrosion kinetics parameters was obtained using a fitting by Stern–Geary equation. To do so, the overall current density values, i, were considered as the sum of two contributions, anodic and cathodic current i_a and i_c, respectively. For the potential domain not too far from the open circuit potential, it may be considered that both processes followed the Tafel law (Stern & Geary, 1957). Thus, it can be derived from equation (3):

$$i = i_a + i_c = i_{corr}\left\{\exp\left[b_a \times \left(E - E_{corr}\right)\right] - \exp\left[b_c \times \left(E - E_{corr}\right)\right]\right\} \tag{3}$$

where i_{corr} is the corrosion current density (A cm^{-2}), b_a and b_c are the Tafel constants of anodic and cathodic reactions (V^{-1}), respectively. These constants are linked to the Tafel slopes β (V/dec) in usual logarithmic scale given by equation (4):

$$\beta = \frac{\ln 10}{b} = \frac{2.303}{b} \tag{4}$$

The corrosion parameters were then evaluated by means of nonlinear least square method by applying equation (3) using Origin software. However, for this calculation, the potential range applied was limited to ± 0.100V around E_{corr}, else a significant systematic divergence was sometimes observed for both anodic and cathodic branches.

The inhibition efficiency was evaluated from the measured i_{corr} values using the relationship:

$$\eta_{PP}\% = \frac{i^0_{corr} - i_{corr}}{i^0_{corr}} \times 100 \tag{4}$$

where i^0_{corr} and i_{corr} are the corrosion current densities values without and with TSC, respectively.

e. EIS Measurements

The electrochemical impedance spectroscopy measurements were carried out using a transfer function analyzer (VoltaLab PGZ 100), with a small amplitude ac. signal (10 mV_{rms}), over a frequency domain from 100 kHz to 10 mHz. The results were then analyzed in terms of equivalent electrical circuit using Z-view program. The polarization resistance R_p, is obtained from the diameter of the semicircle in the Nyquist representation. The inhibition efficiency of the studied inhibitor has been found from the relationship:

$$\eta_{EIS}\% = \frac{R_p - R^0_p}{R_p} \times 100 \tag{5}$$

where R^0_p and R_p are the polarization resistance values in the absence and the presence of TSC, respectively.

The values of surface coverage (θ) have been obtained from electrochemical impedance spectroscopy measurements at various concentrations of TSC. Here, θ can be given as:

$$\theta = \frac{R_p - R_p^0}{R_p} \tag{6}$$

where R_p^0 and R_p are the values of polarization resistance after immersion in solution without and with TSC, respectively.

f. Scanning Electron Microscopy (SEM)

The examination of mild steel specimens after immersion in synthetic cooling water solution, without and with 10^{-3} M of TSC after 15 days immersion time, was carried out using scanning electron microscopy (SEM; JOEL JSM-5500).

3. RESULTS AND DISCUSSION

a. Effect of Concentration of Inhibitor on Scale Inhibition

Table 3 shows scale inhibition efficiency of TSC at different temperature ranges in the synthetic cooling water system. For the blank solution, the results indicated that the solution became more scale with temperature. It is noted also that the inhibition efficiency increases with the TSC addition for all temperature ranges which becomes more important with temperature at great concentration (10^{-2} and 10^{-3} M). These results showed that higher temperature demands higher concentration to keep the scale forming ions into solution. Plummer and Busenberg showed the general behavior of $CaCO_3$ solubility as a function of temperature (Moghadasi et al., 2004). They founded that the calcium carbonate becomes less soluble as temperature increases. So, the hot water the more likely the $CaCO_3$ precipitation. Hence, water, which is no scaling at the surface, may result in scale formation in the injection well if the down hole temperature is sufficiently high. Therefore, a higher concentration of TSC was required at higher temperature to prevent super saturation of the test solution.

In addition, two common calcite scale inhibitors such as PPCA and BHPMP has been investigated and inferred that thermodynamic inhibitors decrease super saturation by lowering the ion activity product through either the metal ions chelating or by decreasing solution pH (Yean et al., 2008). The TSC can avoid scaling by forming complex with Ca^{2+} ions and avoids their precipitations in the salts form. The shift may indicate linking of carboxylate groups with calcium ion via electrovalent linkages as demonstrated by T. Kumar and al. using FTIR and XRD analysis (Kumar et al., 2010).

b. Weight Loss Study

Table 4 gives values of the inhibition efficiency of TSC obtained for mild steel in synthetic cooling water solution at 305 K after 24 h of immersion. Two tests of corrosion weight loss are realized and apparent

Table 3. Inhibition efficiency of scale in the presence of different concentration of TSC for mild steel after 24 h of immersion at different temperature ($\Omega = 0$ rpm).

T (K)	Conc. (M)	NACE Solution		Synthetic Solution (SS)	
Before test		$[Ca^{2+}]$ = 154 ppm		$[Ca^{2+}]$ = 159 ppm	
		$[Ca^{2+}]$ (ppm)	η_{Ca}%	$[Ca^{2+}]$ (ppm)	η_{Ca}%
328	Blank solution	68	-	59	-
	10^{-5}	100	37.2	105	46.0
	5×10^{-5}	108	46.5	110	51.0
	10^{-4}	111	50.0	112	53.0
	5×10^{-4}	128	69.7	135	76.0
	10^{-3}	150	95.0	153	94.0
	5×10^{-3}	148	93.0	148	89.0
338	Blank solution	52	-	42	-
	10^{-5}	111	57.8	110	58.1
	5×10^{-5}	113	59.8	112	59.8
	10^{-4}	122	68.8	114	61.5
	5×10^{-4}	136	82.4	125	70.9
	10^{-3}	142	88.2	140	83.7
	5×10^{-3}	138	84.3	135	79.5
348	Blank solution	35	-	32	-
	10^{-5}	84	41.2	84	40.9
	5×10^{-5}	88	44.5	86	42.5
	10^{-4}	100	54.6	98	51.9
	5×10^{-4}	123	73.9	116	66.4
	10^{-3}	130	79.8	132	78.7
	5×10^{-3}	128	78.1	120	69.3

precision, resulting from used material and reproducibility was esteemed to 8% such as indicated above. It is noted that the inhibition efficiency increases with inhibitor concentration and reaches a maximum at 10^{-3} M. This behavior could be attributed to the increase of the surface area covered by the adsorbed molecules of TSC with increasing its concentration. It is generally assumed that the adsorption of the inhibitor at the metal/solution interface is the first step in the mechanism of inhibition in aggressive media. Four types of adsorption may take place by organic molecules at the metallic surface: (1) electrostatic attraction between the charged metal and the charged molecules, (2) interaction of uncharged electron pairs in the molecule with the metal, (3) interaction of π-electrons with the metal and (4) combination of (1 and 3) (Kumar et al., 2010). So, the chemisorption involves the share or transfer of charge from the molecules to the metallic surface to form a coordinate type bond. Electron transfer is typical of transition metals having vacant low energy electron orbital. As for inhibitors, the electron transfer can be expected with compounds having relatively loosely bound electrons (Schweinsberg et al., 1988).

c. Potentiodynamic Polarization Curves

The potentiodynamic polarization curves obtained for mild steel in synthetic cooling water solution in the absence and the presence of different concentration of TSC at 305 K are presented in Figure 2. It is remarked, in the cathodic range, that the current are less sensitive to the TSC concentration and an increasing of current plateau with TSC addition. This can be explained by the change of molecule inhibitor orientation, which is released from the parts surface metal (Cenoui et al., 2010). It is noted, in

Table 4. Weight loss measurements for mild steel in synthetic cooling water without and with addition of TSC at various concentrations after 24 h of immersion ($T = 305$ K; $\Omega = 0$ rpm).

	Conc. (M)	Corrosion Rate, w (g cm^{-2} h^{-1})	η_w (%)
Blank solution	00	0.1998	-
TSC	10^{-5}	0.0479	76.0
	5×10^{-5}	0.0374	81.2
	10^{-4}	0.0322	83.3
	5×10^{-4}	0.0285	85.7
	10^{-3}	0.0201	89.9
	5×10^{-3}	0.0254	87.2

the anodic range, that that the anodic current densities decrease significantly with inhibitor addition. In addition, a shift of the corrosion potential (E_{corr}) towards more anodic potential are observed when the inhibitor concentration increase. This suggests that TSC acts only on the anodic sites and classifies as an anodic-type inhibitor. This can be explained on the basis that specific adsorption of the TSC anions and on the basis of the competitive adsorption between the TSC anions and Cl^- and SO_4^{2-} ions on the electrode surface, and thus retards the Cl^- destructive action (Touir et al., 2008). Apparently, the bond between the surface atoms of metal and the adsorbed inhibitor anions is strong which can inhibit oxidation (Touir et al., 2008).

Figure 2.

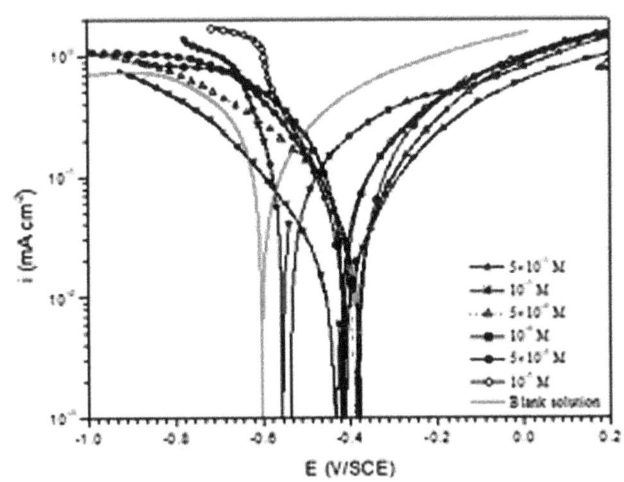

Table 5 shows that the addition of inhibitor has a beneficial effect even at low concentrations. The inhibition effect of TSC decreases gradually with increasing concentration up to the critical concentration value (10^{-3} M). This suggests the formation of soluble TSC complexes (Touir et al., 2008; Mansfeld, 1987).

Table 5. Electrochemical parameters for mild steel in synthetic cooling water without and with addition of TSC at various concentrations (T = 305 K; Ω = 1000 rpm).

	Conc. (M)	E_{corr} (mV/SCE)	i_{corr} (μA cm^{-2})	b_a (mV dec^{-1})	η_{PP} (%)
Blank solution	00	-605	102.00	305	-
TSC	10^{-5}	-395	49.14	241	51.8
	5×10^{-5}	-378	39.12	214	61.6
	10^{-4}	-412	20.71	212	79.7
	5×10^{-4}	-385	14.15	197	86.1
	10^{-3}	-420	8.78	178	91.4
	5×10^{-3}	-548	13.05	291	87.2

d. Electrochemical Impedance Spectroscopy

Figure 3 shows the Nyquist plots of mild steel in synthetic cooling water in the presence of different concentration of TSC after 1 h of immersion time at 305 K and at E_{OCP}. For the free solution, a clear separation of two loops was observed. This behavior may be attributed to the formation of corrosion and/or scale product layer. Similar finding is recently reported by other researchers (Roti & Thomas, 1984). Taking into account the following elementary reactions of the corrosion process:

$$Fe \leftrightarrow Fe^{2+} + 2e^{-} \tag{7}$$

$$O_2 + 2H_2O + 4e^{-} \leftrightarrow 4HO^{-} \tag{8}$$

Figure 3.

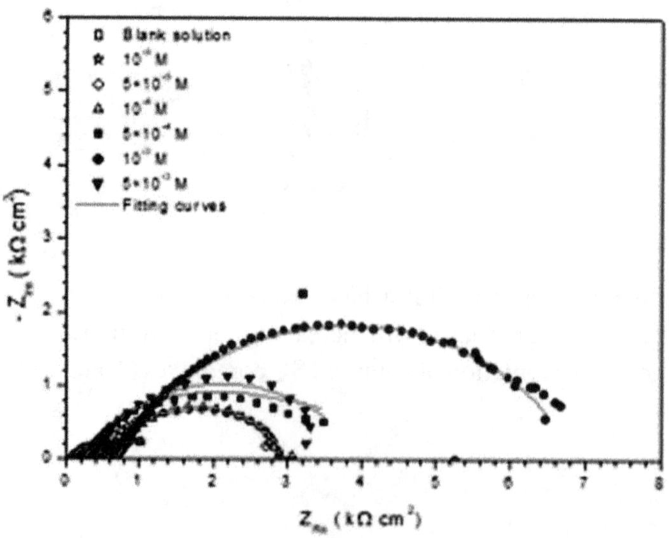

The dissolution reaction of mild steel (5) will generate corrosion products, whereas the oxygen reduction reaction will produce HO^- ions, giving rise to a local modification of pH at the interface. Both reactions promote the formation of Fe(II), and Fe(III) compounds on the surface, including iron hydroxides ($Fe(OH)_2$, $Fe(OH)_3$), oxides (Fe_3O_4, γ-Fe_2O_3), sulfates ($FeSO_4$) and chlorides ($FeCl_3$) of iron. The oxygen reduction reaction additionally gives rise to scaling due to the presence of Ca(II) and HCO_3^- ions in the medium, according to the reaction as follows (Chen et al., 200; Marín-Cruz et al., 2006; Cachet & Wiart, 1994):

$$Ca^{2+} + HCO_3^- + HO^- \leftrightarrow CaCO_3(s) + H_2O \tag{9}$$

Given the reactions (7) and (8) which occur simultaneously in different zones on the same surface, the scale process equation (9) will be coupled to the corrosion process.

Figure 4.

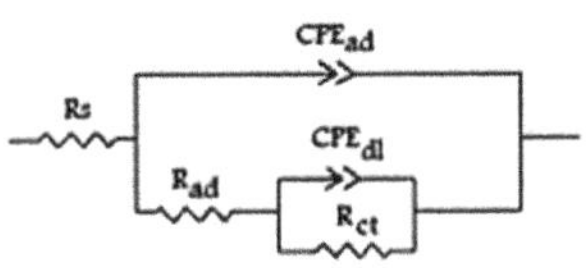

In the presence of TSC, the obtained diagrams are significantly changed with its addition, suggesting a change in the corrosion process of the mild steel/electrolyte interface. These diagrams were composed of two separated loops. The first was attributed to the layer of the formed products and the second can be related to the charge transfer resistance. A simple electrical equivalent circuit (Figure 8) has been proposed to model the experimental data. Where, R_s represents the resistance electrolyte; R_{ct} is the charge transfer resistance and R_{ad} is the resistance of the formed product layer. The Q_{ad} and Q_{dl} are the constant phase elements used to replace the formed product layer and the double layer capacitances, respectively. Constant phase elements have widely been used to account for deviations brought about by surface roughness (Felhosi et al., 2002; Georges et al., 2008). The impedance of CPE is given by equation:

$$Z_{CPE} = (\omega) = Q^{-1}(j\omega)^{-n} \tag{10}$$

where Q is the constant phase element (*CPE*), ω is the sine wave modulation angular frequency (in rad s^{-1}), $j^2 = -1$ is the imaginary number and n is the *CPE* exponent. According to n value, the CPE can be resistance [$Z_{CPE} = R$, $n = 0$], pure capacity [$Z_{CPE} = C$, $n = 1$], inductance [$Z_{CPE} = L$, $n = -1$] or Warburg impedance ($n = 0.5$) (Tsuchiya et al., 2002). The use of these parameters, similar to the constant phase element (*CPE*), allowed the depressed feature of the Nyquist plot to be reproduced readily.

However, the effective calculated double layer capacitance (C) derived from the *CPE* parameters according to the following equation (Brug et al., 1984):

$$C = Q^{\frac{1}{n}} \times R^{\frac{(1-n)}{n}} \tag{11}$$

The impedance parameters derived from the Nyquist plots as well as the polarization resistance R_p are given in Table 6. R_p is determined from the extrapolation of the impedance spectra towards the low frequency limit (the sum of R_{ad} and R_{ct}). It is seen that that the inhibition efficiency increases with TSC concentration reaching a maximum 96.8% for 10^{-3} M of TSC. This result is in good agreement with this obtained from potentiodynamic polarization curves. Additionally, the charge transfer resistance R_{ct} and the electrolytic resistance R_s values increase while the double layer C_{dl} and the products layer C_{ad} capacities decrease with increasing of inhibitor concentration until an optimum concentration (10^{-3} M). These can be explained by the increasing portion of the blocking electrode surface due to the TSC molecules adsorption (Touir, 2016).

Table 6. Electrochemical parameters for mild steel in synthetic cooling water solution at different concentrations of TSC (T = 305 K; Ω = 1000 rpm).

C (M)	R_s (Ω cm²)	C_{ad} (μF cm⁻²)	R_{ad} (Ω cm²)	n_{ad}	C_{dl} (μF cm⁻²)	R_{ct} (Ω cm²)	n_{ct}	R_p (Ω cm²)	η_{EIS} (%)	θ
00	140.0	1.2	13.0	0.56	1 741	125.0	0.62	210	-	-
10^{-5}	76.81	1.06	305.0	0.79	133.54	1167.3	0.86	1398.2	84.9	0.8498
5×10^{-5}	387.8	0.34	367.1	0.85	85.44	2200.5	0.88	2184.3	90.4	0.9038
10^{-4}	136.5	0.16	572.6	0.86	81.81	2455.5	0.89	2891.1	92.7	0.9274
5×10^{-4}	692.0	0.09	851.7	0.89	74.33	4403.5	0.87	4567.7	95.4	0.9540
10^{-3}	270.4	0.07	880.7	0.91	40.46	5988.1	0.91	6599.3	96.8	0.9682
5×10^{-3}	231.4	0.74	412.2	0.83	74.83	3640.7	0.88	3827.8	94.5	0.9451

However, it is noted that the n_{ad} and n_{dl} values are well below 1 and reflect the very flat shape of the impedance diagrams. So, the increase n_{ad} and n_{dl} with TSC concentration reveals the reduction of heterogeneity of mild steel surface which is the result of the adsorption of TSC molecules on the metallic surface (Rochdi et al., 2015).

On the other hand, it is found that the solution resistance R_s increased generally with TSC addition. This increase may be attributed to the slight decreases in the ionic concentration of the solution taking place with TSC molecules presence.

e. Effect of Solution Temperature and Thermodynamic Parameters

To clarify the adsorption nature of TSC, the influence of temperature at the range from 305 K to 335 K on the corrosion rate of mild steel in synthetic cooling water without and with inhibitor was investigated using potentiodynamic polarization curves. The obtained results are given in Figures 5 and 6 and their corresponding data are summarized in Table 7. It is seen that the corrosion current density values increase with temperature for both in uninhibited and inhibited solutions. In addition, the inhibition efficiency decreases slightly from 87.4% to 84.5% when the temperature increases from 305 K to 335 K. These results confirm that TSC acts as an efficient inhibitor at the high temperature.

Figure 5.

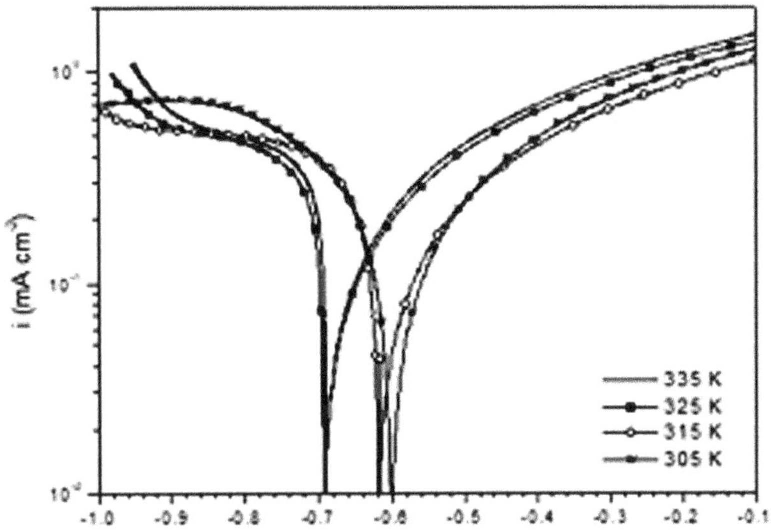

Figure 6.

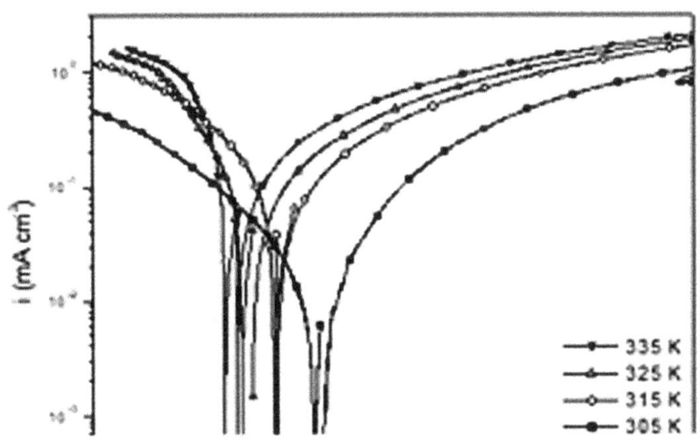

However, the effect of temperature on the corrosion parameter can be deduced by comparing the activation energy with and without TSC. The Arrhenius plots were used to determine the activation energy (E_a) values according to the equation:

$$i_{corr} = A\exp\left(\frac{-E_a}{RT}\right) \tag{12}$$

where i_{corr} is corrosion current density, A is the pr-exponential constant, E_a is the activation energy of the metal dissolution reaction, R is the gas constant and T is the absolute temperature. Figure 6 presents the Arrhenius plots obtained for mild steel in synthetic cooling water without and with TSC addition.

Figure 7.

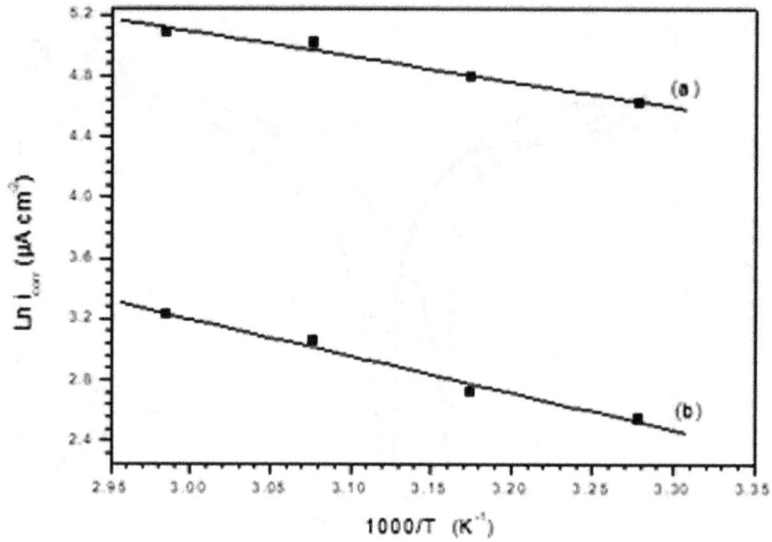

However, the E_a values were calculated from Figure 7 as 14.52 kJ mol^{-1} and 20.03 kJ mol^{-1} in the absence and presence of TSC, respectively. The increase in activation energy in the presence of the inhibitor may suggest the physical adsorption (Xu et al., 2013; Anejjar et al., 2013). However, as it is also discussed in the literature (Solmaz, 2014) the adsorption process could not be classified as purely physical, which is the first stage of adsorption, or chemical. Moreover, the criteria of adsorption type obtained from the change of E_a cannot be taken as a decisive due to competitive adsorption with water molecules, whose removal from the surface requires also some activation energy (Vračar & Dražić 2002). Therefore, it is concluded that, the adsorption of TSC molecules on mild steel surface from synthetic cooling water takes place through both physical and chemical processes simultaneously (Tang et al., 2013).

Other kinetic data are accessible using the alternative formulation of the Arrhenius equation is follow (13) (El Hezzat et al., 20153):

$$Ln\frac{i_{corr}}{T} = \ln\left(\frac{R}{Nh}\right) + \frac{\Delta S_a^*}{R} - \frac{\Delta H_a^*}{RT} \tag{13}$$

where h is Plank's constant, N is Avogadro's number, ΔS_a* and ΔH_a* are the entropy and enthalpy of activation, respectively. Figure 8 represents the plot of ln (i_{corr}/T) versus the reciprocal of temperature in the absence and presence of TSC. The straight lines are obtained with a slope of (- ΔH_a*/R) and an intercept of (ln R/Nh + ΔS_a*/R). The values of ΔH_a* and ΔS_a* are calculated and are listed in Table 7. It is shown that ΔH_a* increase with TSC addition. In the other, literature postulates that the positive sign of the enthalpy (ΔH_a*) is an endothermic nature of the mild steel dissolution process (Adardour et al., 2013). The entropy of activation ΔS_a* in the absence and presence of TSC is negative, which decreases negatively from -170.44 J mol^{-1} K^{-1} to -167.69 J mol^{-1} K^{-1} for mild steel in synthetic cooling water in the absence and presence of TSC, respectively. This phenomenon suggested that a decrease in randomness occurred on going from reactants to the activated complex. The great negative values of entropies in-

Figure 8.

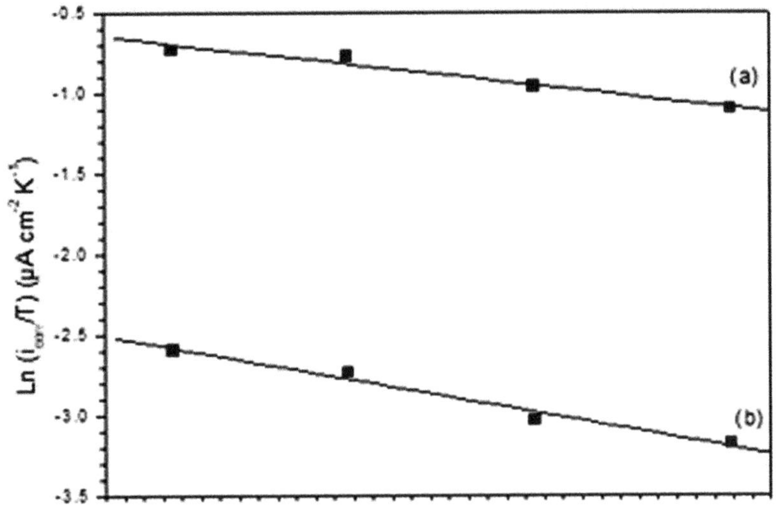

dicate that the activated complex in the rate determining step is an association rather than dissociation step meaning that a decrease in disordering takes place on going from reactants to the activated complex (Fergachi et al., 2018).

f. Adsorption Isotherm

Generally, the adsorption of the organic compound can be described by two main types of interactions: physical adsorption and chemisorption. They are influenced by the metal charge, the chemical structure of inhibitor and the electrolyte type. Several adsorption isotherms, such as Tempkin, Langmuir, Frumkin, Freundlich and Flory–Huggins isotherms, can be used to assess the adsorption behavior of such an inhibitor (Bastidas et al., 2003). So, the coverage surface (θ), which determined from AC impedance, polarization or weight loss measurements by the ratio η/100, can be used. In this study, to make the fitting and select the suitable isotherm, the coverage surface was estimated from AC impedance measurements. So, the following adsorption isotherms and their corresponding parameters are listed in Table 8.

Table 7. Electrochemical and thermodynamic parameters for mild steel in synthetic cooling water without and with addition of 10^{-3} M of TSC at various temperatures.

	T (K)	E_{corr} (mV/SCE)	i_{corr} (μA cm^{-2})	b_a (mV dec^{-1})	η_{pp} (%)	E_a (kJ mol^{-1})	ΔH^*_a (kJ mol^{-1})	ΔS^*_a (J mol^{-1} K^{-1})
Blank solution	305	-605	102	320	-	14.52	11.06	-170.44
	315	-621	121	348	-			
	325	-695	151	339	-			
	335	-698	162	355	-			
10^{-3} M of TSC	305	-420	8.78	178	91.3	20.03	17.38	-167.69
	315	-598	11.23	292	90.7			
	325	-547	17.24	324	85.7			
	335	-573	21.12	356	86.9			

Figure 9.

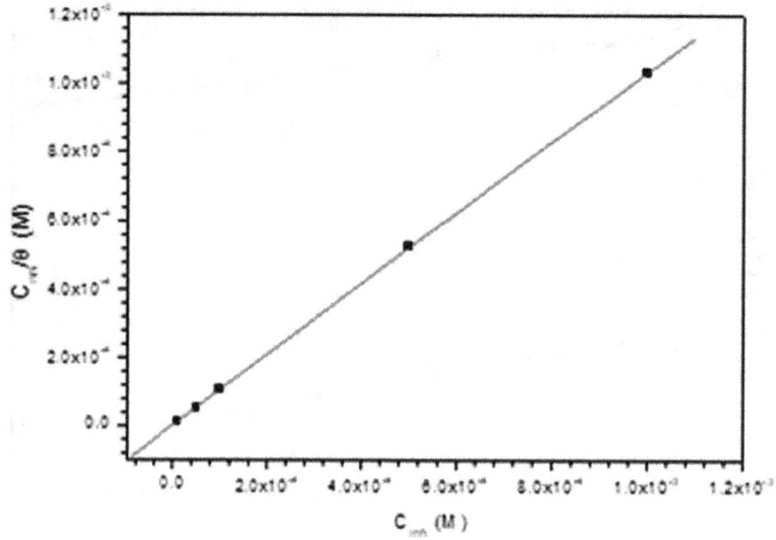

The constant K_{ads} may be determined from the intercept in Figure 9 according to the following assumptions, such as indicated by Langmuir isotherm (Touir et al., 2008):

- That the TSC forms a mono-layer on the mild steel surface at the maximum inhibitor concentration;
- That there is no interaction between the adsorbed species on the electrode surface;
- That TSC is adsorbed on the mild steel surface according to the equation:

Fe⌝ + TSC ® Fe--- MQ=S (14)

Therefore, the obtained value of K_{ads} is about 2.46×10^{7} L mol^{-1}. This high value reflects the high adsorption ability of TSC on mild steel surface. A plot of C_{inh}/θ against C_{inh} shows a straight line (R^2 = 0.9999) indicating that adsorption follows the Langmuir adsorption isotherm as shown in Figure 9. The equilibrium constant K_{ads} and the free energy of adsorption (ΔG^*_{ads}) are related to the following equation (Kliškic et al., 1997):

$$K_{ads} = \frac{1}{55.55}\exp\left(-\frac{\Delta G^*_{ads}}{RT}\right) \tag{15}$$

where 55.55 value represents the water concentration in solution by mol L^{-1}, R is the universal gas constant and T is the absolute temperature.

The negative value of ΔG^*_{ads} calculated from Equation (15), are consistent with the spontaneity of the adsorption process and the stability of the adsorbed layer on the mild steel surface. Generally, values of ΔG^*_{ads} up to -20 kJ mol^{-1} are consistent with physisorption, while those around -40 kJ mol^{-1} or

Table 8. Curve fitting results of TSC adsorption for different adsorption isotherms.

Isotherms	Linear Forms	Curves	Parameters	Values
Langmuir	$\frac{C_{inh}}{\theta}=\frac{1}{K_{ads}}+C_{inh}$	$\frac{C_{inh}}{\theta}=f(C_{inh})$	R^2	0.99998
			K_{ads} (L mol^{-1})	2.46×10^7
			Slope	1.03107
Freundlich	$\ln(\theta)=\ln(K_{ads})+\frac{1}{n}\ln(C_{inh})$	$\ln(\theta)=f(\ln(C_{inh}))$	R^2	0.98135
			K_{ads} (L mol^{-1})	1.1787
			Slope (1/n)	0.0275
			n	36.36
Temkin	$\theta=\frac{1}{f}\ln(K_{ads})+\frac{1}{f}\ln(C_{inh})$	$\theta=f(\ln(C_{inh}))$	R^2	0.98472
			K_{ads} (L mol^{-1})	7.96×10^{19}
			Pente (1/f)	0.02502
			f	39.96
Frumkin	$\ln\left(C_{inh}\times\frac{1-\theta}{\theta}\right)=-\ln(K_{ads})-2f\theta$	$\ln\left(C_{inh}\times\frac{1-\theta}{\theta}\right)=f(\theta)$	R^2	0.98553
			K_{ads} (L mol^{-1})	1.06×10^{15}
			Slope (-2f)	24.95287
			f	-12.47643
Flory-huggins	$\ln\left(\frac{\theta}{C_{inh}}\right)=\ln(K_{ads})+a\ln(1-\theta)$	$\ln\left(\frac{\theta}{C_{inh}}\right)=\ln(1-\theta)$	R^2	0.99561
			K_{ads} (L mol^{-1})	1.91×10^7
			Slope (a)	2.92257

where K_{ads} is the equilibrium constant of the adsorption process, C_{inh} is the inhibitor concentration, f is the factor of energetic inhomogeneity and the parameter" a" is the water molecules number replaced by inhibitor molecules on metallic surface.

higher are associated with chemisorption as a result of the sharing or transfer of electrons from organic molecules to the metal surface to form a coordinate bond (Refay et al., 2004). In the present work the value of ΔG^*_{ads} is equal to ΔG^*_{ads} = - 41.75 kJ mol^{-1}. The large value of ΔG^*_{ads} and its negative sign is usually characteristic of strong interaction and a highly efficient adsorption (Abdallah et al., 2002). Therefore, it is concluded that, the adsorption of TSC molecules on mild steel surface is a mixed type of chemical and physical adsorptions with predominantly the first one (Li et al., 2011; Deng et al., 2011; Ahamad et al., 2010). So, TSC molecules are adsorbed on the metal surface due to the free electron pairs of the oxygen atoms. It exists in the form of ionic species. So, the ionic oxygen atoms may adsorb through electrostatic interaction between the positively charged metal (Touir et al., 2008) and the negatively charged TSC molecules.

g. Surface Analysis

Figure 10a-b shows the scanning electron microscopy (SEM) micrographs of mild steel surface that have been exposed to the synthetic cooling water solution for 15 days in the absence of TSC. It is observed the subtract surface was greatly damaged during the immersion tests, as the corrosion products show a multi-layer morphology. The outer layer of the corrosion products appears uneven and rough, which is compactly adsorbed to the inner layer. A homogeneous attack was observed on the entire surface. In

Figure 10.

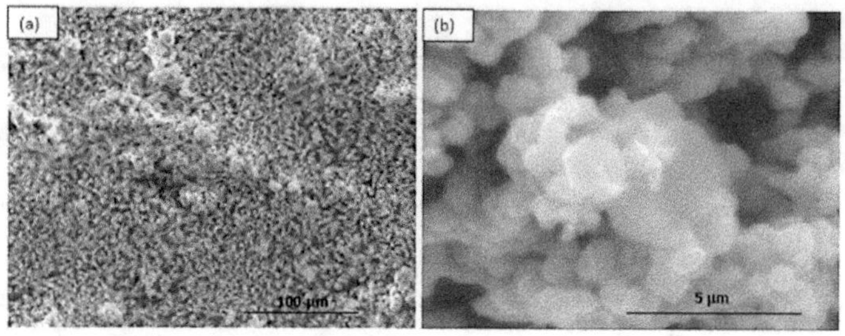

addition, it is remarked that these images show heterogeneous layer of scale products (Figure 10b). In addition, Figure 10b shows sharp and pointed crystal structures. These sharp and pointed crystal tips could have increased the attachment capability of the deposit to adhere on the heat transfer surface more strongly than blunt crystals for the physical water treatment cases (Tijing et al., 2010). However, in our previous works, the EDAX analysis further identified characteristic corrosion product elements (Fe, O, S and Cl) as well as the presence of calcium on this layer (Belakhmima et al., 2015).

In contrast, the metal surface immersed in simulated solution in the presence of TSC (Figure 11) shows a large area free of corrosion and scale products and reveals the formation of a protective layer on the metal surface. This result shows that TSC acts as scale and corrosion inhibitor and it is in accordance with the inhibition efficiency values obtained by electrochemical measurements.

Figure 11.

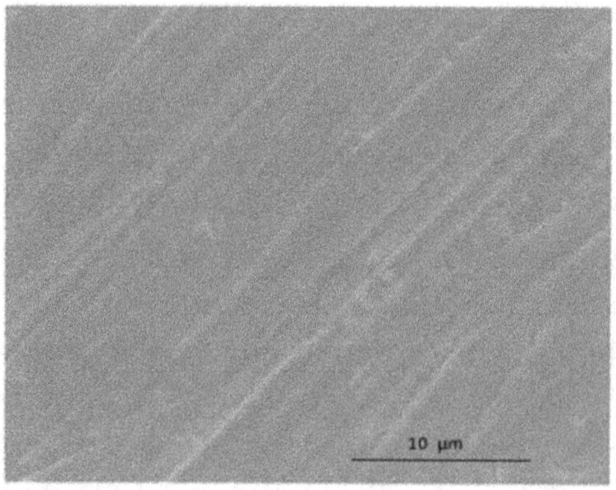

h. Proposed Mechanism

Inhibitors usually act on the metallic surface by a formatting an adsorption film or by involving a layer of corrosion products. So, their adsorptions depend on the metal surface charge, on the charge (or on the dipole moment) of inhibitor molecule and on the adsorption of the other ionic species present in the electrolyte (Nady et al., 2017). In our previous work, it is found by electrochemical impedance spectroscopy that the mild steel surface electrode was positively charged such as mentioned in Figure 12 (Touir, 2016). This figure represents a minimum at – 650 mV/SCE which is more negative than the corrosion

Figure 12.

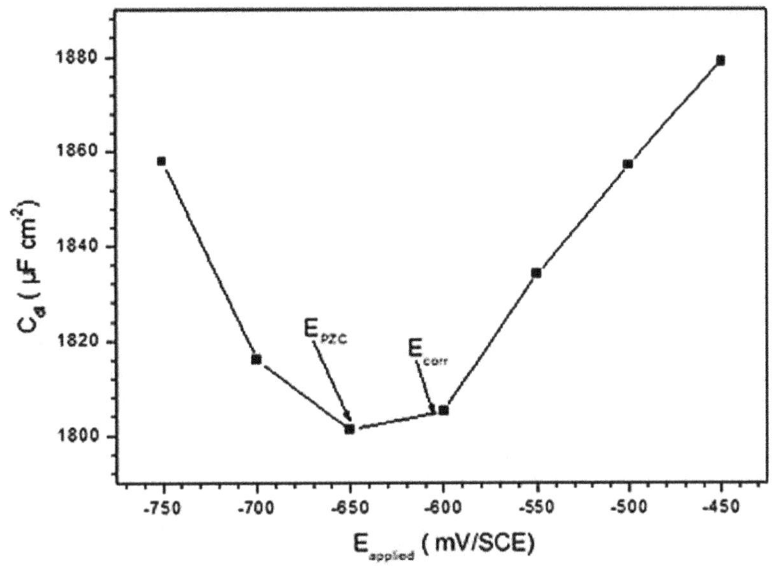

potential (- 605 mV/SCE) indicating that the potential φ is positive (($\varphi = E_{corr} - E_{q=0} = -605 - (-650) =$ 45 mV/SCE).

On the other hand, the inhibition efficiency depends on several factors, such as the adsorption site number and their charge density, molecular size, heat of hydrogenation, mode interaction with the metal surface, and the formation of metal complexes (Wang et al., 2001).

However, the most organic inhibitors contain at least one polar group with a nitrogen, sulfur or oxygen atom, each of which enhances the adsorption center. This compound interferes in the reaction of dissolution by metal surface adsorption in two different ways. Firstly, the inhibitor competes with the

Figure 13.

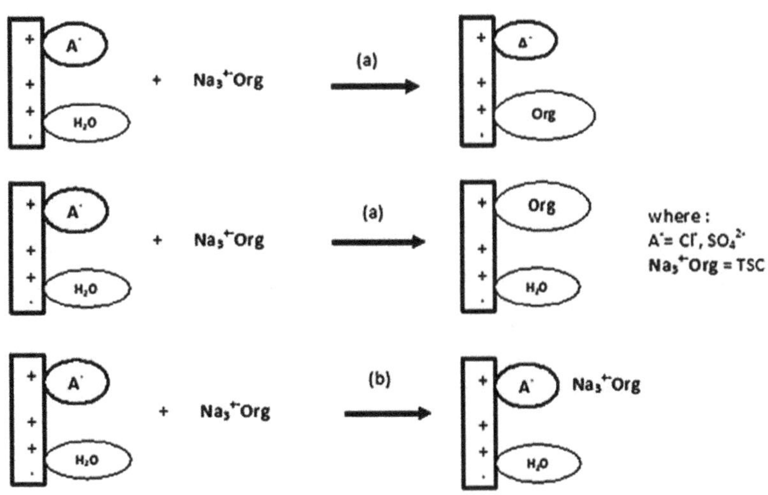

water molecules and the Cl^- and SO_4^{2-} ions adsorbed on the anodic sites covered by water as shown in Figure 12. In addition, the inhibitors are electro-statically adsorbed on the surface covered by the above cited anions. They thus play a dynamic role at the interface and interfere with the dissolution reaction by participation, in a certain number of adsorption stages - desorption, rather than only by blocking the locations (Hackerman et al., 1966; Murakawa & Hackerman, 1964). The two adsorption modes are schematized in Figure 13.

CONCLUSION

The results showed that a layer of corrosion and scale products forms naturally on the mild steel surface and that the corrosion and scale processes occurring in these systems evolve over time. In addition, it was possible to establish that the corrosion and scale processes are closely related and occur simultaneously at different sites on the metallic surface. Moreover, it is noted that the corrosion and scale rate increase with temperature.

However, it was established that the corrosion inhibitor used for this study, generate a layer with resistive properties that provides good protection against corrosion and scale. In addition, it is found that TSC acts as an anodic–type inhibitor for mild steel in synthetic cooling water solution and its inhibition efficiency values increase with its concentration and reaches a maximum at 10^{-3} M. This effectiveness is confirmed by electrochemical impedance spectra. It is found also that TSC takes its performance at high temperature. On the other hand, the adsorption of TSC on the mild steel surface in the corrosive medium obeys a Langmuir adsorption isotherm, indicating that the main inhibition process occurs via adsorption. The negative value of ΔG^*_{ads} obtained from this study indicates that it was strongly adsorbed on the electrode surface (chemisorptions process). Indeed, the SEM observations showed that the tested inhibitor keeps also its efficiency after 15 days of immersion and it's a very good inhibitor for scale process. The AC impedance study and polarization curves confirm the inhibition character of TSC obtained with weight loss.

The good inhibition efficiencies of TSC in synthetic solution makes it a suitable alternative as environmentally friendly corrosion and scale inhibitor in deicing solutions in which Tri-sodium citarte is a main component. The advantages of this inhibitor consist in biodegradability and zero impact on the environment.

REFERENCES

Abdallah, M. (2002). Rhodanine azosulpha drugs as corrosion inhibitors for corrosion of 304 stainless steel in hydrochloric acid solution. *Corrosion Science*, *44*(4), 717–728. doi:10.1016/S0010-938X(01)00100-7

Adardour, K., Touir, R., Elbakri, M., Ramli, Y., Ebn Touhami, M., El Kafsaoui, H., Kalonji Mubengayi, C., & Essassi, E. M. (2013). Thermodynamic properties and comparative studies of quinoxaline derivatives as a corrosion inhibitor for mild steel in 1 M H_2SO_4. *Research on Chemical Intermediates*, *39*, 4175–4188. doi:10.100711164-012-0934-x

Ahamad, I., Prasad, R., & Quraishi, M. A. (2010). Inhibition of mild steel corrosion in acid solution by Pheniramine drug: Experimental and theoretical study. *Corrosion Science*, *52*(9), 3033–3041. doi:10.1016/j.corsci.2010.05.022

Amjad, Z. (1995). *Mineral Scale Formation and inhibition*. Plenum Press. doi:10.1007/978-1-4899-1400-2

Anejjar, A., Zarrouk, A., Salghi, R., Zarrok, H., Ben Hmamou, D., Hammouti, B., Elmahi, B., & Al-Deyab, S. S. (2013). Studies on the Inhibitive Effect of the Ammonium Iron (II) Sulphate on the Corrosion of Carbon Steel in HCl Solution. *Journal of Materials and Environmental Science*, *4*(5), 583–592.

Ashita, J., Tolwani, M. D., & Ruth, C. (2001). Simplified citrate anticoagulation for continuous renal replacement therapy. *Kidney International*, *60*(1), 370–374. doi:10.1046/j.1523-1755.2001.00809.x PMID:11422774

ASTM G1-03. (2011). *Standard practice for preparing, cleaning, and evaluating corrosion test specimens*. http://www.astm.org/Standards/G1

Bastidas, J. M., Pinilla, P., Cano, E., Polo, J. L., & Miguel, S. (2003). Copper corrosion inhibition by triphenylmethane derivatives in sulphuric acid media. *Corrosion Science*, *45*(2), 427–449. doi:10.1016/S0010-938X(02)00123-3

Belakhmima, R. A., Dkhireche, N., Touir, R., & Ebn Touhami, M. (2015). Development of a multicomponent SG with CTAB as corrosion, scale, and microorganism inhibitor for cooling water systems. *Materials Chemistry and Physics*, *152*, 85–94. doi:10.1016/j.matchemphys.2014.12.018

Brug, G. J., Van Den Eeden, A. L. G., Sluyters-Rehbach, M., & Sluyters, J. H. (1984). The analysis of electrode impedances complicated by the presence of a constant phase element. *Journal of Electroanalytical Chemistry and Interfacial Electrochemistry*, *176*(1-2), 275–295. doi:10.1016/S0022-0728(84)80324-1

Cachet, C., & Wiart, R. (1994). Zinc Electrowinning in Acidic Sulfate Electrolytes: Impedance Analysis and Modeling of the Influence on Nickel Impurities. *Journal of the Electrochemical Society*, *141*(1), 131–140. doi:10.1149/1.2054672

Cenoui, M., Dkhireche, N., Kassou, O., Ebn Touhami, M., Touir, R., Dermaj, A., & Hajjaji, N. (2010). Synergistic influence of molybdate ions with TDMTAA on corrosion inhibition of ordinary steel in cooling water system. *Journal of Materials Environmental Science. Journal Mater. Environ. Sci.*, *1*(2), 84–95.

Chen, Y., Hong, T., Gopal, M., & Jepson, W. P. (2000). EIS studies of a corrosion inhibitor behavior under multiphase flow conditions. *Corrosion Science*, *42*(6), 979–990. doi:10.1016/S0010-938X(99)00127-4

Choi, D. J., You, S.J., & Kim, J. G. (2002). Development of an environmentally safe corrosion, scale, and microorganism inhibitor for open recirculating cooling systems. *Journal of Materials Science and Engineering: A*, *335*(1-2), 228–236.

Deng, S., Li, X., & Fu, H. (2011). Two pyrazine derivatives as inhibitors of the cold rolled steel corrosion in hydrochloric acid solution. *Corrosion Science*, *53*(2), 822–828. doi:10.1016/j.corsci.2010.11.019

Dkhireche, N., Abdelhadi, R., Ebn Touhami, M., Oudda, H., Touir, R., Elbakri, M., Saira, M., Hammouti, B., Senhaji, O., & Taouil, R. (2012). Elucidation of Dimethyldodecylphosphonate and CTAB Synergism on Corrosion and Scale Inhibition of Mild Steel in Simulated Cooling Water System. *International Journal of Electrochemical Science*, *7*, 5314–5330.

El Hezzat, M., Assouag, M., Zarrok, H., Benzekri, Z., El Assyry, A., Boukhris, S., Souizi, A., Galai, M., Touir, R., Ebn Touhami, M., Oudda, H., & Zarrouk, A. (2015). Electrochemical and theoretical evaluation of ethyl 6-amino-5-cyano-2-methyl-4-phenyl-4H-pyran-3-carboxylate as corrosion inhibitor for low carbon steel in 1.0 M HCl. *Der Pharma Chemica*, *7*(10), 77–88.

Farr, J. P. G., & Saremi, M. (1982). Molybdate in aqueous corrosion inhibition part II: Effects of molybdate, citrate and phosphate on the anodic behaviour of steel. *Surface Technology*, *17*(1), 19–27. doi:10.1016/0376-4583(82)90057-7

Felhosi, I., Telegdi, J., Palinkas, G., & Kalman, E. (2002). Kinetics of self-assembled layer formation on iron. *Electrochimica Acta*, *47*(13-14), 2335–2340. doi:10.1016/S0013-4686(02)00084-1

Fergachi, O., Benhiba, F., Rbaa, M., Touir, R., Ouakki, M., Galai, M., Lakhrissi, B., Oudda, H., & Ebn Touhami, M. (2018). Experimental and Theoretical Study of Corrosion Inhibition of Mild Steel in 1.0 M HCl Medium by 2(-4(hloro phenyl-1H- benzo[d]imidazol)-1-yl)phenyl)methanone. *Materials Research*, *21*(6), e20171038. doi:10.1590/1980-5373-mr-2017-1038

Garimella Purna, S. K., Pollard, A., & Metzger, L. E. (2006). Effect of Formulation and Manufacturing Parameters on Process Cheese Food Functionality - I. Trisodium Citrate. *Journal of Dairy Science*, *89*(7), 2386–2396. doi:10.3168/jds.S0022-0302(06)72311-6 PMID:16772554

Georges, C., Rocca, E., & Steinmetz, P. (2008). Synergistic effect of tolutriazol and sodium carboxylates on zinc corrosion in atmospheric conditions. *Electrochimica Acta*, *53*(14), 4839–4845. doi:10.1016/j.electacta.2008.01.073

Hackerman, N., Snavely, E., & Payne, J. S. (1966). Effects of Anions on Corrosion Inhibition by Organic Compounds. *Journal of the Electrochemical Society*, *113*(7), 677–681. doi:10.1149/1.2424089

Kliškic, M., Radoševic, J., & Gudic, S. (1997). Pyridine and its derivatives as inhibitors of aluminium corrosion in chloride solution. *Journal of Applied Electrochemistry*, *27*(8), 947–952. doi:10.1023/A:1018405803182

Konstantinos & Mavredaki. (2007). Inhibition and growth control of colloidal silica: Designed chemical approaches. *Journal of Science Direct.*, *213*, 38–46.

Kumar, T., Vishwanatham, S., & Kundu, S. S. (2010). A laboratory study on pteroyl-l-glutamic acid as a scale prevention inhibitor of calcium carbonate in aqueous solution of synthetic produced water. *Journal of Petroleum Science Engineering*, *71*(1-2), 1–7. doi:10.1016/j.petrol.2009.11.014

Labriti, B., Dkhireche, N., Touir, R., Ebn Touhami, M., Sfaira, M., El Hallaoui, A., Hammouti, B., & Alami, A. (2012). Synergism in Mild Steel Corrosion and Scale Inhibition by a New Oxazoline in Synthetic Cooling Water. *Arabian Journal for Science and Engineering*, *37*(5), 1293–1303. doi:10.100713369-012-0257-7

Lake, D. L. (1988). Approching environmental acceptability in cooling water corrosion inhibitor. *Corrosion Prevention & Control*, 113.

Li, X., Deng, S., & Fu, H. (2011). Triazolyl blue tetrazolium bromide as a novel corrosion inhibitor for steel in HCl and H_2SO_4 solutions. *Corrosion Science*, *53*(1), 302–309. doi:10.1016/j.corsci.2010.09.036

Maliekkal, R., Kakkassery, J. T., & Palayoor, V. R. (2018). Efficacies of sodium nitrite and sodium citrate–zinc acetate mixture to inhibit steel rebar corrosion in simulated concrete interstitial solution contaminated with NaCl. *International Journal of Industrial Chemistry.*, *9*(2), 105–114. doi:10.100740090-018-0142-7

Mansfeld, F. (1987). *Corrosion Mechanisms*. Marcel Dekker.

Marín-Cruz, J., Cabrera-Sierra, R., Pech-Canul, M. A., & González, I. (2006). EIS study on corrosion and scale processes and their inhibition in cooling system media. *Electrochimica Acta*, *51*(8-9), 1847–1854. doi:10.1016/j.electacta.2005.02.104

Moghadasi, J., Jamiala hmadi, M., Müller-Steinhagen, H., & Sharif, A. (2004). Formation damage due to scale formation in porous media resulting from water injection. Society of Petroleum Engineers.

Murakawa, T., & Hackerman, N. (1964). The double layer capacity at the interface between iron and acid solutions with and without organic materials. *Corrosion Science*, *4*(1-4), 387–396. doi:10.1016/0010-938X(64)90040-X

NACE Standard TM0374-95. (1997). Laboratory screening tests to determine the ability of scale inhibitors to prevent the precipitation of calcium sulfate and calcium carbonate from solution. *NACE.*

Nady, H., El-Rabiei, M. M., & Samy, M. (2017). Corrosion behavior and electrochemical properties of carbon steel, commercial pure titanium, copper and copper–aluminum–nickel alloy in 3.5% sodium chloride containing sulfide ions. *Egyptian Journal of Petroleum*, *26*(1), 79–94. doi:10.1016/j.ejpe.2016.02.008

Ochoa, N., Baril, G., Moran, F., & Pébère, N. (2002). Study of the properties of a multi-component inhibitor used for water treatment in cooling circuits. *Journal of Applied Electrochemistry*, *32*(5), 497–504. doi:10.1023/A:1016500722497

Refay, S. A., Taha, F., & Abd El-Malak, A. M. (2004). Inhibition of stainless steel pitting corrosion in acidic medium by 2-mercaptobenzoxazole. *Applied Surface Science*, *236*(1–4), 175–185. doi:10.1016/j.apsusc.2004.04.016

Rochdi, A., Dkhireche, N., Kassou, O., Elbakri, M., Ebn Touhami, M., Touir, R., Forsal, I., & Mernari, B. (2011). Protection d'un acier doux et du laiton dans les eaux des circuits de refroidissement par une formulation contenant un oxadiazole et un biocide non oxydant. *Proceeding of 23ème Forum sur les Impédances Electrochimiques*.

Rochdi, A., Touir, R., El Bakri, M., Ebn Touhami, M., Bakkali, S., & Mernari, B. (2015). Protection of low carbon steel by oxadiazole derivatives and biocide against corrosion in simulated cooling water system. *Journal of Environmental Chemical Engineering*, *3*(1), 233–242. doi:10.1016/j.jece.2014.11.020

Schweinsberg, D. P., George, G. A., Nanayakkara, A. K., & Steinert, D. A. (1988). The protective action of epoxy resins and curing agents—Inhibitive effects on the aqueous acid corrosion of iron and steel. *Corrosion Science*, *28*(1), 33–42. doi:10.1016/0010-938X(88)90004-2

Solmaz, R. (2014). Investigation of adsorption and corrosion inhibition of mild steel in hydrochloric acid solution by 5-(4-Dimethylaminobenzylidene)rhodanine. *Corrosion Science*, *79*, 169–176. doi:10.1016/j.corsci.2013.11.001

Stern, M., & Geary, A. L. (1957). Electrochemical Polarization: A Theoretical Analysis of the Shape of Polarization Curves. *Journal of the Electrochemical Society*, *104*(1), 56–63. doi:10.1149/1.2428496

Tang, Y., Zhang, F., Huc, S., Cao, Z., Wu, Z., & Jing, W. (2013). Novel benzimidazole derivatives as corrosion inhibitors of mild steel in the acidic media. Part I: Gravimetric, electrochemical, SEM and XPS studies. *Corrosion Science*, *74*, 271–282. doi:10.1016/j.corsci.2013.04.053

Tijing, L. D., Kim, H. Y., Lee, D. H., Kim, C. S., & Cho, Y. I. (2010). Physical water treatment using RF electric fields for the mitigation of CaCO3 fouling in cooling water. *International Journal of Heat and Mass Transfer*, *53*(7-8), 1426–1437. doi:10.1016/j.ijheatmasstransfer.2009.12.009

Touir, R. (2016). *Nouvelles formulations pour la protection des circuits de refroidissement*. Éditions Universitaires Européennes.

Touir, R., Cenoui, M., El Bakri, M., & Ebn Touhami, M. (2008). Sodium gluconate as corrosion and scale inhibitor of ordinary steel in simulated cooling water. *Corrosion Science*, *50*(6), 1530–1537. doi:10.1016/j.corsci.2008.02.011

Touir, R., Dkhireche, N., Ebn Touhami, M., Lakhrissi, M., Lakhrissi, B., & Sfaira, M. (2009). Corrosion and scale processes and their inhibition in simulated cooling water systems by monosaccharides derivatives, Part I: EIS study. *Desalination*, *249*(3), 922–928. doi:10.1016/j.desal.2009.06.068

Touir, R., Dkhireche, N., Ebn Touhami, M., Sfaira, M., Senhaji, O., Robin, J. J., Boutevin, B., & Cherkaoui, M. (2010, July). corrosion inhibition in simulated cooling water. *Materials Chemistry and Physics*, *122*(1), 1–9. doi:10.1016/j.matchemphys.2010.02.063

Touir, R., Dkhireche, N., Ebn Touhami, M., Sfaira, M., Senhaji, O., Robin, J.J., Boutevin, B., & Cherkaoui, M. (2010). *Study of phosphonate addition and hydrodynamic conditions on ordinary steel*. Academic Press.

Touir, R., El Bakri, M., Dkhireche, N., Ebn Touhami, M., & Rochdi, A. (2010). Operational parameters effect of sodium gluconate inhibition used in the protection of cooling water system. *Journal of Materials Environmental Science*, *1*(S1), 317–328.

Tsuchiya, H., Fujimoto, S., Chihara, O., & Shibata, T. (2002). Semiconductive behavior of passive films formed on pure Cr and Fe–Cr alloys in sulfuric acid solution. *Electrochimica Acta*, *47*(27), 4357–4366. doi:10.1016/S0013-4686(02)00508-X

Vračar, Lj. M., & Dražić, D. M. (2002). Adsorption and corrosion inhibitive properties of some organic molecules on iron electrode in sulfuric acid. *Corrosion Science*, *44*(8), 1669–1680. doi:10.1016/S0010-938X(01)00166-4

Wang, L., Yin, G.-J., & Yin, J.-G. (2001). 2-Mercaptothiazoline and Cetyl Pyridinium Chloride as Inhibitors for the Corrosion of a Low Carbon Steel in Phosphoric Acid. *Corrosion Science*, *43*(6), 1197–1202. doi:10.1016/S0010-938X(00)00138-4

Xu, B., Liu, Y., Yin, X., Yang, W., & Chen, Y. (2013). Experimental and theoretical study of corrosion inhibition of 3-pyridinecarbozalde thiosemicarbazone for mild steel in hydrochloric acid. *Corrosion Science*, *74*, 206–213. doi:10.1016/j.corsci.2013.04.044

Yean, S., Al Saiari, H., Kan, A. T., & Tomson, M. B. (2008). Ferrous carbonate nucleation and inhibition. In *The SPE International Oilfield Scale Conference*. Society of Petroleum Engineers. 10.2118/114124-MS

Zheng, J., & Li, Y. Q. (2011). One-pot synthesis of tetrahydrobenzo[b]pyran and dihydropyrano [c] chromene derivatives in aqueous media by using trisodium citrate as a green catalyst. *Archives of Applied Science Research*, *3*(2), 381–388.

Section 2
Biomaterials Corrosion and Protection

Chapter 3
Corrosion of Biomaterials:
Electrochemical Techniques, in Vitro Test, and Methods to Mitigate Corrosion

Oscar Sotelo-Mazon
CIATEQ A. C., Mexico

John Henao
CONACYT-CIATEQ A. C., Mexico

Astrid Lorena Giraldo-Betancur
https://orcid.org/0000-0002-5056-7270
CONACYT-CINVESTAV, Mexico

Carlos A. Poblano-Salas
CIATEQ A. C., Mexico

Jorge Corona-Castuera
CIATEQ A. C., Mexico

ABSTRACT

The goal of this chapter is to define concepts and methods currently used to study the corrosion behavior of biomaterials in physiological conditions. One of the interesting points of corrosion of biomaterials is that they must be designed to fulfill different physical and chemical requirements within the human body. For instance, ceramic biomaterials are designed, sometimes, to accomplish a bone growth task. Alternatively, bioceramics are also employed to provide high wear resistance to implant surfaces with the lowest corrosion activity in biological environment. Depending on the part of the human body where the biomaterial is implanted, the corrosion process and chemical interaction with body fluids can be accelerated, and consequently, the product of these reactions can have a negative effect in the health of the patients. For this reason, the chapter is also focused on explaining how to study the interaction between biomaterial surface and body fluids as well as the existing methods to prevent corrosion phenomena that could lead to affect human health.

DOI: 10.4018/978-1-7998-2775-7.ch003

INTRODUCTION

The diseases and damages produced in the human body have had a high incidence in the development of society over the years. Multidisciplinary research in different areas such as physical, biological, materials, medical, clinical, and applied science, have been grouped to address this problem (Bose et al., 2016). Biomaterials is a field within materials science that has been established to attend the development of materials to treat the diseases of the body. A biomaterial has been defined as "any substance or a combination obtained from natural or synthetic sources used for any period to treat or replace parts of a living system of the body or to be and function in close contact with the living tissue in a physiologically acceptable manner" (Balani et al., 2015; Park & Lakes, 2007). Therefore, the biomaterials area need to have a transdisciplinary approach to develop and get closer to innovative and useful products that mimic the nature of the part that is being replaced (Park & Lakes, 2007).

A variety of applications have been developed to treat specific injuries. Bone cement, joint replacements, prosthesis, dental implants, among others are some examples of the devices/applications focus in to treat diseases of the musculoskeletal system; otherwise, stents, catheters, artificial heart, and kidney are part of the demands of the circulatory and urinary systems (Balani et al., 2015).

Active research and development have been carried out around biomaterials since their discovery. Due to the necessity for having more biofunctions in these materials, they have evolved from simple bioinert materials to accomplish different functions when implanted in the human body. Currently, biomaterials can be classified into four generations according to the biological response to attend the issues around the diseases. Figure 1 shows a schematic classification of biomaterials according to some authors (Ficai et al. 2011; Hernández-Montes, et al. 2017; Yang et al. 2017; Marchi, 2016):

Figure 1. Biomaterials evolution by generation
(Adapted from Ficai et al. 2011; Hernández-Montes, et al. 2017; Allo et al. 2012)

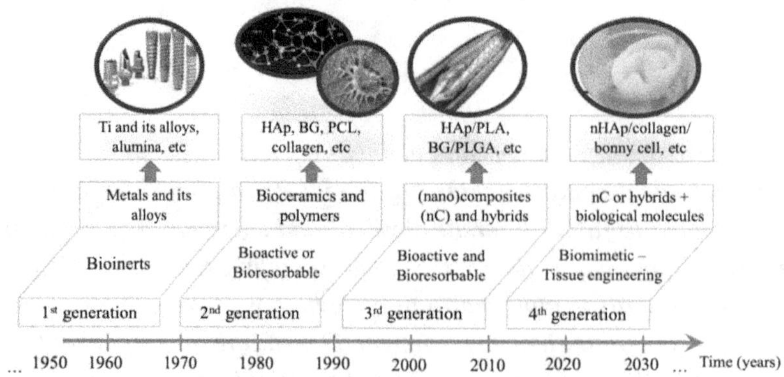

The elements from the first generation of the biomaterials, present biological "inertness" or bioinert behavior; neither bioactive nor bioresorbable. The advantage of this generation of materials is the mechanical resistance, especially to the applications that demand load-bearing. Metals and alloys are part of this generation. More common materials of this generation are Titanium (Ti) and its alloys, Stainless-steel (SS), Cobalt-Chromium (Co-Cr) alloy, titania (TiO_2), zirconia (ZrO_2), alumina (Al_2O_3), among others. Due to the failure of the implants from the first generation of biomaterials associated with the inertness of its surface, in the second generation, a bioactive or bioresorbable behavior was introduced in biomaterials with the aim of favoring the interaction of the implant surface with the surrounding tissue. Bioceramics and polymers are part of this generation, including calcium phosphates such as hydroxyapatite (HAp), bioactive glasses (BG), natural and synthetic polymers, such as poly (ε-caprolactone)-PCL and collagen. Despite the biological response induced for the second generation of biomaterials, some of them have shown limited durability in the body over time. Further research was focused on the necessity to find materials able to promote a natural integration with the surrounding tissue, which led to the development of the third-generation of biomaterials. The third generation of biomaterials synergically combines bioactive and bioresorbable behavior, which is achieved by the design of either composite or hybrid materials. Interestingly, the third generation of biomaterials present superior properties compared to the previous generations. Combinations of bioceramics and polymers are common in the third generation, such as (nano)HAp/collagen, HAp/polylactic acid-PLA, bioactive glass-BG/poly lactic-co-glycolic acid-PLGA, etc. In the last decade, tissue regeneration has been a target property to include in biomaterials, giving rise to the fourth generation of biomaterials. These new biomaterials combine the bioactive and bioresorbable behavior from the third generation but also favor tissue regeneration thanks to the presence of a biological molecule like a bone morphogenetic protein (BMP), bonny and stem cells, among others (Ficai et al. 2011; Yang et al. 2017; Allo et al. 2012).

Each generation has been crucial for the development of the biomaterials, taking into account that before the first generation of biomaterials, materials used to replace parts of the body presented failures mainly due to the mechanical and corrosion resistance mismatch with bone. These issues opened the research opportunity in this area, and currently, it is exponentially growing (Bose et al., 2016).

Another form to categorize biomaterials takes into account the interaction between the material/device surface and the host tissue. Hence, the biomaterials can be classified as bioinert, biotolerant, bioactive, and bioresorbable. Bioinert materials have minimal interaction with the surrounding tissue and look like an isolated part in the body; this kind of materials creates a mechanical interlocking between implant and tissue. Biotolerant materials are separated from the implant by a formed thin connective tissue capsule without any adherence to the implant surface. This kind of material does not present a toxic reaction in the body but could show rejection and failure of the implant. Unlike bioinert and biotolerant materials, bioactive materials exhibit the ability to interact with the tissue around them. The interaction capability of these materials is related with the material composition because of the ion exchange in the physiological environment to promote the formation of bone-like apatite layer between the material/device surface and the tissue generating a stable biological union. Finally, bioresorbable materials have an *in-vivo* gradual dissolution being replaced by the autologous tissue (Bose et al., 2016; Aherwar et al. 2015). The above-explained behaviors can be seen in Figure 2.

On the other hand, biomaterials are also classified in a generic way considering its origin, according to Vallet-Regí. Artificial origin group is divided into ceramics and glasses (TiO_2, apatites, calcium phosphates, glass-ceramics, among others), metals (Ti and its alloys, stainless steel, Co-Cr alloys), polymers (biostable and biodegradables), and composites (ceramic matrix, metallic matrix, organic matrix). In the natural group, collagen, elastin, chitin, etc., are include [10]. This classification is in agreement with other reports. A resume of this classification is presented in Table 1 (Bose et al., 2016; Park & Lakes, 2007; Vallet-Regí, 2011; Wei et al., 2005). Regardless of the origin or application of biomaterial, its success depends on the interaction with the surrounding tissue; i.e., it is necessary the biocompatibility

Figure 2. Classification by the interaction between the material/device surface and host tissue

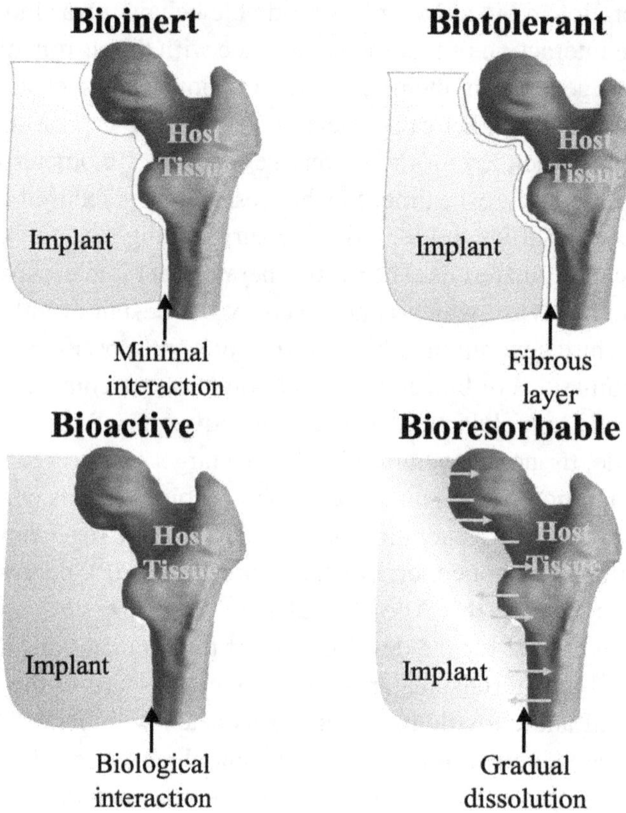

property, which is related with nontoxicity, nonallergenicity, nonimmunogenicity, noncarcinogenicity performance (Park & Lakes, 2007).

MEDICAL DEVICES AND THEIR APPLICATIONS

When an organ or a part of the human body presents a decrease in its functionality or fail, it should be replaced. Then, it is crucial to establish an appropriate design and materials that can replace the part of the body according to the disease type, developing a custom solution. This kind of solution are known as biomedical devices, which are necessary for the prevention, diagnosis, treatment, or rehabilitation of the damages caused by an illness or a disease. The instruments used to attend problems related to the body need to have not only medical and technological attributes but also the materials to manufacture them need to meet some criteria related to biological, physical-chemical and mechanical specifications. Specifications can be summarized as follows (Bose et al., 2016; Haschek et al., 2013; Ratner et al. 2006):

1. The most critical feature that needs to be accomplished by a medical device is safety; i.e., the material's behavior might not present any toxicity; for instance, corrosion resistance materials should be used for production.

Table 1. Generic classification of Biomaterials

Origin	Material		Advantages	Disadvantages	Examples
Artificial	**Ceramics**	Apatites, Calcium phosphates (HAp), glass ceramics, TiO_2, Al_2O_3, ZrO_2,	• Biocompatible • Composition tailored for bioactivity and bioresorption • Good corrosion resistance • High compressive strength	• Brittle • Low fatigue resistance • Weak in tension • Anisotropic mechanical properties	• Dental parts • Orthopedic implants • Coatings for load-bearing and dental implants • Bone fillings and cements
	Metals	Titanium (Ti) and its alloys, cobalt-chromium (Co-Cr) and molybdenum (Mo) and magnesium (Mg) alloys, stainless steel (SS)	• Strong • Tough • Ductile • High mechanical strength • High fatigue strength • Good corrosion resistance • Cost effective in some cases	• Poor biocompatibility • High elastic modulus compared to bone • Dense • Possibility of corrode in an in vivo environment • Wear resistance	• Load-bearing implants • Screws • Articulation prosthesis and join replacements • Dental implants • Skull plaques
	Polymers	Poly (ε-caprolactone)-PCL, poly lactic-co-glycolic acid-PLGA, sodium alginate	• Easy to fabricate in different forms • Bioresorption rate can be tailored	• Poor mechanical strength • Deformation over the time • Generation of toxic products from the degradation process	• Sutures • Spacer in intervertebral implants • Carrier for drug delivery • Blood vessels • Soft tissues
	Composites or hybrids	Metals with ceramic	• Composition tailored for the application • Biocompatible • Good corrosion resistance	• Lack of homogeneity • Difficult to fabricate	• Artificial articulations • Hip and knee implants • Bone cements • Medical resins
Natural	Collagen, chitin, elastin, hyaluronic acid, etc		• Biocompatibility • Tissue regeneration	• Availability • Processing procedures	• Tendons • Ligaments • Soft tissue

(Adapted from Bose et al., 2016; Balani et al., 2015; Park & Lakes, 2007; Vallet-Regí, 2001; Wei et al., 2005).

2. The durability and wear resistance to increase the lifetime and reducing the fatigue problems after their use.
3. To reduce the mismatch between the implant and the organ or site, it is necessary to have a mechanical match. Although the requirements mentioned above are met, there is always the possibility that the implanted device reacts in the body generating a risk of infection or failure and compromising the device performance and the health of the patient.

The tools, instruments, or elements that fulfill the requirements mentioned above are known as medical or biomedical device (Ratner et al., 2006). According to *Pecquet M.E. and Goad D.L.*, a biomedical device is defined as a system used in a permanent or temporally way, inside or outside the body to evaluate the medical condition, preserve, restore or augment the function of an organ, tissue or part of a specific system (Haschek et al., 2013). There are a wide variety of devices and materials used in the treatment of disease or injury; a resume of some applications/devices depending on the organ or part/site on action and system are shown in Table 2 (Balani et al., 2015; Park & Lakes, 2007).

Table 2. Biomaterial applications for organ replacement or sites of action

System/Organ or Site	Application
Musculoskeletal/Muscles, bones	Join replacements (hip, knee, shoulder, elbow, etc), tendons and ligaments, artificial limb, bone plates, spacers, dental implants, bone cements, sutures.
Cardiovascular/Heart, arteries, veins	Artificial heart, heart valve, cardiac pacemaker, heart-lung machine, stent, vessels (arteries and veins).
Respiratory/Lung, trachea, larynx, bronchus	Heart-lung machine, tracheotomy tubes, chest wall, lungs, diaphragm, implants: trachea, larynx, bronchus.
Digestive/Liver, esophagus	Tooth fillings, bile ducts, artificial liver, esophagus.
Genitourinary/Kidney, bladder, urethra	Artificial kidney, kidney dialysis machine, catheters, stent, ureter, bladder, urethra.
Nervous and sensory/Brain, spinal cord, eyes.	Hydrocephalus drain, nerve stimulator, corneal and lens prosthesis, cochlear implant, carotid pacemaker, contact lens, corneal bandage
NA/Other soft tissue and cosmetic	Cosmetic prosthesis: breast, maxillofacial.

(Adapted from Balani et al., 2015; Park & Lakes, 2007).

The fabrication of medical devices requires the selection of materials in a careful way. The choice needs to focus on mechanical demand, chemical and structural properties and the biological necessities (Goldberg & Kuhn, 2018).

It is important to remark that among the medical devices; the prosthetic ones are structures to replace parts of the body. As summarized in table 2, prosthetic pieces are more common to treat diseases of the cardiovascular and musculoskeletal systems. Hip fracture and heart valve disease are the most common problem from these systems, respectively and present high incidence in the world population (Aherwar et al., 2015; Ratner et al., 2006).

Heart valve disease is a medical problem, and the patients who suffer it, have short life expectancy without surgical intervention. On the other hand, different reasons cause hip trauma, but the treatment is the same. Total replacement is the procedure to treat these diseases. It consists of a cup and femoral element; the head of the femoral component fits in the cup to give the articulation movement to the device (Aherwar et al., 2015; Ratner et al., 2006; Poitout, 2004).

In the last trauma mentioned, metal, metal alloys, ceramics, ceramic-metallic, composite, and polymers biomaterials have been used to manufacture or to give a surface treatment of the device to attend orthopedic diseases (Poitout, 2004). Implants fabricated using metal-type material have been widely used. The metallic devices combine its biological inertness with the mechanical strength to achieve the body demand in that part (Yang et al. 2017).

Metals employed for the manufacture of biomedical device have high strength and high resistance to fracture and are designed to resist corrosion. The last property is the result of the nature of metallic bonds. Metals are strong, tough and ductile (deformable comparing with ceramics) compared with ceramics or polymers (Goldberg & Kuhn, 2018). Most metals are used in alloy form, but in some cases, the pure element can be used. The main metallic materials used in biomedical applications are (Balani et al., 2015; Aherwar et al., 2015); (Goldberg & Kuhn, 2018):

1. *Ti and its alloys:* these materials present high strength and corrosion resistance, lightness, good mechanochemical behavior, among other properties, compared with other metallic biomaterials. These materials are better for implants compared with the other groups. Ti and its alloys are often

used to manufacture joins replacement (knee and hip), surgical instruments, coronary stents, dental bridges, among others.

2. *SS316L:* It presents low carbon content that enhances the corrosion resistance. This biomaterial has low cost, availability, and good strength, and it is usually used in the fabrication of artificial joints, stents, bone fracture fixation, artificial valves, heart valves, among others.
3. *Co-Cr alloys:* It has a high wear resistance and has been used in dentistry, bone fracture fixation, total hip arthroplasty components, and stems.

In the last decades, some efforts have been made to improve the current metallic materials used in biomedical application. One of these efforts is related to their biological behavior. As mentioned before, the metallic implants employed currently present inertness, but they are shifting to have bioactive behavior. This behavior can be achieved, considering the biomechanical design, a porous structure, biological activation, and biodegradable design. The last issue is the new trend in tissue regeneration. Some applications of four generation of implant devices are related to orthopedic and cardiovascular implants, such as metallic bone scaffolds and stents demanding good mechanical strength and excellent integration in the host tissue (Bose et al., 2016; Yang et al., 2017).

BACKGROUND

Although biomaterials have been used for long time in the human history, biomaterials science have only been developed during the last century. In the 1950s, most of materials used for biomedical applications had a low probability of success due to the poor understanding of healthcare professionals about biocompatible behavior and sterilization of available materials at that time. In fact, about the year 600 ad, Mayan civilization used to repair teeth by employing nacre teeth from seashells; Interestingly, there is some evidence about successful osseointegration by using materials from seawater. Moreover, early evidence of biomaterials usage is found in the Greek and Egyptian civilizations, where some materials were often implanted in humans such as animal's bones and wood (Bobbio et al., 1972). Vestiges of using various types of materials as implants through the human history have all in common that no materials science, no advanced medicine, or no complex biological understanding existed at those times. The born of the modern era of synthetic biomaterials date from 1546 with the development of the gold-plate, which was employed to repair a cleft palate (Ratner et al., 1996). Similarly, the born of the modern era of biomaterials science is associated with the first reports of in-vivo studies involving the analysis of materials such as silver, platinum, and gold in dogs about 1829. Afterward, in early 1900s, some metals such as vanadium, copper, nickel, silver, and Cr-Ni alloys were tested for biocompatibility and for bone fracture applications. First electrochemical studies focused on corrosion resistance of metals under physiological conditions resulted in the development of some famous alloys such as the stainless steel 316. Later, after the second world war, the knowledge developed at that time about metallic alloys drove to the use of titanium and its alloys for orthopedic applications. Titanium alloys used to be employed for aircraft applications and showed excellent corrosion resistance in seawater; this fact was correlated to the extreme environmental conditions within the human being and was then proposed for medical applications. In 1947, some orthopedic devices such as plates and screws were implanted and studied in patients. In the 1960´s, the basis of the modern era biomaterial science was stablished by the introduction of some important developments such as total hip implants, hydroxyapatite scaffolds, and heart valves. This fact was also favored by the increasingly spread of successful surgical operations in developed countries (Ratner et al., 1996; Eliaz et al., 2019). Nowadays, biomaterials have progressed not only for the variety of diseases requiring different type of materials including pure metals and alloys, ceramics, polymers and composites, but also due to the understanding gained by the scientific community, especially, trying to comprehend how, for instance, corrosion may influence biomechanical behavior of materials in physiological environments.

MAIN FOCUS OF THE CHAPTER

The main focus of this chapters is to explore some concepts of corrosion from the point of view of biomaterials. The present book chapter explains how biomedicine has been related to the world of materials and how such interaction has led to the development of different types of materials such as metallic alloys, ceramics, polymers and composites. The purpose of the chapter is also to provide information about how corrosion is presented when biomaterials are implanted in human body and gives information about some available methods to study the corrosion behavior of biomaterials under physiological conditions. Some examples are included in the present chapter by using simulated body fluids to evaluate corrosion resistance of biomaterials. Considering factors such as corporal temperature, physiological medium, and in some cases the interaction of the surface with cells or bacteria, it can be possible to study a biomaterial under in-vitro conditions. Studying the corrosion resistance of biomaterials relies on understanding the interaction between the surface of the biomaterial and corporal fluids. For instance, most of metallic alloys undergo degradation in the human body, affecting somehow the natural health state of the patient. Another purpose of this book chapter is to understand the relationship between corrosion/degradation of biomaterials and their biocompatibility and bioactivity. The chapter presents some examples of bioactive materials such as hydroxyapatite coatings, which are currently an alternative to improve biocompatibility and bioactivity of implants. Moreover, the perception of the authors about current and future developments in the field of metallic biomaterials and coatings is captured in the last part of this book chapter. In particular, the importance of manufacturing processes such as additive manufacturing and thermal spray techniques are mentioned as alternatives to produce medical devices to mitigate corrosion. Additive manufacturing processes are increasingly adopted in the biomedical industry to produce implants due to their versatility in terms of design. Similarly, thermal spray processes have been widely used to fabricate coatings to protect metallic implants and to provide them of bioactivity. Overall, the present book chapter highlights the main concepts about biocorrosion and presents some approaches to control corrosion in metallic biomaterials.

IN-VITRO BEHAVIOR AND CORROSION ISSUES IN BIOMATERIALS

Currently, different organizations around the world have developed testing procedures for studying the biocompatibility of materials. One of these procedures is contained in the 20 parts of the ISO 10993: “biological evaluation of medical devices” standard, which shows one testing methodology to evaluate the response of the medical devices within the human body. In addition, different countries have their own guideline such as those prescribed by the European Pharmacopedia or by the Food and Drug Administration or National Institute of Health in the United States. *In-vitro* test are critical for the evaluation of biocompatibility in metallic materials. The term *in vitro*, comes from Latin: in glass, which means outside the living organisms; the type of tests is designed to study the cytotoxic effects of biomaterials, and sometimes can include the presence of cell lines to evaluate cell viability and proliferation. A very important factor that must be considered for studying the biocompatibility of biomaterials is the choice of cells. Since there are different species of cells, the researcher should have a criterion to select the right combination of cells in function of the biomaterial and its biomedical application. Table 3 shows individual standards of the ISO 10993. For instance, the ISO 10993-12 standard is related to the preparation of the biomaterial before *in-vitro* testing (Balani et al., 2015). On the other hand, the corrosion resistance of materials must be measured in simulated physiological electrolytes at 37±0.5°C according to ASTM 59-97 (ASTM standards No.G59-97, 2003).

Table 3. Individual Standards of the ISO 10993: Biological Evaluation of Medical Devices

No.	Standard
1	Evaluation and testing in the risk management process
2	Animal welfare requirements
3	Tests for genotoxicity, carcinogenicity and reproductive toxicity
4	Selection of test for interactions with blood
5	Test for in vitro cytotoxicity
6	Tests for local effects after implantation
7	Ethylene oxide sterilization residuals
8	Selection of reference materials
9	Framework for identification and quantification of potential degradation products
10	Tests for irritation and delayed-type hypersensitivity
11	Tests for systemic toxicity
12	Sample preparation and reference materials
13	Identification and quantification of degradation products from polymeric medical devices
14	Identification and quantification of degradation products from ceramics
15	Identification and quantification of degradation products from metals and alloys
16	Toxicokinetic study design for degradation products and leachables
17	Establishment of allowable limits for leachable substances
18	Chemical characterization of materials
19	Physicochemical, morphological and topographical characterization of materials
20	Principles and methods for immunotoxicology testing of medical devices

Corrosion Issues in Biomaterials

Metallic materials such as stainless steels, cobalt-chromium alloy, noble metal alloys, titanium, and titanium alloys are suitable biomaterials used as implants or prosthesis in different parts of the body. It is well-known that the titanium or titanium alloys are the most popular materials used to fabricate dental implants and prosthesis due to its mechanical properties, excellent corrosion resistance, and biocompatibility. It has been reported that titanium has a corrosion resistance superior to those of stainless steels and cobalt-chromium alloys (Koizumi et al., 2019). The spontaneous ability of Ti to form a stable and dense oxide layer (mainly TiO_2) is responsible for its corrosion resistance, and for that reason, is well accepted in biomedical applications (Rocha et al., 2013). On the other hand, the stainless steels have also been used in many applications in odontology, such as, elaboration of instruments and wire for orthodontics (orthodontic archwires, metallic fillings, or temporomandibular joints). The high content of chromium (^{3} 12% wt.) present in austenitic stainless steels promotes high corrosion resistance due to the formation of a passive chromium oxide film (Cr_2O_3) at the metal surface, which also prevents the ions release (Fe^{2+}, Cr^{3+} and Ni^{2+}). It is worth mentioning that some studies carried out in simulated body fluid have shown that ions release is too low that it has no toxic effects on human healt (Hwang et al., 2001; Mikulewicza et al., 2015). Some studies confirm that stainless steel and alloys with the addition of Cr are materials often used in dental applications due to its excellent corrosion resistance (Al Jabbari et al., 2016; Upadhyay et al.,2006). Also, noble metals (Au, Pt, Pd and Ag) haven been added to the base Ti metals in order to enhance the corrosion resistance of these nonnoble alloy. In consequence, noble metals like gold or silver alloys and alloys containing chromium are used for the fabrication of crowns, inlays, and denture bases. In addition, silver inhibit the colonization of oral bacteria, which is important factor in the development of biomaterials (Salgado et al., 2016).

One of the biological environments where corrosion resistance is one of the most important properties of biomaterials is the oral cavity. In materials as Cr alloy, Titanium and Co alloys can be observed an important cause of degradation promoted by a phenomenon known as bio-tribocorrosion, which leads to functional and/or biological detrimental effects due to an increased release of metallic ions and wear debris. This type of corrosion is caused by several simultaneous factors presents in oral cavity, such as mechanical, electrochemical corrosion, and microbiological. For instance, during mastication, implants are subjected to different forces, which may create micromovements at several interfaces, such as implant/crown, implant/abutment, abutment/crown and implant/bone (Vieira et al. 2016). These micromovements can lead to a fretting corrosion process, which mean that wear and corrosion of the system can occur simultaneously. In consequence, the protective passive film is removed due to the contact by rub between the two surfaces, which lead to release of corrosion products into the system. Also, it has been reported that fretting corrosion is one of the most important factors of the failure of prosthesis made of titanium and titanium alloys under *in-vivo* conditions (Kumar et al, 2010).

In term of corrosion within the oral cavity, the metals are exposed at different corrosive environments, where the saliva is often the corrosion medium. However, the soft-drinks, toothpastes and mouthwashes are commercial products that contain chloride, fluoride and high acidity which dissolve the passive oxide film. The carbonated drinks, and specially cola soft-drink has highest acidity (pH = 2.6) among the carbonated drinks (Peralta et al., 2017). In addition, changes in temperature and pH, and bacterial biofilm also influence in the degradation of dental devices promoting the release of inflammatory mediators from various cells that contribute to bone resorption or the encapsulation of the implant by fibrous tissue with subsequent failure of the implant (Rocha et al., 2013).

Different issues have been found in the case of the Co-Cr alloys-based implants, for instance, it have lower biocompatibility and osseointegration capacity than the Ti-based implants. Therefore, Ti-based implants are mainly used for the elements that make direct contact with the bone, whereas element such as the rods in spinal fixation that do not interfere with the bone directly are made using Co-Cr based alloys. However, it has also been revealed that metal corrosion and shredding, may result at the site contacting Ti or Co implant, due to increasing frictional load. The tissues near the metallic implant experience metallosis (corrosion due to erosion of metallic components, that produces particles which induce hipersensibility), as in the case of total hip arthoroplasty, spinal fixation, and knee implants (Zindani et al., 2019).

In vitro tests using Hank´s solution at corporal temperature (36.7-37.2 °C) have revealed that the 316L stainless steel, which is a commercial material used as implant, can undergo pitting corrosion, due to high Cl^- concentration in the environment. This fact leads to the release of metal ions of Fe, Cr and specially Ni, which can cause toxicity and impair the biocompatibility (Yang et al., 2007). Pitting manifests as a breakdown of the passive film at local areas giving raise to the formation of microgalvanic cells in which the anode area is represented by pits and the cathode is the large surrounding impervious passive surface (Frangini & De Cristofaro, 2003).

TECHNIQUES FOR THE EVALUATION OF CORROSION AND BIOACTIVITY IN BIOMATERIALS

Because corrosion is one of the main issues related to failure of implants, a criterion for selection of alternative or new materials are the characteristics of the passive films formed, its mechanical properties and the bioactivity of the surface (Arrieta et al. 2011). Corrosion behavior of biomaterials can be studied by means of *in vitro* test using artificial physiological fluids at 37 °C, which is equivalent to human body temperature. Hank´s solution, artificial saliva and artificial sweat are some examples that replicate different corrosive media in the human body (Karthega et al., 2007; Huang et al., 2003; Song et al., 2007). Nowadays, the evaluation of corrosion is carried out using electrochemical techniques. The more common techniques are polarization curves or also known as Tafel curves, measurement of open circuit potential (OCP), linear polarization resistance (LPR), electrochemical impedance spectroscopy (EIS) and electrochemical noise (EN).

In the present subchapter, a case-study related with the electrochemical evaluation of three metallic materials is discussed. The three metallic materials studied were Verabond®, Titanium and SS-316L (see chemical composition in table 4), which are considered as metallic biomaterials and are used as implant devices in the oral cavity. These materials have been evaluated in artificial saliva during 48 h at 37 °C. The chemical composition of artificial saliva used in the present study is shown in the Table 5. Electrochemical test were performance using a Gamry interface 1000 Potentiostat/Galvanostat/ZRA analyzer controlled by a personal computer. A typical three-electrode cell was used with a saturated calomel electrode (SCE) as reference electrode, and a platinum wire was the auxiliary counter electrode (CE). Corrosion resistance of materials was obtained by polarization curves from -400mV a 1500 mV with respect to the corrosion potential, $E_{corr,}$ a scanning rate of 60 mV/min. To evaluate the ability of the different materials to form a protective oxide layer on their surface upon immersion in the artificial saliva environment free corrosion potential respect to the time of the working electrode was measured versus a saturated calomel reference electrode (SCE). LPR tests were performance by polarizing the specimen from ±20 mV with respect to the free corrosion potential value, E_{corr}, at a scanning rate of 10 mV/min. EIS test were obtained applying an amplitude of input sine wave of ±10 mV in a range of frequency from 10000 Hz to 0.01 Hz. It is worth mentioning that many studies use the cyclic polarization technique to evaluate pitting and crevice corrosion resistances of the different materials according to the ASTM F2129-01 and ASTM G3-89 or ASTM G61-86. The procedure to develop this technique consists in stabilizing the free corrosion potential of the working electrode for an hour, then it is applied a sweep (0.166 mV/s) in the anodic direction until the potential reaches a predetermined value. After that, the scan direction is reversed until the hysteresis loop closes. At the point where the current increases sharply is known as the potential of pitting nucleation, *Enp*, and when the current is zero (during the reverse sweep) are called the protection potential, *Epp* (Arrieta et al., 2011).

Table 4. Elemental chemical composition of Verabond® and Stainless Steel 316 (wt. %)

Material	Ni	Cr	Mo	Al	Be	Fe	C	Mn	Si
Verabond*	77.9	12.6	5.0	2.9	1.9				
SS-316L	12	17	2.5			65.49	0.01	2.0	1.0

*composition provided by the manufacturer AalbaDent
(ASTM A213.04, 2004)

It is worth mentioning that, before evaluating a sample in a corrosive medium, it must be prepared according to ASTM E3-11 (Preparation of Metallographic Specimens). Since corrosion is a surface phenomenon and imperfections in the surface of the material often promote an increase in the corrosion rate, polishing is a recommendable step to avoid misunderstandings in the results of corrosion tests.

Table 5. Chemical composition of artificial saliva (pH=6.5)

Compound	Content (g/L)
NaCl	0.600
KCl	0.720
$CaCl_2 \cdot 2H_2O$	0.220
KH_2PO_4	0.680
$Na_2HPO_4 \cdot 12H_2O$	0.856
KSCN	0.060
$NaHCO_3$	1.500
Citric acid	0.030

(Duffo & Quezada 2004).

Electrochemical Techniques Evaluation in Metallic Biomaterials

Polarization Curves

The polarization curve technique is defined for the anodic and cathodic branches. The anodic branch is used to determine the behavior to material corrosion. In this branch is possible to observe phenomena such as passivation, transpassivation, pitting and uniform corrosion. Figure 3 shows the polarization curves obtained for the Verabond®, pure Titanium and Stainless Steel 316L in artificial saliva after 48

Figure 3. Polarization Curves evaluation for the Verabond®, Titanium, SS-316 in artificial saliva at 37 °C.

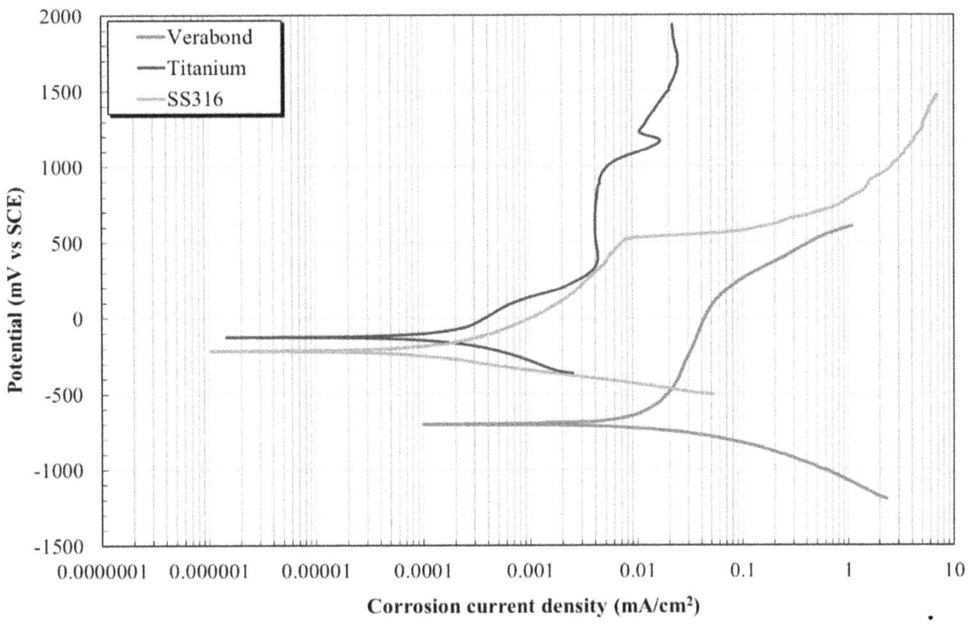

h of evaluation at 37°C. Ti presented the noblest corrosion potential (E_{corr} = -125 mV), while Verabond presented the most active corrosion potential with E_{corr} equal to -700 mV. Besides, Ti presented the lowest corrosion current density (i_{corr} = 0.0014 mA/cm^2). It is worth to mention that the corrosion current density can be interpreted as the corrosion rate. However, it is possible to observe that the current density value and corrosion potential of the Ti are very close at obtained to SS-316L. In both materials, the protection of corrosion is attributed to a protective film of corrosion products formed on the metal surface. Ti element form a protective film of oxide titanium known as "titania" (TiO_2), which provides a high corrosion resistance (Cui et al., 2009). On the other hand, the SS-316L form a protective film of oxide chromium known as "chromia" (Cr_2O_3). The highest corrosion current density and more active corrosion potential for the Verabond is related with the minor content of chromium. A higher content of chromium provide more stable protective films (Kamimura & Stratmann, 2001).

All materials presented passive zones in their anodic branches, which is a consequence of the formation of a protective film formed on the metal surface. However, when the SS-316L reached a corrosion potential of 526 mV the current density increase sharply. This behavior is associated with a pitting corrosion process, and has been observed in corrosive environment with high Cl^- concentration and temperature of 37 °C (Yang et al., 2007). A similar behavior is observed in the anodic branch of Titanium at a value of 1010 mV, but when it reached a value of E_{corr} equal to1223 mV, a secondary passivation region is observed. The values of E_{corr}, i_{corr}, B_a (anodic slope) and B_c (catodic slope) were determined using the Tafel extrapolation method (see Table 6).

Open Circuit Potential (OCP)

Open circuit potential measurements are shown in the Figure 4. The corrosion potential (E_{corr}) is a simple way to study the formation of protective films and the passivation of metals in a corrosive environment

Table 6. Electrochemical parameters of the materials evaluated in artificial saliva at 37 °C.

Material	E_{corr} (mV)	ba (mV/Dec)	bc (mV/Dec)	I_{corr} (mA/cm²)
Titanium	-125	287	172	0.00014
SS-316	-214	331.7	188	0.00027
Verabond	-700	1086	129.6	0.00601

in function of time. That is, the increase of the potential in the positive direction is associates with the formation of a passive film, while a stable potential is related to the formation of a stable and protective film. A potential that fluctuates between the positive and negative sides can indicate that the alloy is not forming a stable and protective oxide layer on its surface. A decrease in the potential toward the negative direction indicates a rupture, dissolution or non-formation of the protective film. In this sense, the potential value itself shows how stable is the material in a given environment. For instance, a high potential in stands is associated with a high corrosion resistance. Nevertheless, when the material has a low potential at rest, it has either the ability to oxidize, precipitating soluble compounds, and/or the formation of non-soluble products (i.e. oxide layer that decreases the corrosion resistance) (Gurrapa & Reddy, 2005; Manarache & Hornberger, 2007; Peralta et al, 2017). In the present study, Titanium obtained the most noble corrosion potential value (E_{corr} = -60.4 mV), which is very close to the obtained to SS-316 (E_{corr} = -80.13 mV). On the other hand, Verabond® presented the most active E_{corr} =-290.1 mV. The three material presented a corrosion potential almost stable, which indicates that the film remains

Figure 4. OCP measurements for the Verabond®, Titanium and SS-316 after 48 h immersed in artificial saliva at 37 °C.

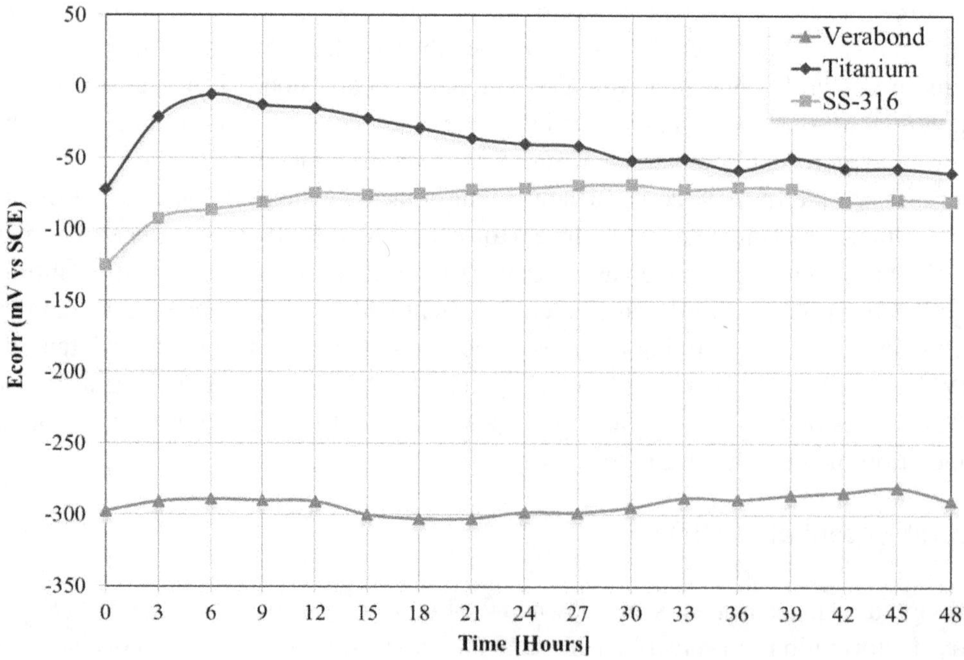

intact and protective. In the case of Titanium, in the first hours of evaluation, presented a potential drop in the negative direction, but after 30 h of test the corrosion potential was stable.

Linear Polarization Resistance (LPR)

Figure 5 shows the Linear Polarization Resistance curves for Titanium, SS-316L and Verabond® after 48 h immersed in artificial saliva at 37 °C. It is observed that, all materials displayed an increase in their polarization resistance after the first hours of test. This corrosion behavior is related to the growth and formation of a passive layer. After 12 h of testing, Titanium and SS-316L showed a behavior almost stable until end the test. However, at 36 h the SS-316L increased its polarization resistance until reaching the highest corrosion resistance. The Verabond® presented the lowest corrosion resistance, which is due to its minor content of Cr with respect to the SS-316L.

Figure 5. LPR measurements for the Verabond®, Titanium and SS-316 after 48 h immersed in artificial saliva at 37 °C.

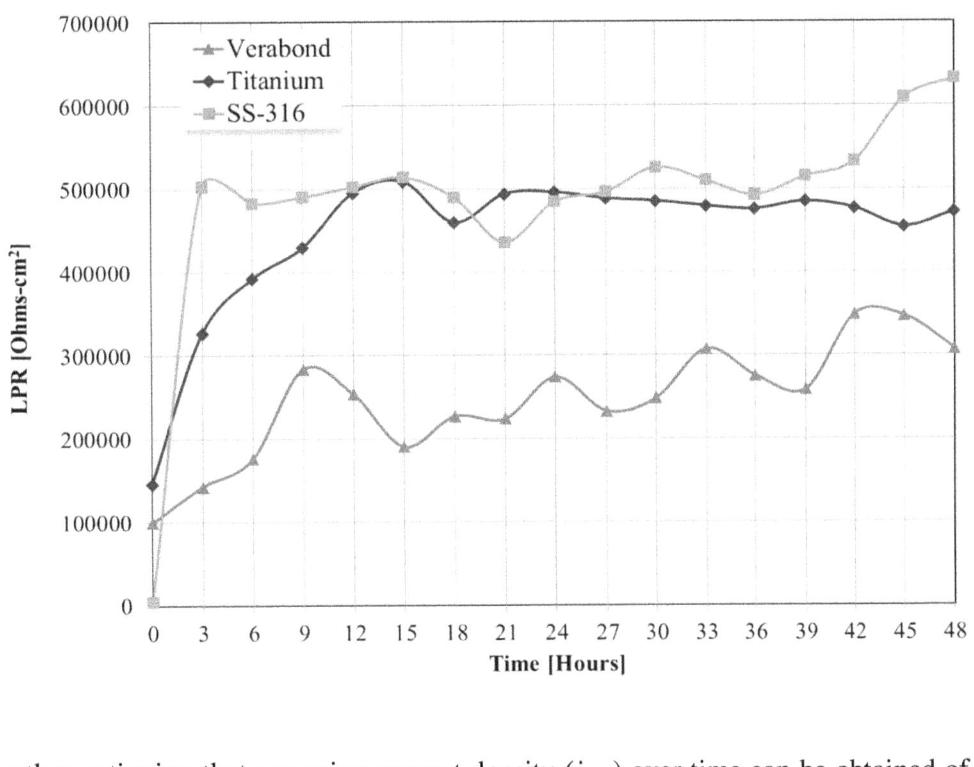

It is worth mentioning that corrosion current density (i_{corr}) over time can be obtained of data from linear polarization resistance measurements using Stern-Geary expression.

$$i_{corr} = \frac{b_a b_c}{2.303 R_p \left(b_a + b_c\right)} \quad (1)$$

Where the *ba* and *bc* value were those reported in Table 6.

The electrochemical reactions of the main elements in the Verabond® and Stainless Steel are shown below, the anodic reactions (oxidation reactions) are (Yaya et al., 2011):

$Fe \rightarrow Fe^{2+} + 2e$-

$Cr \rightarrow Cr^{3+} + 3e$-

$Ni \rightarrow Ni^{2+} + 2e$-

$Mo \rightarrow Mo(VI) + 6e$-

These reactions involve the release of ions due to the dissolution of the metal. On the other hand, the cathodic reactions consist of the reduction of the dissolved oxygen in the electrolyte (Nikolopoulou 2006):

$O_2 + 4e^- + 2H_2O\_4OH^-$

$O_2 + 4H^+ + 4e^-\_2H_2O$

The reactions presented above occur on a metal surface exposed to body fluids, as artificial saliva. Therefore, it is possible that chemical reactions involved in the dissolution of Verabond and Stainless Steel are associated with the following ionic species Fe^{2+}, Cr^{3+}, Ni^{2+}, Mo(VI), $CrOH_2^+$, FeOH+, NiOH+, $Cr(OH)_2^+$, $Cr(OH)_3$, $Fe(OH)_2$, $Ni(OH)_2$, $CrCl^{2+}$, $FeCl^+$, $FeCl_2$, $NiCl^+$, MoO_4^{2-}, $HMoO_4^-$, H_2MoO_4, H^+, OH^-, Cl^- in function of chemical reactions presented below.

$Cr^{3+} + H_2O \leftrightarrow Cr\ (OH)^{2+} + H^+$

$Cr(OH)_2 + H_2O \leftrightarrow Cr(OH)^+_2 + H^+$

$Cr(OH)^{2+} + H_2O \leftrightarrow Cr(OH)_{3<aq>} + H^+$

$Cr^{3+} + Cl^- \leftrightarrow CrCl^{2+}$

$Cr^{3+}\ 2Cl^- \leftrightarrow CrCl^{2+}$

$Fe^{2+} + H_2O \leftrightarrow Fe(OH)^+ + H^+$

$Fe(OH)^+ + H_2O \leftrightarrow Fe(OH)_{2<aq>} + H^+$

$Fe^{2+} + Cl^- \leftrightarrow FeCl^+$

$Fe^{2+} + 2Cl^- \leftrightarrow FeCl_2$

$Ni^{2+} + H_2O \leftrightarrow Ni(OH)^+ + H^+$

$Ni(OH)^+ + H_2O \leftrightarrow Ni(OH)_{2<aq>} + H^+$

$Ni^{2+} + Cl^- \leftrightarrow NiCl^+$

$Mo(VI) + H_2O \leftrightarrow MoO^{2-}_4 + H^+$

$MoO^{2-}_4 + H^+ \leftrightarrow HMoO^-_4$

$HMoO^-_4 + H^+ \leftrightarrow H_2MoO_4$

$H_2O \leftrightarrow H^+ + OH^-$

Electrochemical Impedance Spectroscopy (EIS)

The impedance technique is very useful to understand the corrosion process. It is considered a useful tool to study different areas of corrosion, due to its ability to discern between the solution resistance and/or the protective oxide layers. In other words, the impedance method shows the metal-solution interphase behavior. It is an ideal technique in high resistance systems and offers a great view of the corrosive phenomenon that occur in the system. Nyquist plot shows the different mechanisms that occur in the system, that are mass transport, charge transfer resistance (Rct) and solution resistance value "Rs". The module of impedance lZl shows the charge transfer resistance (Rct) and solution resistance, where "Rct" is measured in the low frequency range and "Rs" in the high frequency range, where Rp + Rs is equivalent to corrosion resistance. On the other hand, phase angle (°) plot presents time constants similar to a gauss bell, each time constant is associate to a layer, for instance, a coating layer or the formation of a corrosion products layer.

Figure 6 shows the impedance plots in the formats, Nyquist, impedance module (lZl) and phase angle (°) obtained in artificial saliva at 37 °C during 48 h. The Nyquist diagram shows the formation of a semicircle that is a charge transfer resistance process for Verabond® and SS-316L. Impedance module |Z| shows the corrosion resistance value, the stainless steel 316L presented the highest corrosion resistance. On the other hand, phase angle plot for SS-316L presented one time constant and the Verabond® obtained two-time constants, which is related to the formation of two layers on the surface.

Bioactivity in Biomaterials

Bioactivity is the ability of a biomaterial to mimic the response in a living system. Bioactive materials must elicit the biological response at the interface to develop a strong bond between the material and the human tissue. Hydroxyapatite (HAp) is a biomaterial (HAp resembles the mineral component of bone and teeth) with good biocompatibility and bioactivity in physiological environment, and consequently, it has been considered the ideal material to construct bone-tissue scaffold (Saber et al, 2016). To increase bioactivity and mechanical properties, some composites of polymer and bioactive ceramics have been developed for bone tissue engineering. HAp has been added to polymer to form polymer/HAp composites that improve the activity and viability of cells cultured on them, or to improve both the mechanical and cell-attachment properties of scaffolds (Kong et al., 2006).

Figure 6. EIS measurements for the Verabond® and SS-316 after 48 h immersed in artificial saliva at 37 °C.

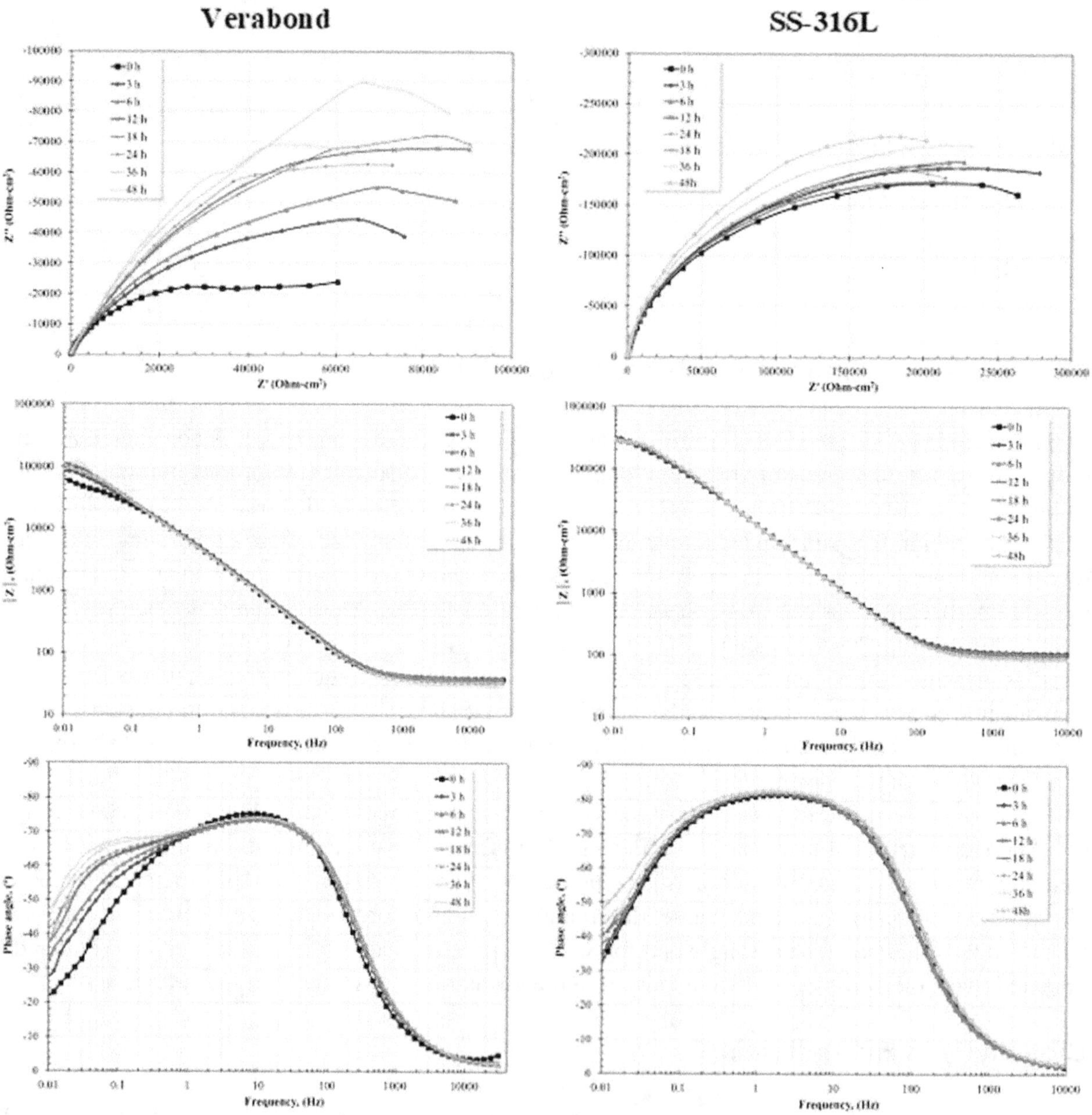

The bioactive ceramics present important characteristics; one of these is a time-dependent kinetic modification of the surface that occurs upon implantation: a biological active carbonate hydroxyapatite layer can form on their surface, which is chemically and structurally equivalent to the mineral phase in bone and provides an interfacial bonding between materials and tissues. Therefore, the bioactivity of these artificial materials can be attributed to the formation of a biologically active bone-like, carbonate-containing apatite layer (Song et al. 2004). In order to predict the process of apatite formation on the bioactive materials in living body, *in vitro* test in simulated body fluid (SBF) has been performed based on the standard ISO/FDIS 23317:2007 (E). It is worth mentioning that apatite coating is another method to prepare a surface scaffold for bone tissue engineering. Some techniques used to make coatings are, for instance, thermal spray techniques (HVOF or plasma spraying) and ion-assisted deposition (Barrere et al., 2002; Barrere et al., 2002; Abe et al., 1990).

Figure 7. Schematic presentations of the origin of the negative charge on the HAp surface, and the process of bone-like formation thereon in SBF.
Reprinted from (Kim et al. 2004). (all rights are preserved by the original source).

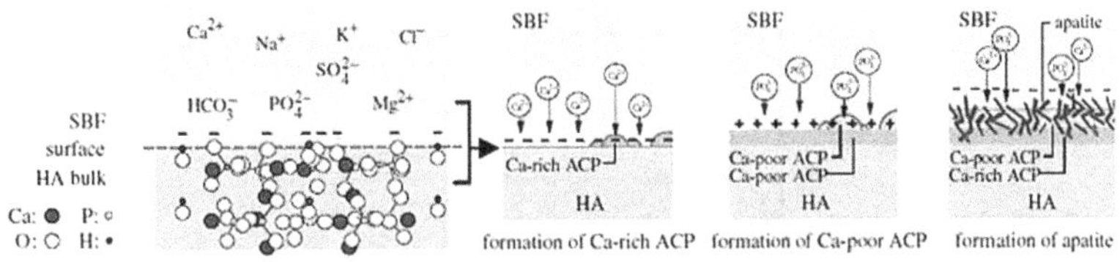

Figure 7 presents a schematically representation of the mechanism of biomineralization of bone-like apatite on synthetic hydroxyapatite in SBF. *In vitro* test was carried out to determine this mechanism, after immersion in SBF, the HAp undergoes three characteristics surface structural changes in the process of apatite formation on its surface. Firstly, surface structural change is the formation of Ca-rich amorphous calcium phosphate (ACP) on the HAp, which take place in the nominal soaking period 0-6 h. In view of the change in the Ca/P ratio, the formation of Ca-rich ACP is assumed to be a consequence of interaction of the HAp surface specifically with the calcium ions in the SBF. Secondly, surface structural change is the formation of Ca-poor ACP on the HAp in the next soaking period 6-9 h. In this soaking period, the HAp appears to use the Ca-rich ACP on its surface to interact with the phosphate ion in the fluid to form the Ca-poor ACP. The third surface structural change is the formation of apatite in the soaking period 9-12h, in which the Ca-poor ACP on the HAp appears to crystallize into apatite with bone mineral-like composition and structure. Thereafter with increased soaking time in SBF, the apatite grows incorporating the calcium and phosphate ions in the SBF. This mechanism was revealed based on its zeta potential, which was measured using laser electrophoresis spectroscopy.

This analysis indicates that the HAp changes the surface potential, which was highly negative immediately after exposure to SBF, in a very consonant way during the above soaking periods of surface structural changes. The HAp was shown to increase the surface potential to reveal a maximum positive charge when it was forming the Ca-rich ACP and to decrease it to again reveal a negative charge when it was forming the Ca-poor ACP. It then continued to decrease the surface potential whilst forming the apatite and then stabilized it at a constant negative charge. Therefore, the changes of bioactive surfaces during in vitro exposures have been documented in term of the Ca/P ratio and surface potential (see figure 8) (Kim et al. 2004).

In vitro cytotoxicity test has been used to determinate the inhibition of cell grow or record the damage or dead of cells. Cytotoxity tests can be by direct contact (Staining with vital dyes and release of radioactive chromium) and by indirect contact (extended in agar, elution of extracts, milipore filter method, among others) (Ríos & Cepero 2006). For instance, a study carried out by W. Xu *et al.* for dental applications

Figure 8. Ca/P atomic ratio and zet potential of the surface of HA as a function of soaking time in SBF.
Reprinted from (Kim et al. 2004) (all rights are preserved by the original source).

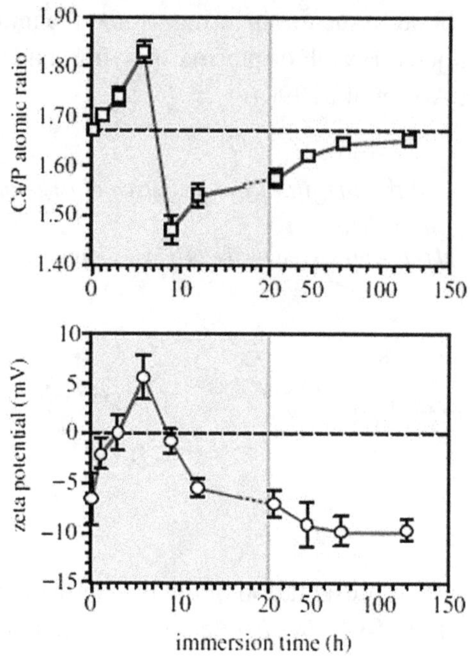

was performed with murine osteoblast cells (MC3T3-E1) to examine the biocompatibility of Ti-12Mo alloy. Pure Ti and Ti6Al-4V were used as comparison materials. In the indirect test, the extracts were obtained based on the international standard ISO 10993-5 (ISO 10993-5, 1999), the cytotoxical level was divided into six groups: 0: ³ 100%; 1: 75%-99%; 2: 50%-74%; 3: 25%-49%; 4: 1%-24%; 5: £ 1%. The cell viability ratio (*CVR*) of cell can be calculated by the formula as following (Xu et al. 20018):

$$CVR = (OD_{570nm} \text{ in experiment extract}/OD_{570nm} \text{ in control extract}) \times 100\% \quad (2)$$

The cytotoxicity grade determinates that Ti-12Mo alloy for MC3T3-E1 cells is 0 grade, while those of Pure Ti and Ti-6Al-4V are 0-1 grade. Therefore, the Ti-12Mo alloy shows good *in vitro* biocompatibility to MC3T3-E1 cells. In turn, corrosion test were performed, such as RPL and EIS, which suggested that the Ti-12Mo alloy is more corrosion resistant that Pure Ti and Ti6Al4V in artificial saliva (FAS), phosphate-buffered saline (PBS), simulated body fluid (SBF), and saline solution (SS). Titanium oxide compounds, present on the surface of commercially titanium implants, are responsible for the favorable biological interaction that occur on the bone-fixation interface. Thus, this oxide layer provides titanium biocompatible features (Schoroeder et al., 1981).

ADDITIVE MANUFACTURING AND THERMAL SPRAY TECHNIQUES IN MEDICAL DEVICES TO MITIGATE CORROSION

Titanium alloys have been used for many years as bulk solid implant materials; bone resorption due to stress shielding is one of the main issues of implants fabricated employing these alloys, which is the result of a different metal stiffness with respect to the bone (an order of magnitude greater young's modulus of the metal than the cortical bone, and two orders of magnitude greater than the trabecular bone (Arabnejad et al., 2016). Another attendant issue is the wear debris production for contacting surfaces (metal/metal), null vascularization, and ion dissolution (Hansen, 2008; Matusiewicz, 2014). However, one of the advantages of Titanium alloys is the formation of a stable passive film on the surface that minimizes the diffusion of metal ions from the bulk material preventing the corrosion of the implant in contact with body fluids (Niinomi & Boehlert, 2005). Alternatively, stainless steel (316L) and Co-Cr (or CoCr-Mo) alloys have been also employed as bulk materials for implants instead of titanium alloys, in particular, when strength, good fatigue properties, and wear resistance are required. Stainless steel (316L) and Co-Cr (or CoCr-Mo) alloys rely on the presence of chromium to support their corrosion resistance in aqueous media (Park & Kim, 2003).

Since surface characteristics and geometric details of metallic implants play an important role in their mechanical and chemical behavior when implanted in the human body, in recent years, the biomedical industry have focused their efforts to build up prosthesis that approaches to the physicochemical properties of the bone (Yilmaz et al. 2018; Al Ketan et al., 2018). However, it is still challenging the fabrication of bulk metallic prosthesis with similar characteristics (i.e. density, roughness) of human bone by employing conventional manufacturing methods. Additive manufacturing (AM) has recently illustrated not only the potential for fabricating complex implant components but also the prospect of fabricating patient-specific implants parts. AM processes are capable of producing parts by depositing materials layer-by-layer and building the part up from a design file (.STL) previously designed by the user according to the medical requirement (Savio et al. 2018). The workpiece (the medical device or prosthesis) is virtually sliced into several two-dimensional layers, which are deposited one on top of the other, consecutively, forming a 3D part. In recent years, AM has been of great interest for the production of medical devices, prosthesis and scaffolds in the field of orthopedy, tissue engineering, dentistry, regenerative medicine, and drug delivery.

AM comprises a group of processes that allows the production of ceramics, polymeric, metallic and composite parts. These processes are divided in the following families: A) Solid based systems, (B) powder-based systems (C) Liquid based systems (Vayre et al., 2012), as shown in Figure 9. Among these AM families, the powder-based deposition processes are usually employed to produce metallic parts for applications in the biomedical industry. The powder-based printing technologies include the direct metal laser sintering (DMLS) and the electron beam melting (EBM) processes. In DMLS, a high-powered laser fully melts each layer of metal powder producing printed parts that are dense (> 90%) and strong (no need of further heat treatment for consolidation). DMLS process involves very high-energy, as each layer of metal powder must be heated above the melting point of the metal. The high temperature gradients that occur during DMLS manufacturing can also lead to stresses and dislocations inside the final part, which can affect its physical properties (Simchi, 2006; Rizzuti et al., 2019). On the other hand, the EBM process employs an electron beam, which is generated to scan uniformly a powder bed layer by layer. EBM process optimization involves beam focus selection and melt scan beam current depending upon the powder properties, especially, the melting point (Galati & Iuliano, 2018).

Figure 9. Classification of AM processes.

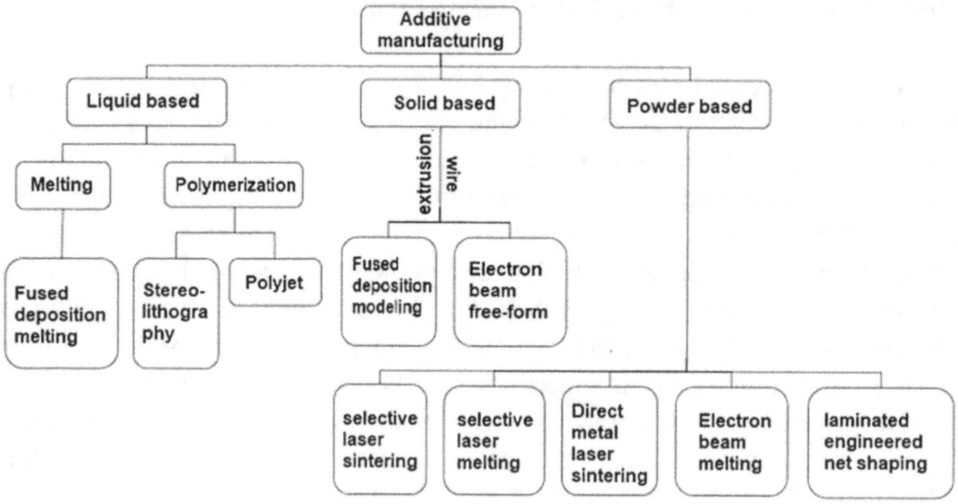

Since customization is very important in the biomedical field, the AM technologies are a good option to produce a wide range of products with specific properties and shapes customized to the patient's needs. For instance, diagnostic platforms can be employed to produce tailored-made orthopedic, dental implants, medical devices, and tissue scaffolds (Bank, 2013; Calignano et al., 2019). In the orthopedic field, for example, hip prosthesis can be designed to meet patient needs by starting from collection of the medical data of the patient. To attain this objective, computed Tomography (CT) and the Magnetic

Figure 10. Customized DMLS printed femur from CT-Scan data obtained from a patient, CT-scan data were processed in a STL file and provided to the DMLS machine to print it employing a Ti-6Al-4V powder bed.

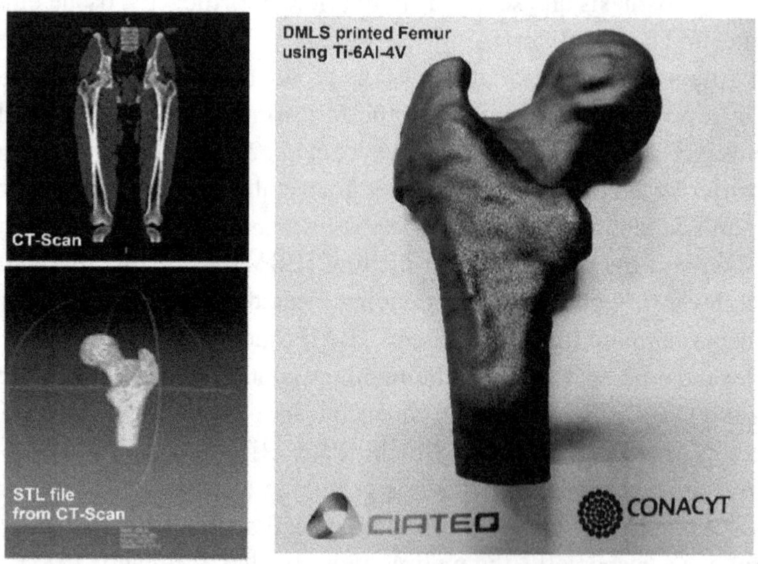

Resonance Imaging (MRI) technologies can be used. These technologies allow the collection of images of the patient affected zone; these images can be converted into computer aided design (CAD) model via Digital Imaging and Communications in Medicine (DICOM) directory. The CAD model obtained from the patient is then converted to an STL file. This file is introduced in the AM machine to produce the 3D implant. Figure 10 shows a customized 3D-printed hip prothesis manufactured by DMLS employing Ti-6Al-4V biomedical grade powder and using CT scan data of a patient.

Due to the advantage of the AM for the manufacture of biomedical devices, issues such as poor surface characteristics, poor dimensional accuracy, low mechanical strength, biocompatibility, corrosion of the implants can be currently addressed by exploiting the capacities of AM processes (Calignano et al., 2019). For instance, porous surface having interconnected pores are recommended for the improvement of bone formation rate and also for the better fixation of the implant. Implant porosity promotes positive results in bone formation *in-vivo* (Bandyopadhyay et al., 2019), as increased contact area between the biomaterials and bone tissue, result into better implant stability. The porosity can be designed in areas based on the patient's need to enhance biological fixation and achieve long-term in-vivo stability (Ghouse et al., 2019). Porosity can be also employed to create implants to adjust the elastic modulus to that of the bone, Figure 11 shows a porous structure printed by DMLS using Ti-6Al-4V alloy with potential applications in the biomedical field.

Figure 11. DMLS printed Ti-6Al-4V porous structure developed by CIATEQ A.C., México.

Corrosion is another phenomenon that may appear during the conventional adaption of implants due to the fact that plastic deformation may break the passive layer on the surface of the implant or by the occurrence of infections in the region around the surface of the implant (Jacobs et al., 1998). Infections can increase the corrosion rate as well as corrosion products. To mitigate corrosion, modifying the surfaces of the implants results as the best option, which at the same time can enhance biocompatibility, and improve osseointegration without compromising the bulk properties. For instance, incorporating passivating elements such as Cr, Ti and Al, modifying the topography and changing the crystalline structure can lead to improve corrosion resistance and implant-bone integration (Manivasagam et al., 2010). Previous studies have studied the effect of surface roughness, porosity and microstructure in the corrosion behavior of AM Ti-6Al-4V implants (Balakrishan et al. 2008; Sivakumar et al. 2017; Bandyopadhyay et al., 2010).

Overall, previous studies have reported that the biphasic beta and alpha microstructure has lower corrosion resistance than fully martensitic or alpha/martensitic phase in Titanium alloys. This fact is attributed to the formation of a homogeneous titanium dioxide film in the martensitic phase. In the biphasic microstructure, vanadium is stabilizer element of the beta phase, which segregates into the beta phase during cooling. Since the properties of the titanium dioxide formed on the surface of the alloy depends on the substrate composition, the oxide formed on beta phase has lower stability in chloride containing solutions, so the native oxide film turns more prone to localized attack in such areas, reducing the passivity. In addition, biphasic microstructure also has more interfacial regions, this fact increases the susceptibility of the alloy to experience corrosion (Choubev et al., 2005; Atapour et al. 2011; Zhou et al. 2019).

Previous studies (Xu et al. 2019; Longhitano et al. 2018) also report the effect of heat treatment on the corrosion behavior of AM Ti-6Al-4V parts in simulated body fluid. For example, Longhitano et al. 2018 (Longhitano et al. 2018) obtained DMLS parts that consisted in alpha and martensitic phases, with beta phase among the contours and lamellar structure of alpha and beta phases. Nucleation and growth of the beta phase were detrimental to corrosion resistance reducing pitting corrosion resistance on heat treated samples. These results highlight the importance of AM parameters control to avoid the formation of undesired phases.

An option to mitigate corrosion in AM implants is the deposition of a protective film on the surface of the AM part. For instance, Longhitano et al. 2018 (Longhitano et al. 2018), also proposed the anodizing technique to enhance the corrosion resistance of a Ti-6Al-4V implant in simulated body fluid. Interestingly, the anodizing process can promote the formation of a homogenous oxide layer, which is reflected in the improvement of the pitting corrosion resistance despite the presence of the undesired beta phase in the surface of the AM implant.

Thermal spray coatings are also employed to provide bioactivity and corrosion resistance to the surface of AM implants. Thermal spray is a group of processes used to apply metallic and non-metallic coatings. These processes employ energy sources to heat the material, often in powder shape, to a molten or a semi-molten state. The heated powder is then accelerated and propelled toward the surface to be coated. The coating is formed by the successive impact of particles (Lima & Marple, 2007). Among the various thermal spray processes, the atmospheric plasma spray (APS) and the high velocity oxygen fuel spray (HVOF) have been mainly used to this purpose in recent years. In APS process, an inert gas (typically a mixture of hydrogen and argon) is superheated by an electric arc forming an ionized gas that achieve temperatures up to 14000°C, which is significantly above the melting point of any material. The APS process is one of the thermal spray techniques currently used in commercial hip prosthesis to protect the implant surface and to provide it with bioactivity. Alternatively, the HVOF process employs a fuel gas (such as propane, acetylene etc.) and oxygen to produce a combustion jet in the range of temperature from 2000°C to 3200°C. The HVOF process is equipped with a de laval nozzle which is in charge of accelerate the particle into the supersonic regime. In this sense, the HVOF process is usually employed when high density and adhesion strength are required. It is also helpful when the initial characteristics of the powder material must be preserved in the final coating (Davis, 2004). HAp ($Ca_{10} PO_4 (OH)$), a calcium phosphate, is one of the most common materials employed in the biomedical industry as a coating material to provide bioactivity and corrosion resistance in implants. This ceramic material is often used since calcium is one of the most abundant elements in the human body, especially, stored in bones and teeth's (bone is a composite of collagen fibers and hydroxyl apatite). The HVOF process is usually preferred over the APS process for the preparation of HAP coatings due to the temperatures achieved during the spraying process, which favor the preservation of great number of initial phases in the final coating (Gadow & Stiegler., 2010; Berndt et al., 2014). Figure 12 shows an example of a HAp coating deposited by HVOF on a Ti-6Al-4V DMLS printed part with potential application in the biomedical field.

Figure 12. DMLS printed Ti-6Al-4V porous structure with a HAp coating deposited by HVOF at CIATEQ A.C. México.

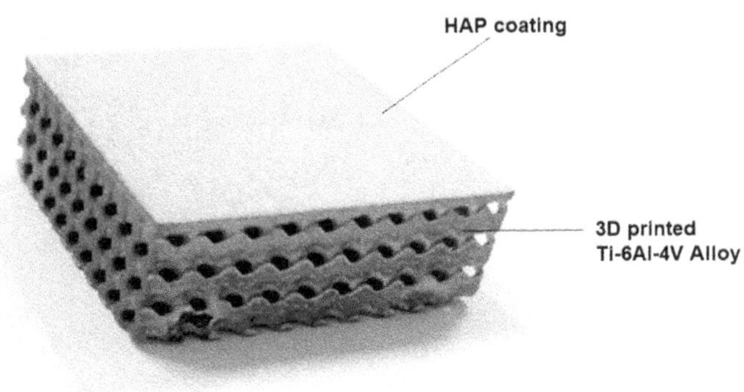

The use of ceramic coatings on metallic implants results advantageous for osteointegration and protection against corrosion. For instance, when HAP coating is immersed in body fluids a dissolution-precipitation process occurs (Berndt et al., 2014; Henao et al. 2018). This process leads to the formation of new bone on the surface of the implant when implanted in-vivo (Brown & Constantz, 2017) In the absence of HAP coating, the metallic implant do not stimulate bone-regeneration, instead, a fibrous tissue appears around the implant surface, which subsequently favor the occurrence of micromovements (Heimman, 2016), see Figure 13. This fact is one of the main causes of implants failure (Hirakawa et al. 2004).

Figure 13. In-vivo tested non-coated titanium implant (left) and coated titanium implant (right) displaying bone formation when bioactive coating is employed (right), while a fibrous tissue is formed in the absence of a bioactive coating (left)
(Adapted from Heimman, 2016).

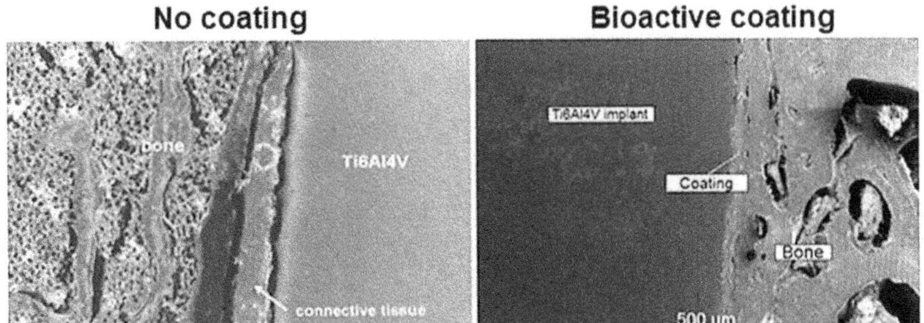

In terms of corrosion, the use of a HAp coating also contributes to the increase in the corrosion resis-

Figure 14. Corrosion potential measurements of Ti-6Al-4V and a HAp coating immersed in simulated body fluid at 37 °C for 48h.

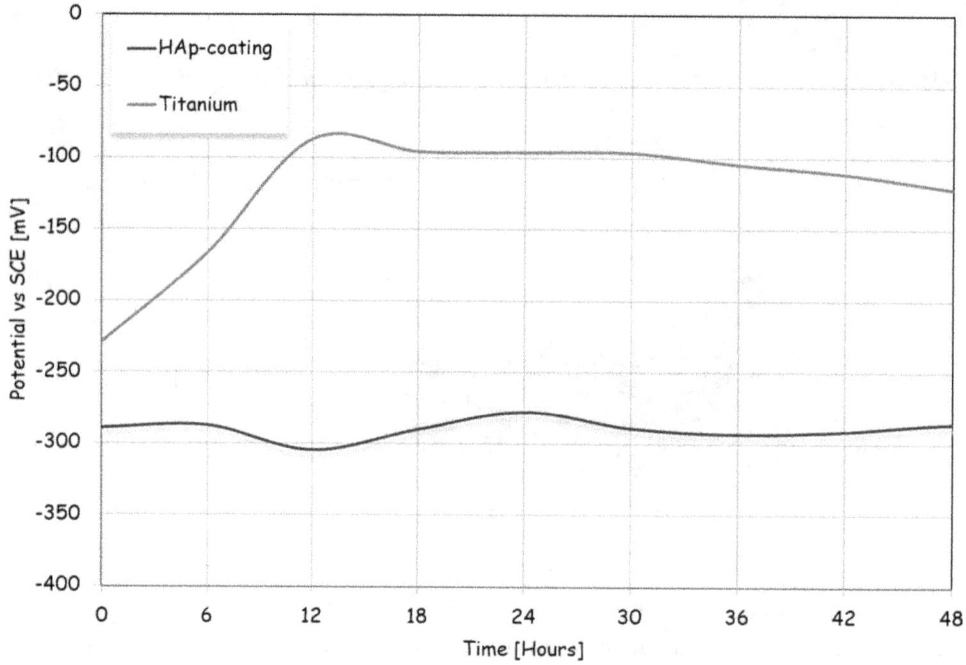

Figure 15. Linear polarization resistance measurements of Ti-6Al-4V and a HAp coating immersed in simulated body fluid at 37 °C for 48h.

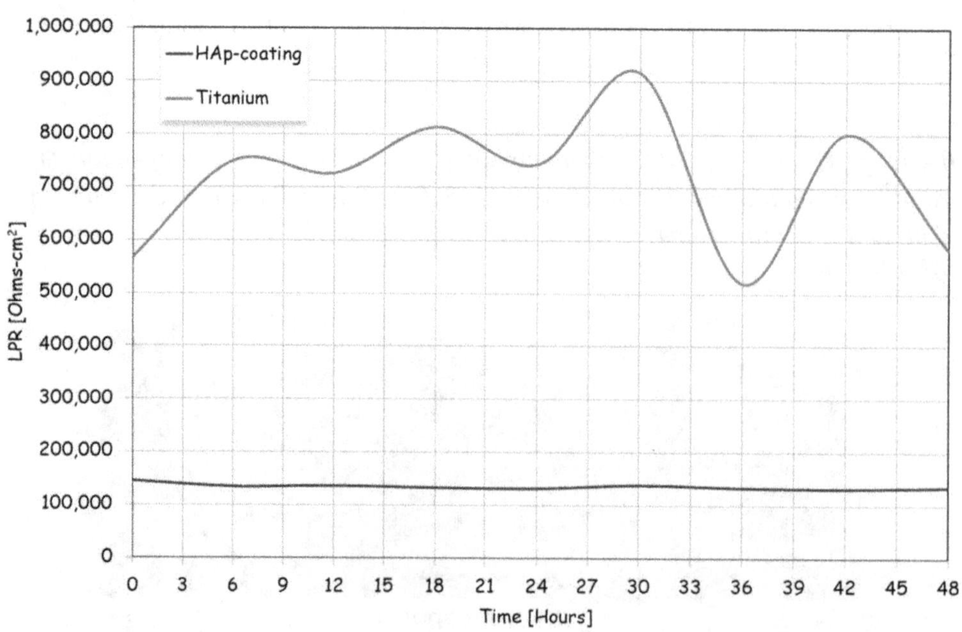

tance in metallic implants. Figure 14 shows the corrosion potential, Figure 15 shows lineal polarization and Figure 16 shows impedance curves of a HAp coating after 48 hours of immersion in simulated body fluid. It is worth mentioning that although the corrosion resistance obtained was higher in the Titanium than in the HAp coating during 48 h of test, the HAp coating provides better protection and bioactive properties. In addition, potential and RPL shows more stability in the HAp coating, which indicates homogeneity and compactness.

Figure 16. Electrochemical Impedance spectroscopy measurements of Ti-6Al-4V and a HAp coating immersed in simulated body fluid at 37 °C for 48h.

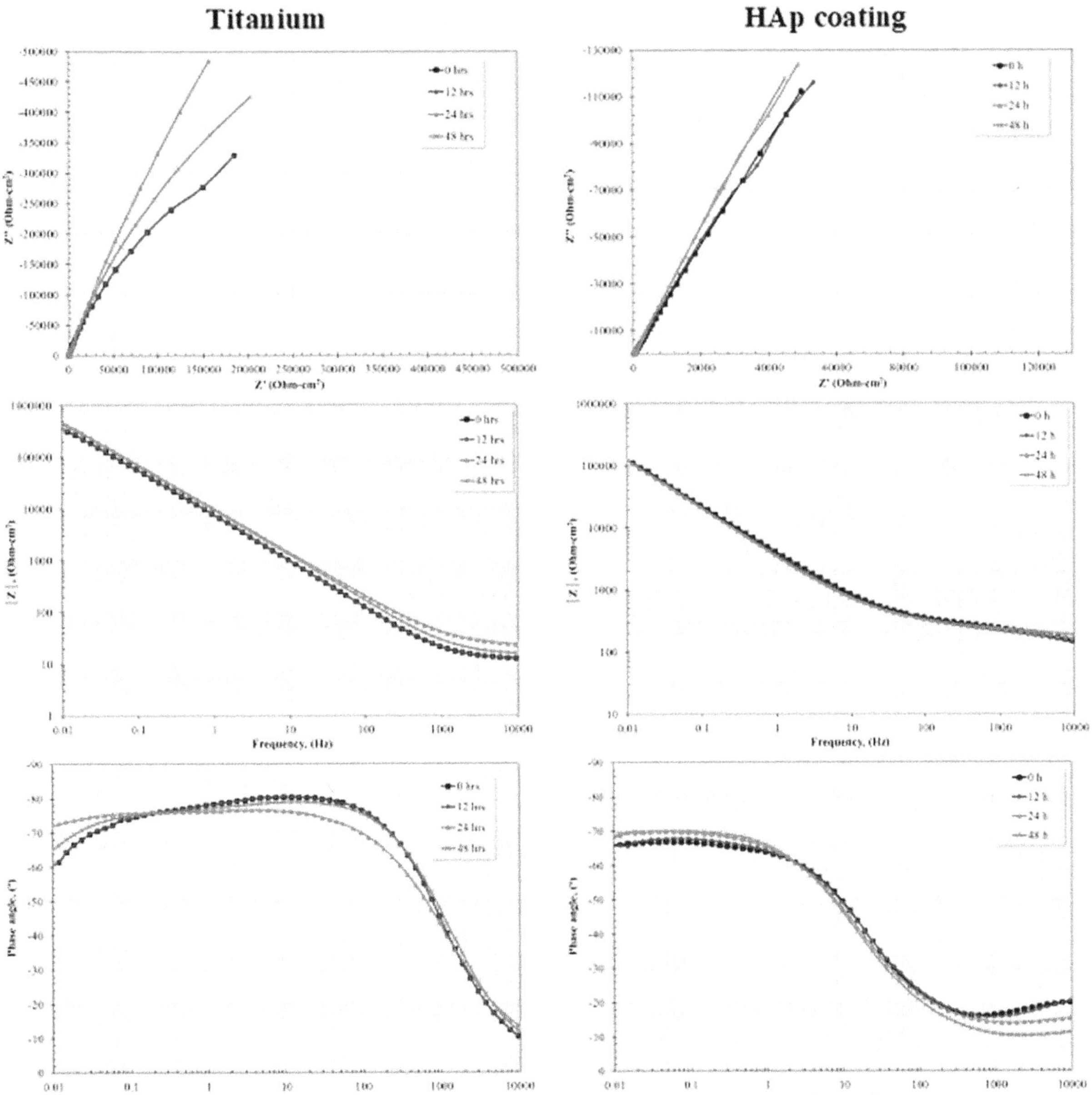

The corrosion resistance of HAp coatings has been tested not only on Titanium substrates but also on other medical alloys such as stainless steel 316L, resulting in the increase of corrosion resistance due to the role of physical barrier between the physiological fluid and the metallic surface (Singh et al., 2013; Singh et al. 2012; Melero et al. 2018). Some studies have also shown that HAp/Titania composite coatings also increase the corrosion resistance of metallic implants in body fluid (Singh et al., 2012; Melero et al., 2018; Kumari & Majumdar, 2017). The advantage of this type of coatings lies on the excellent mechanical behavior before and after immersion in body fluid while keeping biocompatible behavior. However, the bioactive properties are diminished when Titania composite coatings are used with respect to pure HAp coatings. This fact is attributed the limited bioactive behavior of the titanium dioxide with respect to the HAp. In this sense, multilayered graded coatings (*i.e.* coating with multiple layers, starting with a pure Titania coating at the bottom layer, composite layers at the inner layers, and a pure HAp coating at the outer surface) has been recently developed, achieving a compromise between mechanical and chemical response of the coated implant when immersed in simulated body fluid (Henao et al., 2018). In any case, *in-vitro* cell culture studies have shown that HAp-coated 316L SS and HAp-coated Ti-6Al-4V specimens appeared more bioactive and with better corrosion resistance than the uncoated metallic specimens (Sun et al., 2001). An alternative method to improve bioactive behavior in titania coatings consists in the post-spray treatment by using sulfuric acid at room temperature for 24h (Ban et al. 2006; Takeuchi et al., 2003). Acid-treated titania coatings can show bone-like apatite formation on the surface after soaked in simulated body fluid for a period of time. The concentration of sulfuric acid is an important parameter for the promotion of bioactive properties on titania coatings. The bioactivity of plasma sprayed titania coating cannot be improved below 0.01 M sulfuric acid treatment. Interestingly, cell culture test has shown that cells can attach, grow and proliferate well on the surface of titania coatings and shows good corrosion protection when compared with the uncoated metallic surface (Conner et al., 2003). This chemical treatment can be also applied to 3D printed titanium surfaces previously anodized (to promote the formation of a titanium dioxide uniform layer) to provide corrosion resistance and bioactive behavior to the metallic implant.

CHALLENGES AND FUTURE TRENDS OF METALLIC BIOMATERIALS FOR BIOMEDICAL APPLICATIONS

When designed a biomaterial, biocompatibility is the main targeted property. Metallic biomaterials not only must have biocompatibility but also corrosion resistance in biological environment. It is worth mentioning that corrosion is one of the main problems disturbing the biocompatibility of metallic implants in the human body, especially, because *in-vivo* conditions are very aggressive for implanted alloys. *In-vivo* biological environment may involve various interactions between the biomaterial and proteins, cells, fluids, and enzymes, which can induce undesired chemical reactions in the surface of the metallic implant, resulting in allergy and/or toxicity. For instance, metals such as nickel when released into the human body can cause cancer, allergy, and genotoxic and mutagenic activities (accepted rate must be 20 mg per kg each day) (Salnikow & Zhitkovich, 2007; Denkhaus & Salnikow, 2002).

Currently, research efforts are focused on the development of new alloy compositions that can possess excellent biocompatibility but also high corrosion resistance in biological environment. One example of this trend has been the development of nickel-free steels, which are often associated with the FeCrMnMoN systems (Talha & Sinha 2013; Talha & Sinha, 2015). The role of the nitrogen in these alloys is to increase the passive ðlm stability in biological environments. Nickel free steels have also good resistance to localized corrosion. Overall, the formation of a passive ðlm on surfaces of metallic biomaterials makes them physiologically inert and increases their corrosion resistance, these are aspect currently studied when developed new biomedical alloys.

In recent years, niobium (Nb) and its alloys have shown a great potential for biomedical applications, for example, the biocompatibility of niobium in both hard and soft tissue environments have reported to be as good as titanium alloys (Rubitschek et la., 2012; Wang et al. 2016). This metallic material also present high corrosion resistance in physiological environment with the formation of a passive film, which has a superior breakdown potential (10V) than that of Ti-6Al-4V (4V), as an alloy element, Nb has also demonstrated beneficial effects for improving biocompatibility; it is the case of the Ti-6Al-7 Nb alloy, which has shown better corrosion resistance, enhanced cell proliferation and superior mitochondrial activity than conventional Ti-6Al-4V alloy (Walkowiak et al. 2012; Contu et al., 2004). These properties have been attributed to the uniform passive layer, with a smaller number of electronic defects, developed on the surface of the alloy in the presence of Nb. In fact, one of the advantages of using Nb (pure or alloyed) is related to its tendency for rapidly forming a stable and protective oxide layer. Comparing Nb with respect to other alloys such as stainless steel, cobalt-chromium and titanium, the corrosion resistance is remarkably better, which is a crucial factor that has called the attention of researchers for its use in the biomedical field (Krasicka & Amin, 2012).

Despite the interesting results obtained with niobium, this material has not been widely used in medical applications to date. This fact is mainly promoted by the poor mechanical properties of pure niobium with respect to the conventional medical grade alloys and bone (Xiao et al., 1991; Matsuno et al., 2001). In this sense, niobium alloys have been only limited to small devices such as stents for cardiovascular applications (O'Brien, 2015). However, current needs within the biomedical field include the replacement of conventional alloys that have shown good mechanical properties, moderate corrosion resistance and biocompatibility in physiological environment, but also have shown some unfavorable effects for the health of the patients. For instance, popular Ti-6Al-4V alloy will be in the incoming years one of the candidates to disappear progressively of orthopedic applications such as several years ago occurred with stainless steels. This situation is supported by the presence of aluminum and vanadium, which are elements associated with the occurrence of Alzheimer's and Parkinson's diseases and toxicity in the organism (Yegambaram et al., 2015; Wilk et al., 2017). Nevertheless, if the presence of aluminum and vanadium are considered as negative point in the use of Ti-6Al-4V, it is not less important the fact that this alloy presents a high elastic modulus (110 GPa), which is the main cause of stress shielding and failure of orthopedic implants (Au et al., 2007; Ridzwan et al., 2007). Thus, considering that implants are more and more being placed in younger people and likely remain in the human body for longer periods of time, implant materials with better biocompatibility and without a harmful effect on the health are a requirement to improve the quality of life of patients. In this sense, most titanium alloys of the new generation will be focused on employing alloy elements that can provide a compromise between biocompatibility in longer periods of time and mechanical properties that can approach to that of the human bone. In that order of ideas, Ti – Nb systems have risen as one candidate for the next generation of biomedical alloys (Karre et al., 2019; Ehtemam eet al. 2019). In fact, various studies have devoted to the to these systems in the last years, and expectations suggests that in the following years new investigations can stablish a composition that can perform adequately in the human body. Other promising options for orthopedic and long-term implantation are ternary or quaternary systems involving niobium, zirconium and tantalum additions to titanium. Tantalum (Ta) as Nb is also considered as another promising metal for biomedical applications. This metallic material has been studied in physiological environment, presenting good corrosion resistance and biocompatibility. However, due to similar reasons than Nb, Ta is also being employed as an alloy element for the design of next generation of medical grade alloys. Alloy systems containing these elements such as Ti15Zr4Nb4Ta alloy have presented better corrosion resistance and enhanced biocompatibility than Ti6Al7Nb alloys (Niinomi & Nakai, 2019; Ijaz et al., 2018).

Opposite to the tendency of developing a durable, biocompatible, and non-harmful metallic alloys, there is also a well stablished need for the development of biodegradable metals for certain biomedical applications by controlling corrosion and mechanical properties of the alloy. These materials degrade in the physiological environment once their purpose is accomplished (Drelich, 2019; Dehghanghadikolaei et al. 2019). This concept is remarkably wider than the conventional concept of permanent biomedical devices. However, compared against conventional biomedical alloys, research on biodegradable metals is in an early stage with enough field for innovation. Promising biodegradable metals are based on elements that exist in the human body; for example, iron and magnesium alloys are included among the biodegradable alloys. Especially, magnesium alloys have shown low thrombogenicity and excellent biocompatibility, which is attributed to its essential presence in the human body and high permissible content in body fluid. Another advantage of magnesium alloys is associated to the elastic modulus, which is similar to that of the human bone, preventing stress shielding and allowing natural bone remodeling. Mg-Zn and Mg-Ca alloys have shown to have shown to stimulate tissue healing while degradation occurs (Gu et al. 2018; Roche et al., 2019).

Alternatively, Fe-based alloys (crystalline and metallic glasses) are also suitable as biodegradable materials. Interestingly, when implanted, OH–and Fe2+ions are produced in the region of the prosthesis leading to the oxidation of Fe2+ions. This corrosion process is biocompatible; however, this fact delays oxygen transport required for the continuous degradation of iron (Cimpoeşu et al., 2018). Consequently, the degradation rate of the iron alloy is slower than the rate of tissue healing. Currently, several attempts are focused on accelerate the corrosion rate of Fe-based alloys by alloying with different elements to induce galvanic corrosion that can help to increase degradation rates in physiological environment. One feasible option to improve degradation rate of biodegradable Fe-based alloys is the use of AM to create porous Fe-base structures. This geometrical solution can enhance degradation rates without detrimental of biocompatible properties. However, further investigation is required in the following years since there is currently few available information about this topic. The authors of this chapter are currently working on the processing of Fe-base metallic glass alloys by means of DMLS, having as a target in a near future considering the study of Fe-base metallic glass printed structures as biodegradable alloys in the biomedical field.

Additive manufacturing, surface treatments and the application of coatings are still alternatives to process metallic materials to improve biocompatibility and corrosion resistance of biomedical implants. One future challenge for the AM industry lies on the fact that many of the promising biomedical alloys (non-biodegradable and biodegradable) are still not available, the AM industry should work in widening the number of biomedical alloys available in the market. Otherwise, latest advances in the design of biomedical alloys will last longtime before they can be exploited in AM. In the AM field, customization of implants and miniaturization (development of small medical devices) will be the focus of attention in the incoming years (Chung et al., 2019). In terms of biocompatibility, biodegradation and enhancement of corrosion properties, a lot of parametrization work and 3D design should be performed for evaluating these properties in AM parts.

Figure 17. Fe-base metallic glass powder being processed by DMLS at CIATEQ A.C, México. (Powder is a courtesy of EPSON ATMIX, Japan).

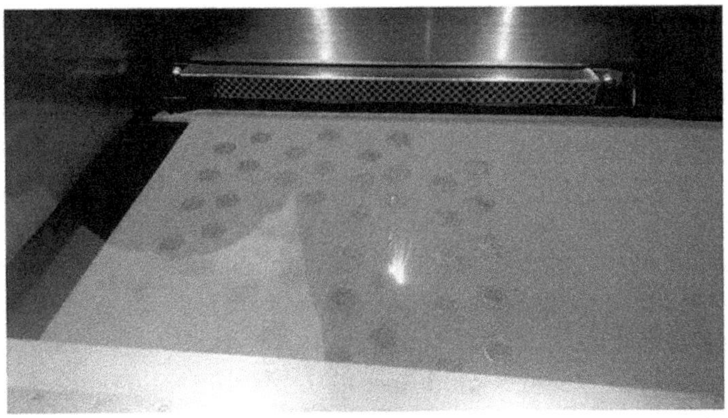

In the field of surface treatments and coatings manufacturing, recent efforts have been focused on applying the latest advances in the design of biomedical alloys. The challenge of biomedical specialists is how to combine the known properties of conventional alloys with the promising properties of new alloys, taking advantage of conventional alloys as supporting material and the new alloy deposited as a coating or formed by a surface treatment on the surface of the implant. For instance, as mentioned before in this chapter, Nb surfaces are known for the formation of a highly stable and inert oxide layer. This fact provides the basis for the excellent corrosion resistance and biocompatibility when applied in the surface of an implant. Up to date, several technologies have been employed for applying niobium or niobium oxides as coating materials, with the aim of developing an implant with attractive bulk mechanical properties with ideal biocompatible surfaces. Physical vapor deposition (PVD) techniques and sol-gel methods have been employed for the deposition of Nb and Nb nitride coatings on 316l SS, Ti and Ti-6Al-4V alloys (Biswas et al., 2017; Guzman et al., 2018; Ramirez et al., 2011). Similarly, Ta metal, Ta oxide, Ta nitride and Ta Carbides has been explored recently by magnetron sputtering and by electrochemical methods, and the tendency exploring new coating deposition methods that can offer a compromise between biological response and mechanical stability of these coatings in physiological environment. One of the promising coating deposition methods to obtain this type of coating surfaces in the future is the cold gas spray process (CGS) (Moridi et al., 2014). This thermal spray technique is suggested for the application of new biomedical alloys because the processing conditions involve lower temperature than conventional HVOF and APS techniques and the mechanism of coating formation is based on the plastic deformation of the alloys, offering the possibility to keep the initial properties of the alloy and/or open the possibility to tailor microstructures in the final coating on the surface of the prosthesis. The CGS technique also offer strong bonding with the coated surface and no significant thermal and mechanical changes are promoted on the substrate, being ideal for processing implants in the biomedical industry.

CONCLUSION

The present book chapter focused on corrosion behavior of metallic and coating biomaterials in the human body, since the increasing use of orthopedic and dental implants has boosted the interest of the biomedical field in improving the long-term behavior of these type of materials. An introduction to biomaterials and main techniques for the evaluation of corrosion and ionic interaction between biomaterials and body fluids are presented. A number of metallic biomaterials have been developed so far; however, only a very small number of these have ever been put into practical applications. Greater numbers of such alloys will be considered for practical applications in the future given recent advances in metallurgy and understanding of interaction of some alloy elements with the physiological environment. Currently, titanium alloys have been developed to provide advanced functionalities such as a high biological biocompatibility; low Young's modulus, high fracture toughness, and high wear resistance. Other advanced biological functionalities are currently the focus of attention such as long-term stability and/or biodegradation; these new functionalities have been created in titanium alloys by means of designing new compositions including elements such as niobium, tantalum and zirconium. The use of electrochemical techniques and *in-vitro* tests are the first approach to comprehend the behavior of new biomaterials in physiological environment and currently, ISO and ASTM standards provide a guide for the evaluation of these materials in *in-vitro* conditions.

ACKNOWLEDGMENT

The authors acknowledge the support of the National Council of Science and Technology (CONACYT) México through its "Cátedras CONACYT" program, specifically projects 848 and 881. The support from CENAPROT (national laboratory of thermal spray) and from Institute of Physical Science at the national autonomous university of Mexico (ICF-UNAM) is also acknowledged. The present work is derived from scientific advancements obtained in the basic science project (253365) funded by CONACYT.

REFERENCES

Abe, Y., Kokubo, T., & Yamamuro, T. (1990). Apatite coating on ceramics, metals and polymers utilizing a biological process. *Journal of Materials Science. Materials in Medicine*, *1*(4), 233–238. doi:10.1007/BF00701082

Aherwar, A., Singh, A. K., & Patnaik, A. (2015). Current and future biocompatibility aspects of biomaterials for hip prosthesis. *AIMS Bioengineering*, *3*(1), 23–43. doi:10.3934/bioeng.2016.1.23

Al-Jabbari, Y. S., Ntasi, A., Gaintatzopoulou, M., Mueller, W. D., Eliades, G., Sherif, E. S. M., & Zinelis, S. (2016). Elemental, Morphological, and Corrosion Characterization of Different Surface States of Co-Cr Alloy for Prosthodontic Applications. Academic Press.

Al-Ketan, O., Rowshan, R., & Al-Rub, R. K. A. (2018). Topology-mechanical property relationship of 3D printed strut, skeletal, and sheet based periodic metallic cellular materials. *Additive Manufacturing*, *19*, 167–183. doi:10.1016/j.addma.2017.12.006

Allo, B. A., Costa, D. O., Dixon, S. J., Mequanint, K., & Rizkalla, A. S. (2012). Bioactive and biodegradable nanocomposites and hybrid biomaterials for bone regeneration. *Journal of Functional Biomaterials*, *3*(2), 432–463. doi:10.3390/jfb3020432 PMID:24955542

Arabnejad, S., Johnston, R. B., Pura, J. A., Singh, B., Tanzer, M., & Pasini, D. (2016). High-strength porous biomaterials for bone replacement: A strategy to assess the interplay between cell morphology, mechanical properties, bone ingrowth and manufacturing constraints. *Acta Biomaterialia*, *30*, 345–356. doi:10.1016/j.actbio.2015.10.048 PMID:26523335

Arrieta-González, C. D., Porcayo-Calderon, J., Salinas-Bravo, V. M., Chacon-Nava, J. G., Martinez-Villafañe, A., & Gonzalez-Rodriguez, J. G. (2011). Corrosion behavior of Ni-Cr based coatings in simulated human body fluid environment. *International Journal of Electrochemical Science*, *6*, 3644–3655.

Arrieta-Gonzalez, C. D., Porcayo-Calderon, J., Salinas-Bravo, V. M., Gonzalez-Rodriguez, J. G., & Chacon-Nava, J. G. (2011). Electrochemical behavior of Fe3Al modified with Ni in Hank's solution. *International Journal of Electrochemical Science*, *6*(9), 4016–4031.

Atapour, M., Pilchak, A. L., Frankel, G. S., & Williams, J. C. (2011). Corrosion behavior of β titanium alloys for biomedical applications. *Materials Science and Engineering: C, 31*(5), 885-891.

Au, A. G., Raso, V. J., Liggins, A. B., & Amirfazli, A. (2007). Contribution of loading conditions and material properties to stress shielding near the tibial component of total knee replacements. *Journal of Biomechanics*, *40*(6), 1410–1416. doi:10.1016/j.jbiomech.2006.05.020 PMID:16846605

Balakrishnan, A., Lee, B. C., Kim, T. N., & Panigrahi, B. B. (2008). Corrosion behaviour of ultra fine grained titanium in simulated body fluid for implant application. *Trends in Biomaterials & Artificial Organs*, *22*(1), 58–64.

Balani, K., Verma, V., Agarwal, A., & Narayan, R. (2015). *A Materials Science and Engineering Perspective*. John Wiley & Sons, The American Ceramic Society.

Ban, S., Iwaya, Y., Kono, H., & Sato, H. (2006). Surface modification of titanium by etching in concentrated sulfuric acid. *Dental Materials*, *22*(12), 1115–1120.

Bandyopadhyay, A., Espana, F., Balla, V. K., Bose, S., Ohgami, Y., & Davies, N. M. (2010). Influence of porosity on mechanical properties and in vivo response of Ti6Al4V implants. *Acta Biomaterialia*, *6*(4), 1640–1648. doi:10.1016/j.actbio.2009.11.011 PMID:19913643

Bandyopadhyay, A., Mitra, I., Shivaram, A., Dasgupta, N., & Bose, S. (2019). Direct comparison of additively manufactured porous titanium and tantalum implants towards in vivo osseointegration. *Additive Manufacturing*, *28*, 259–266. doi:10.1016/j.addma.2019.04.025 PMID:31406683

Banks, J. (2013). Adding value in additive manufacturing: Researchers in the United Kingdom and Europe look to 3D printing for customization. *IEEE Pulse*, *4*(6), 22–26. doi:10.1109/MPUL.2013.2279617 PMID:24233187

Barrere, F., van Blitterswijk, C. D., De Groot, K., & Layrolle, P. (2002). Influence of ionic strength and carbonate on the Ca-P coating formation from SBF× 5 solution. *Biomaterials*, *23*(9), 1921–1930. doi:10.1016/S0142-9612(01)00318-0 PMID:11996032

Barrere, F. V., Van Blitterswijk, C. A., De Groot, K., & Layrolle, P. (2002). Nucleation of biomimetic Ca–P coatings on Ti6Al4V from a SBF× 5 solution: Influence of magnesium. *Biomaterials*, *23*(10), 2211–2220. doi:10.1016/S0142-9612(01)00354-4 PMID:11962662

Berndt, C. C., Hasan, F., Tietz, U., & Schmitz, K. P. (2014). A review of HAP coatings manufactured by thermal spray. In *Advances in calcium phosphate biomaterials* (pp. 267–329). Springer. doi:10.1007/978-3-642-53980-0_9

Biswas, B., Purandare, Y., Sugumaran, A. A., Loch, D. A., Creasey, S., Khan, I., Ehiasarian, A. P., & Hovsepian, P. E. (2017). Defect growth in multilayer chromium nitride/niobium nitride coatings produced by combined high power impulse magnetron sputtering and unbalance magnetron sputtering technique. *Thin Solid Films*, *636*, 558–566. doi:10.1016/j.tsf.2017.06.027

Bobbio, A. (1972). The first endosseous alloplastic implant in the history of man. *Bulletin of the History of Dentistry*, *20*, 1–6. PMID:4505221

Bose, S., Banerjee, D., & Bandyopadhyay, A. (2017). Introduction to biomaterials and devices for bone disorders. In *Materials for Bone Disorders* (pp. 1–27). Academic Press. doi:10.1016/B978-0-12-802792-9.00001-X

Brown, P. W., & Constantz, B. (2017). *HAP and related materials*. CRC press.

Calignano, F., Galati, M., Iuliano, L., & Minetola, P. (2019). Design of Additively Manufactured Structures for Biomedical Applications: A Review of the Additive Manufacturing Processes Applied to the Biomedical Sector. *Journal of Healthcare Engineering*, *2019*, 2019. doi:10.1155/2019/9748212 PMID:30992744

Choubey, A., Basu, B., & Balasubramaniam, R. (2005). Electrochemical behavior of Ti-based alloys in simulated human body fluid environment. *Trends in Biomaterials & Artificial Organs*, *18*(2), 64–72.

Chung, K. S., Shin, D. A., Kim, K. N., Ha, Y., Yoon, H., & Yi, S. (2019). Vertebral reconstruction with a customized 3D-printed spine implant replacing large vertebral defect with 3-year follow-up: A case report. *World Neurosurgery*.

Cimpoeşu, N., Săndulache, F., Istrate, B., Cimpoeşu, R., & Zegan, G. (2018). Electrochemical Behavior of Biodegradable FeMnSi–MgCa Alloy. *Metals*, *8*(7), 541. doi:10.3390/met8070541

Conner, K. A., Sabatini, R., Mealey, B. L., Takacs, V. J., Mills, M. P., & Cochrann, D. L. (2003). Guided bone regeneration around titanium plasma-sprayed, acid-etched, and HAP-coated implants in the canine model. *Journal of Periodontology*, *74*(5), 658–668. doi:10.1902/jop.2003.74.5.658 PMID:12816298

Contu, F., Elsener, B., & Böhni, H. (2004). Serum effect on the electrochemical behaviour of titanium, Ti6Al4V and Ti6Al7Nb alloys in sulphuric acid and sodium hydroxide. *Corrosion Science*, *46*(9), 2241–2254. doi:10.1016/j.corsci.2004.01.005

Cui, X., Kim, H. M., Kawashita, M., Wang, L., Xiong, T., Kokubo, T., & Nakamura, T. (2009). Preparation of bioactive titania films on titanium metal via anodic oxidation. *Dental Materials*, *25*(1), 80–86. doi:10.1016/j.dental.2008.04.012 PMID:18599115

Davis, J. R. (Ed.). (2004). *Handbook of thermal spray technology*. ASM International.

Dehghanghadikolaei, A., Ibrahim, H., Amerinatanzi, A., & Elahinia, M. (2019). Biodegradable magnesium alloys. In *Metals for Biomedical Devices* (pp. 265–289). Woodhead Publishing. doi:10.1016/B978-0-08-102666-3.00009-2

Denkhaus, E., & Salnikow, K. (2002). Nickel essentiality, toxicity, and carcinogenicity. *Critical Reviews in Oncology/Hematology*, *42*(1), 35–56. doi:10.1016/S1040-8428(01)00214-1 PMID:11923067

Drelich, J. W. (2019). Characterization of Biodegradable Medical Materials. *JOM*, *71*(4), 1404–1405. doi:10.100711837-019-03381-3 PMID:31693017

Duffo, G. S., & Castillo, E. Q. (2004). Development of an artificial saliva solution for studying the corrosion behavior of dental alloys. *Corrosion*, *60*(6), 594–602. doi:10.5006/1.3287764

Ehtemam-Haghighi, S., Attar, H., Dargusch, M. S., & Kent, D. (2019). Microstructure, phase composition and mechanical properties of new, low cost Ti-Mn-Nb alloys for biomedical applications. *Journal of Alloys and Compounds*, *787*, 570–577. doi:10.1016/j.jallcom.2019.02.116

Eliaz, N. (2019). Corrosion of metallic biomaterials: A review. *Materials (Basel)*, *12*(3), 407. doi:10.3390/ma12030407 PMID:30696087

Ficai, A., Andronescu, E., Voicu, G., & Ficai, D. (2011). Article. In B. Attaf (Ed.), Advances in Collagen/Hydroxyapatite Composite Materials (pp. 3–32). Advances in Composite Materials for Medicine and Nanotechnology.

Frangini, S., & De Cristofaro, N. (2003). Analysis of the galvanostatic polarization method for determining reliable pitting potentials on stainless steels in crevice-free conditions. *Corrosion Science*, *45*(12), 2769–2786. doi:10.1016/S0010-938X(03)00102-1

Gadow, R., Killinger, A., & Stiegler, N. (2010). HAP coatings for biomedical applications deposited by different thermal spray techniques. *Surface and Coatings Technology*, *205*(4), 1157–1164. doi:10.1016/j.surfcoat.2010.03.059

Galati, M., & Iuliano, L. (2018). A literature review of powder-based electron beam melting focusing on numerical simulations. *Additive Manufacturing*, *19*, 1–20. doi:10.1016/j.addma.2017.11.001

Ghouse, S., Reznikov, N., Boughton, O. R., Babu, S., Ng, K. G., Blunn, G., Cobb, J. P., Stevens, M. M., & Jeffers, J. R. (2019). The design and in vivo testing of a locally stiffness-matched porous scaffold. *Applied Materials Today*, *15*, 377–388. doi:10.1016/j.apmt.2019.02.017 PMID:31281871

Goldberg, A. J., & Kuhn, L. T. (2018). Biomaterials. In Regenerative Engineering. CRC Press.

Gu, X., Wang, F., Xie, X., Zheng, M., Li, P., Zheng, Y., Qin, L., & Fan, Y. (2018). In vitro and in vivo studies on as-extruded Mg-5.25 wt.% Zn-0.6 wt.% Ca alloy as biodegradable metal. *Science China Materials*, *61*(4), 619–628. doi:10.100740843-017-9205-x

Gurrappa, I., & Reddy, D. V. (2005). Characterisation of titanium alloy, IMI-834 for corrosion resistance under different environmental conditions. *Journal of Alloys and Compounds*, *390*(1-2), 270–274. doi:10.1016/j.jallcom.2004.08.040

Guzman, P., Caballero, J. L., Orozco-Hernández, G., Aperador, W., & Caicedo, J. C. (2018). Tribocorrosion Behavior of Niobium-Based Thin Films for Biomedical Applications. *Tribology in Industry, 40*(4).

Hansen, D. C. (2008). Metal corrosion in the human body: The ultimate bio-corrosion scenario. *The Electrochemical Society Interface, 17*(2), 31.

Haschek, W. M., Rousseaux, C. G., Wallig, M. A., Bolon, B., & Ochoa, R. (Eds.). (2013). *Haschek and Rousseaux's handbook of toxicologic pathology*. Academic Press.

Heimann, R. B. (2016). Plasma-sprayed hydroxylapatite-based coatings: Chemical, mechanical, microstructural, and biomedical properties. *Journal of Thermal Spray Technology, 25*(5), 827–850. doi:10.100711666-016-0421-9

Henao, J., Cruz-bautista, M., Hincapie-Bedoya, J., Ortega-Bautista, B., Corona-Castuera, J., Giraldo-Betancur, A. L., Espinosa-Arbelaez, D. G., Alvarado-Orozco, J. M., Clavijo-Mejía, G. A., Trapaga-Martínez, L. G., & Poblano-Salas, C. A. (2018). HVOF HAP/Titania-Graded Coatings: Microstructural, Mechanical, and In Vitro Characterization. *Journal of Thermal Spray Technology, 27*(8), 1302–1321. doi:10.100711666-018-0811-2

Hernández, M. R., & Cañas, J. C. (2006). Citotoxicidad in vitro: Sistema para la evaluación de biomateriales y equipos médicos implantables en Cuba. *Revista CENIC. Ciencias Biológicas, 37*(3), 173–176.

Hernández-Montes, V., Betancur-Henao, C. P., & Santa-Marín, J. F. (2017). Titanium dioxide coatings on magnesium alloys for biomaterials: A review. *Dyna (Bilbao), 84*(200), 261–270. doi:10.15446/dyna.v84n200.59664

Hirakawa, K., Jacobs, J. J., Urban, R., & Saito, T. (2004). Mechanisms of failure of total hip replacements: Lessons learned from retrieval studies. *Clinical Orthopaedics and Related Research, 420*, 10–17. doi:10.1097/00003086-200403000-00003 PMID:15057073

Huang, H. H., Chiu, Y. H., Lee, T. H., Wu, S. C., Yang, H. W., Su, K. H., & Hsu, C. C. (2003). Ion release from NiTi orthodontic wires in artificial saliva with various acidities. *Biomaterials, 24*(20), 3585–3592. doi:10.1016/S0142-9612(03)00188-1 PMID:12809787

Hwang, C. J., Shin, J. S., & Cha, J. Y. (2001). Metal release from simulated fixed orthodontic appliances. *American Journal of Orthodontics and Dentofacial Orthopedics, 120*(4), 383–391. doi:10.1067/mod.2001.117911 PMID:11606963

Ijaz, M. F., Zhukova, Y., Konopatsky, A., Dubinskiy, S., Korobkova, A., Pustov, Y., Brailovski, V., & Prokoshkin, S. (2018). Effect of Ta addition on the electrochemical behavior and functional fatigue life of metastable Ti-Zr-Nb based alloy for indwelling implant applications. *Journal of Alloys and Compounds, 748*, 51–56. doi:10.1016/j.jallcom.2018.03.033

Jacobs, J. J., Gilbert, J. L., & Urban, R. M. (1998). Current concepts review-corrosion of metal orthopaedic implants. *Jbjs, 80*(2), 268–282. doi:10.2106/00004623-199802000-00015

Kamimura, T., & Stratmann, M. (2001). The influence of chromium on the atmospheric corrosion of steel. *Corrosion Science, 43*(3), 429–447. doi:10.1016/S0010-938X(00)00098-6

Karre, R., Kodli, B. K., Rajendran, A., Nivedhitha, J., Pattanayak, D. K., Ameyama, K., & Dey, S. R. (2019). Comparative study on Ti-Nb binary alloys fabricated through spark plasma sintering and conventional P/M routes for biomedical application. *Materials Science and Engineering C*, *94*, 619–627. doi:10.1016/j.msec.2018.10.006 PMID:30423747

Karthega, M., Raman, V., & Rajendran, N. (2007). Influence of potential on the electrochemical behaviour of β titanium alloys in Hank's solution. *Acta Biomaterialia*, *3*(6), 1019–1023. doi:10.1016/j.actbio.2007.02.009 PMID:17478134

Kim, H. M., Himeno, T., Kawashita, M., Kokubo, T., & Nakamura, T. (2004). The mechanism of biomineralization of bone-like apatite on synthetic hydroxyapatite: An in vitro assessment. *Journal of the Royal Society, Interface*, *1*(1), 17–22. doi:10.1098/rsif.2004.0003 PMID:16849149

Koizumi, H., Takeuchi, Y., Imai, H., Kawai, T., & Yoneyama, T. (2019). Application of titanium and titanium alloys to fixed dental prostheses. *Journal of Prosthodontic Research*, *63*(3), 266–270. doi:10.1016/j.jpor.2019.04.011 PMID:31147298

Kong, L., Gao, Y., Lu, G., Gong, Y., Zhao, N., & Zhang, X. (2006). A study on the bioactivity of chitosan/nano-hydroxyapatite composite scaffolds for bone tissue engineering. *European Polymer Journal*, *42*(12), 3171–3179. doi:10.1016/j.eurpolymj.2006.08.009

Krasicka-Cydzik, E., & Amin, A. N. (2012). Anodic layer formation on titanium and its alloys for biomedical applications. In *Titanium Alloys-Towards Achieving Enhanced Properties for Diversified Applications*. InTech. doi:10.5772/34395

Kumar, S., Narayanan, T. S., Raman, S. G. S., & Seshadri, S. K. (2010). Surface modification of CP-Ti to improve the fretting-corrosion resistance: Thermal oxidation vs. anodizing. *Materials Science and Engineering C*, *30*(6), 921–927. doi:10.1016/j.msec.2010.03.024

Kumari, R., & Majumdar, J. D. (2017). Studies on corrosion resistance and bio-activity of plasma spray deposited hydroxylapatite (HA) based Titania and ZrO2 dispersed composite coatings on titanium alloy (Ti-6Al-4V) and the same after post spray heat treatment. *Applied Surface Science*, *420*, 935–943. doi:10.1016/j.apsusc.2017.05.208

Li, M., & Christofides, P. D. (2009). Modeling and control of high-velocity oxygen-fuel (HVOF) thermal spray: A tutorial review. *Journal of Thermal Spray Technology*, *18*(5-6), 753.

Lima, R. S., & Marple, B. R. (2007). Thermal spray coatings engineered from nanostructured ceramic agglomerated powders for structural, thermal barrier and biomedical applications: A review. *Journal of Thermal Spray Technology*, *16*(1), 40–63. doi:10.100711666-006-9010-7

Longhitano, G. A., Arenas, M. A., Conde, A., Larosa, M. A., Jardini, A. L., de Carvalho Zavaglia, C. A., & Damborenea, J. J. (2018). Heat treatments effects on functionalization and corrosion behavior of Ti-6Al-4V ELI alloy made by additive manufacturing. *Journal of Alloys and Compounds*, *765*, 961–968. doi:10.1016/j.jallcom.2018.06.319

Manaranche, C., & Hornberger, H. (2007). A proposal for the classification of dental alloys according to their resistance to corrosion. *Dental Materials*, *23*(11), 1428–1437. doi:10.1016/j.dental.2006.11.030 PMID:17466365

Manivasagam, G., Dhinasekaran, D., & Rajamanickam, A. (2010). Biomedical implants: Corrosion and its prevention-a review. *Recent Patents on Corrosion Science*, *2*(1), 40–54. doi:10.2174/1877610801002010040

Marchi, J. (Ed.). (2016). *Biocompatible glasses: from bone regeneration to cancer treatment* (Vol. 53). Springer. doi:10.1007/978-3-319-44249-5

Matsuno, H., Yokoyama, A., Watari, F., Uo, M., & Kawasaki, T. (2001). Biocompatibility and osteogenesis of refractory metal implants, titanium, hafnium, niobium, tantalum and rhenium. *Biomaterials*, *22*(11), 1253–1262. doi:10.1016/S0142-9612(00)00275-1 PMID:11336297

Matusiewicz, H. (2014). Potential release of in vivo trace metals from metallic medical implants in the human body: From ions to nanoparticles–a systematic analytical review. *Acta Biomaterialia*, *10*(6), 2379–2403. doi:10.1016/j.actbio.2014.02.027 PMID:24565531

Melero, H. C., Sakai, R. T., Vignatti, C. A., Benedetti, A. V., Fernández, J., Guilemany, J. M., & Suegama, P. H. (2018). Corrosion Resistance Evaluation of HVOF Produced HAP and Titania-HAP Coatings in Hanks' Solution. *Materials Research*, *21*(2). Advance online publication. doi:10.1590/1980-5373-mr-2017-0210

Mikulewicz, M., Wołowiec, P., Loster, B. W., & Chojnacka, K. (2015). Do soft drinks affect metal ions release from orthodontic appliances? *Journal of Trace Elements in Medicine and Biology*, *31*, 74–77. doi:10.1016/j.jtemb.2015.03.007 PMID:26004895

Moridi, A., Hassani-Gangaraj, S. M., Guagliano, M., & Dao, M. (2014). Cold spray coating: Review of material systems and future perspectives. *Surface Engineering*, *30*(6), 369–395. doi:10.1179/1743294414Y.0000000270

Niinomi, M., & Boehlert, C. J. (2015). Titanium alloys for biomedical applications. In *Advances in Metallic Biomaterials* (pp. 179–213). Springer. doi:10.1007/978-3-662-46836-4_8

Niinomi, M., & Nakai, M. (2019). Ti-Based Biomedical Alloys. In *Novel Structured Metallic and Inorganic Materials* (pp. 61–76). Springer. doi:10.1007/978-981-13-7611-5_3

Nikolopoulou, F. (2006). Saliva and dental implants. *Implant Dentistry*, *15*(4), 372–376. doi:10.1097/01.id.0000239320.57403.8d PMID:17172955

O'Brien, B. (2015). Niobium biomaterials. In *Advances in Metallic Biomaterials* (pp. 245–272). Springer. doi:10.1007/978-3-662-46836-4_11

Park, B. J., & Kim, Y. K. (2003). Metallic biomaterials. *Balance (Alexandria, Va.)*, *1*, 50.

Park, J., & Lakes, R. S. (2007). *Biomaterials: an introduction*. Springer Science & Business Media.

Peralta-Lopez, D., Sotelo-Mazon, O., Henao, J., Porcayo-Calderon, J., Valdez, S., Salinas-Solano, G., & Martinez-Gomez, L. (2017). Electrochemical Study of Three Stainless Steel Alloys and Titanium Metal in Cola Soft Drinks. *Journal of Electrochemical Science and Technology*, *8*(4), 294–306. doi:10.33961/JECST.2017.8.4.294

Poitout, D. G. (Ed.). (2004). *Biomechanics and biomaterials in orthopedics*. doi:10.1007/978-1-4471-3774-0

Ramírez, G., Rodil, S. E., Arzate, H., Muhl, S., & Olaya, J. J. (2011). Niobium based coatings for dental implants. *Applied Surface Science*, *257*(7), 2555–2559. doi:10.1016/j.apsusc.2010.10.021

Ratner, B. D., Hoffman, A. S., Schoen, F. J., & Lemons, J. E. (2006). Biomaterials science: An introduction to materials in medicine. *MRS Bulletin*, *31*, 59.

Ridzwan, M. I. Z., Shuib, S., Hassan, A. Y., Shokri, A. A., & Ibrahim, M. M. (2007). Problem of stress shielding and improvement to the hip implant designs: A review. *J. Med. Sci*, *7*(3), 460–467. doi:10.3923/jms.2007.460.467

Rizzuti, S., De Napoli, L., & Ventra, S. (2019). The Influence of Build Orientation on the Flatness Error in Artifact Produced by Direct Metal Laser Sintering (DMLS) Process. In *Advances on Mechanics, Design Engineering and Manufacturing II* (pp. 463–472). Springer. doi:10.1007/978-3-030-12346-8_45

Rocha, L. A., Oliveira, F., Cruz, H. V., Sukotjo, C., & Mathew, M. T. (2013). Bio-tribocorrosion in dental applications. In *Bio-tribocorrosion in biomaterials and medical implants* (pp. 223–249). Woodhead Publishing. doi:10.1533/9780857098603.3.223

Roche, V., Koga, G. Y., Matias, T. B., Kiminami, C. S., Bolfarini, C., Botta, W. J., ... Junior, A. J. (2019). Degradation of biodegradable implants: The influence of microstructure and composition of Mg-Zn-Ca alloys. *Journal of Alloys and Compounds*, *774*, 168–181. doi:10.1016/j.jallcom.2018.09.346

Rubitschek, F., Niendorf, T., Karaman, I., & Maier, H. J. (2012). Corrosion fatigue behavior of a biocompatible ultrafine-grained niobium alloy in simulated body fluid. *Journal of the Mechanical Behavior of Biomedical Materials*, *5*(1), 181–192. doi:10.1016/j.jmbbm.2011.08.023 PMID:22100093

Saber-Samandari, S., Saber-Samandari, S., Ghonjizade-Samani, F., Aghazadeh, J., & Sadeghi, A. (2016). Bioactivity evaluation of novel nanocomposite scaffolds for bone tissue engineering: The impact of hydroxyapatite. *Ceramics International*, *42*(9), 11055–11062. doi:10.1016/j.ceramint.2016.04.002

Salgado-Salgado, R. J., Porcayo-Calderon, J., Sotelo-Mazon, O., Rodriguez-Diaz, R. A., Salinas-Solano, G., Salinas-Bravo, V. M., & Martinez-Gomez, L. (2016). Effect of Ag addition on the electrochemical performance of Cu10Al in artificial saliva. *Bioinorganic Chemistry and Applications*, *2016*, 2016. doi:10.1155/2016/4792583 PMID:27660601

Salnikow, K., & Zhitkovich, A. (2007). Genetic and epigenetic mechanisms in metal carcinogenesis and cocarcinogenesis: Nickel, arsenic, and chromium. *Chemical Research in Toxicology*, *21*(1), 28–44. doi:10.1021/tx700198a PMID:17970581

Savio, G., Rosso, S., Meneghello, R., & Concheri, G. (2018). Geometric modeling of cellular materials for additive manufacturing in biomedical field: A review. *Applied Bionics and Biomechanics*, *2018*, 2018. doi:10.1155/2018/1654782 PMID:29487626

Schroeder, A., van der Zypen, E., Stich, H., & Sutter, F. (1981). The reactions of bone, connective tissue, and epithelium to endosteal implants with titanium-sprayed surfaces. *Journal of Maxillofacial Surgery*, *9*, 15–25. doi:10.1016/S0301-0503(81)80007-0 PMID:6939769

Simchi, A. (2006). Direct laser sintering of metal powders: Mechanism, kinetics and microstructural features. *Materials Science and Engineering A*, *428*(1-2), 148–158. doi:10.1016/j.msea.2006.04.117

Singh, G., Singh, H., & Sidhu, B. S. (2013). Corrosion behavior of plasma sprayed HAP and HAP-silicon oxide coatings on AISI 304 for biomedical application. *Applied Surface Science*, *284*, 811–818. doi:10.1016/j.apsusc.2013.08.013

Singh, T. P., Singh, H., & Singh, H. (2012). Characterization, corrosion resistance, and cell response of high-velocity flame-sprayed HA and HA/TiO 2 coatings on 316L SS. *Journal of Thermal Spray Technology, 21*(5), 917-927

Sivakumar, B., Pathak, L. C., & Singh, R. (2017). Role of surface roughness on corrosion and fretting corrosion behaviour of commercially pure titanium in Ringer's solution for bio-implant application. *Applied Surface Science*, *401*, 385–398. doi:10.1016/j.apsusc.2017.01.033

Song, W. H., Jun, Y. K., Han, Y., & Hong, S. H. (2004). Biomimetic apatite coatings on micro-arc oxidized titania. *Biomaterials*, *25*(17), 3341–3349. doi:10.1016/j.biomaterials.2003.09.103 PMID:15020106

Song, Y. W., Shan, D. Y., & Han, E. H. (2007). Corrosion behaviors of electroless plating Ni–P coatings deposited on magnesium alloys in artificial sweat solution. *Electrochimica Acta*, *53*(4), 2009–2015. doi:10.1016/j.electacta.2007.08.062

Sun, L., Berndt, C. C., Gross, K. A., & Kucuk, A. (2001). Material fundamentals and clinical performance of plasma-sprayed HAP coatings: A review. *Journal of Biomedical Materials Research: An Official Journal of The Society for Biomaterials, The Japanese Society for Biomaterials, and The Australian Society for Biomaterials and the Korean Society for Biomaterials*, *58*(5), 570–592. doi:10.1002/jbm.1056 PMID:11505433

Takeuchi, M., Abe, Y., Yoshida, Y., Nakayama, Y., Okazaki, M., & Akagawa, Y. (2003). Acid pretreatment of titanium implants. *Biomaterials*, *24*(10), 1821–1827.

Talha, M., Behera, C. K., & Sinha, O. P. (2013). A review on nickel-free nitrogen containing austenitic stainless steels for biomedical applications. *Materials Science and Engineering C*, *33*(7), 3563–3575. doi:10.1016/j.msec.2013.06.002 PMID:23910251

Talha, M., Behera, C. K., & Sinha, O. P. (2015). Effect of nitrogen and cold working on structural and mechanical behavior of Ni-free nitrogen containing austenitic stainless steels for biomedical applications. *Materials Science and Engineering C*, *47*, 196–203. doi:10.1016/j.msec.2014.10.078 PMID:25492189

Upadhyay, D., Panchal, M. A., Dubey, R. S., & Srivastava, V. K. (2006). Corrosion of alloys used in dentistry: A review. *Materials Science and Engineering A*, *432*(1-2), 1–11. doi:10.1016/j.msea.2006.05.003

Vallet-Regí, M. (2001). Ceramics for medical applications. *Journal of the Chemical Society, Dalton Transactions: Inorganic Chemistry*, (2), 97–108. doi:10.1039/b007852m

Vayre, B., Vignat, F., & Villeneuve, F. (2012). Metallic additive manufacturing: State-of-the-art review and prospects. *Mechanics & Industry*, *13*(2), 89–96. doi:10.1051/meca/2012003

Vieira, A. C., Ribeiro, A. R., Rocha, L. A., & Celis, J. P. (2006). Influence of pH and corrosion inhibitors on the tribocorrosion of titanium in artificial saliva. *Wear*, *261*(9), 994–1001. doi:10.1016/j.wear.2006.03.031

Walkowiak-Przybyło, M., Klimek, L., Okrój, W., Jakubowski, W., Chwiłka, M., Czajka, A., & Walkowiak, B. (2012). *Adhesion, activation, and aggregation of blood platelets and biofilm formation on the surfaces of titanium alloys Ti6Al4V and Ti6Al7Nb*. Academic Press.

Wang, W., & Alfantazi, A. (2016). Electrochemical Corrosion Behavior of Niobium Alloys as Metallic Bio-implants. In *NACE International Corrosion Conference Proceedings* (p. 1). NACE International.

Wei, K., Wang, Y., Lai, C., Ning, C., Wu, D., Wu, G., & Ye, J. (2005). Synthesis and characterization of hydroxyapatite nanobelts and nanoparticles. *Materials Letters*, *59*(2-3), 220–225. doi:10.1016/j.matlet.2004.08.034

Wilk, A., Szypulska-Koziarska, D., & Wiszniewska, B. (2017). The toxicity of vanadium on gastrointestinal, urinary and reproductive system, and its influence on fertility and fetuses malformations. Advances in Hygiene & Experimental Medicine. *Postepy Higieny i Medycyny Doswiadczalnej*, 71.

Xiao, L., Kim, Y. S., Abbaschian, R., & Hecht, R. J. (1991). Processing and mechanical properties of niobium-reinforced MoSi2 composites. *Materials Science and Engineering A*, *144*(1-2), 277–285. doi:10.1016/0921-5093(91)90234-E

Xu, W., Lu, X., Wang, L. N., Shi, Z. M., Lv, S. M., Qian, M., & Qu, X. H. (2018). Mechanical properties, in vitro corrosion resistance and biocompatibility of metal injection molded Ti-12Mo alloy for dental applications. *Journal of the Mechanical Behavior of Biomedical Materials*, *88*, 534–547. doi:10.1016/j.jmbbm.2018.08.038 PMID:30223215

Xu, Y., Lu, Y., Liang, J., & Sisson, R. D. (2019). Microstructure and corrosion behaviour of additively manufactured Ti–6Al–4V with various post-heat treatments. *Materials Science and Technology*, *35*(1), 89–97. doi:10.1080/02670836.2018.1542052

Yang, H., Yang, K., & Zhang, B. (2007). Pitting corrosion resistance of La added 316L stainless steel in simulated body fluids. *Materials Letters*, *61*(4-5), 1154–1157. doi:10.1016/j.matlet.2006.06.071

Yang, K., Zhou, C., Fan, H., Fan, Y., Jiang, Q., Song, P., & Zhang, X. (2017). Bio-functional design, application and trends in metallic biomaterials. *International Journal of Molecular Sciences*, *19*(1), 24. doi:10.3390/ijms19010024 PMID:29271916

Yaya, K., Khelfaoui, Y., Malki, B., & Kerkar, M. (2011). Numerical simulations study of the localized corrosion resistance of AISI 316L stainless steel and pure titanium in a simulated body fluid environment. *Corrosion Science*, *53*(10), 3309–3314. doi:10.1016/j.corsci.2011.06.006

Yegambaram, M., Manivannan, B., Beach, G., & Halden, R. (2015). Role of environmental contaminants in the etiology of Alzheimer's disease: A review. *Current Alzheimer Research*, *12*(2), 116–146. doi:10.2174/1567205012666150204121719 PMID:25654508

Yılmaz, E., Gökçe, A., Findik, F., Gulsoy, H. O., & İyibilgin, O. (2018). Mechanical properties and electrochemical behavior of porous Ti-Nb biomaterials. *Journal of the Mechanical Behavior of Biomedical Materials*, *87*, 59–67. doi:10.1016/j.jmbbm.2018.07.018 PMID:30041140

Zhou, L., Yuan, T., Tang, J., He, J., & Li, R. (2019). Mechanical and corrosion behavior of titanium alloys additively manufactured by selective laser melting–A comparison between nearly β titanium, α titanium and α+ β titanium. *Optics & Laser Technology*, *119*, 105625. doi:10.1016/j.optlastec.2019.105625

Zindani, D., Kumar, K., & Davim, J. P. (2019). *Metallic biomaterials - A review*. Mechanical Behavior of Biomaterials. doi:10.1016/B978-0-08-102174-3.00004-8

Section 3

Application of Eco-Friendly Corrosion Inhibitors

Chapter 4
Monosaccharide Derivatives as Eco-Friendly Corrosion Inhibitors for Carbon Steel in Neutral and Hydrochloric Acid Media

Mohamed Rbaa
Ibn Tofaïl University, Morocco

Loubna Lakhrissi
Mohamed V University, Morocco

Younes Lakhrissi
Ibn Tofaïl University, Morocco

Parul Dohare
Indian Institute of Technology (BHU), India

Brahim Lakhrissi
https://orcid.org/0000-0002-5188-7831
Ibn Tofail University, Morocco

ABSTRACT

In this chapter, the authors present a synthesis of some works carried out within the framework of the study of the corrosion inhibiting activity of a set of monosaccharide derivatives as new biodegradable, nontoxic corrosion inhibitors, and respectful of the environment. The comparative study that they present reveals the existence of more than 20 compounds derivatives of monosaccharide used as corrosion inhibitors in neutral and acidic mediums, and which have given rise to very good corrosion inhibiting efficiencies. These studies of the corrosion inhibiting effect were studied using potentiodynamic polarization curves and electrochemical impedance spectroscopy methods.

DOI: 10.4018/978-1-7998-2775-7.ch004

1. INTRODUCTION

Most of the objects used in our daily life are made of metals, because these latter have several uses due to their physical properties, such as electrical conduction, thermal conduction and resistance to mechanical shocks, in particular iron and its derivatives (carbon steels) (Nam et al., 2012; Tecuapa-Flores et al., 2019; Cruz-Borbolla et al., 2017). However, a high percentage of the quantity of steel used is damaged by corrosion in unsuitable environments, in particular in acidic environments (Rbaa et al., 2019a). These are the reasons why researchers are still trying to find solutions to combat this phenomenon of inappropriate degradation.

Among the proposed solutions, there is the protection of coatings, protection against oxidation and protection against corrosion (Laabaissi et al., 2019; Rbaa et al., 2019b). Currently, researchers are focusing more on inhibiting corrosion because it has the advantage of being the simplest and most economical technique (Fergachi et al., 2019; Rbaa et al., 2019c). Protection against corrosion is based on the use of organic or inorganic chemical compounds, but unfortunately these later sometimes show signs of toxicity during or after use (Martin, 2001; Schüürmann et al., 2011). Also, the application of synthetic compounds as corrosion inhibitors is limited because of the increasing ecological awareness and strict environmental regulations as most of the traditional corrosion inhibitors are synthesized using expensive and toxic chemicals and solvents. In view of this, recently, glucose and their derivatives are being used as corrosion inhibitors as they are available at the economic and commercial level (Han, 2010). Because of the natural and biological origin they can be regarded as one of the greenest sources of the chemical compounds used as corrosion inhibitors. More so, the presence of numerous polar functional groups such as –OH further enhance their corrosion inhibition effectiveness by increasing their molecular size as well as increase solubility in the polar electrolytes. These polar functional groups can also serve as adsorption centers during the meta-inhibitor interactions. Looking the greenness of glucose and their derivatives along with their high protection efficiency, the synthesis, characterization and utilization in all fields of science and technology, including in the area of metallic corrosion inhibition should be further explored (Zhang et al., 2018).

In order to preserve the health of users of these chemical compounds when applied to corrosion inhibition tests, it is desirable to use non-toxic and biodegradable chemicals. Several international organizations and committees have spoken out against the use of toxic chemical products (Han et al., 2010; Zhang et al., 2018). In particular the World Health Organization (WHO) (Teissedre t al., 2000), which is the most active on this side (Rbaa et al., 2020a).

In order to meet the standards of the World Health Organization (WHO), in this study, we have strategically developed the synthesis of two new compound derivatives of D-glucose biodegradable and non-toxic as corrosion inhibitors for steel in 1.0 M HCl. The anticorrosion effect of these two compounds was studied using electrochemical methods such as Potentiodynamic Polarization and Impedance Spectroscopy methods.

2. BACKGROUND

Corrosion is an unavoidable but a controllable process. Due to the issues of toxicity of substances like chromate inhibitors, there is an increasing interest in exploration and utilization of eco-friendly inhibitors, which are also known as green inhibitors. This review briefly discusses some of the interesting features of the synthesized Glucose derivatives used as corrosion inhibitors of metal in different medium. The evaluation of the corrosion inhibiting effect was studied using Potentiodynamic Polarization and Impedance Spectroscopy methods. This study has shown that these compounds provide good protection for metals surface against corrosion in different medium such as (HCl, NaCl, and cooling water system).

3. MAIN FOCUS OF THE CHAPTER

This review briefly discusses some Glucose derivatives used as eco friendly corrosion inhibitors for metals in different medium. Our laboratory has developed several synthesized D-glucose derivatives which have been tested and gives a very good performance against corrosion of metals in different environments. These compounds were synthesized by easy and selective methods, and characterized by the usual spectroscopic methods. The anticorrosion effect of these compounds was studied using gravimetric and electrochemical methods such as Potentiodynamic Polarization and Impedance Spectroscopy methods.

4. GREEN INHIBITORS

In recent years, owing to the growing interest and attention of the world towards the protection of the environment and the hazardous effects of using chemicals on the ecological balance, the traditional approach on corrosion inhibition has gradually changed. As mentioned before, for an inhibitor to be an effective protector against metal corrosion, it should be readily adsorbed on the metal surface through either physisorption or chemisorption processes. Either of these adsorption processes depends primarily on the physicochemical properties of the inhibitor group such as functional groups, electronic density at the donor atom, molecular structure, etc. For instance, organic molecules, which have had a wide applicability and that have been extensively studied and used as corrosion inhibitors, often contain nitrogen, oxygen, and sulfur atoms, as well as multiple bonds in their molecules.

5. MONOSACCHARIDES INHIBITORS

A monosaccharide is the most basic form of carbohydrates. Monosaccharides can by combined through glycosidic bonds to form larger carbohydrates, known as oligosaccharides or polysaccharides. An oligosaccharide with only two monosaccharides is known as a disaccharide. When more than 20 monosaccharides are combined with glycosidic bonds, a oligosaccharide becomes a polysaccharide. Some polysaccharides, like cellulose, contain thousands of monosaccharides. A monosaccharide is a type of monomer, or molecule that can combine with like molecules to create a larger polymer.

Monosaccharides have many functions within cells. First and foremost, monosaccharides are used to produce and store energy. Most organisms create energy by breaking down the monosaccharide glucose, and harvesting the energy released from the bonds. Other monosaccharides are used to form long fibers, which can be used as a form of cellular structure. Plants create cellulose to serve this function, while some bacteria can produce a similar cell wall from slightly different polysaccharides. Even animal cells surround themselves with a complex matrix of polysaccharides, all made from smaller monosaccharides.

Recently some Monosaccharides derivatives are used as corrosion inhibitors. A list of various Monosaccharides compounds that have been reviewed in this book chapter is given in Table 1.

6. RESULTS AND DISCUSSION

The Monosaccharid inhibitors cited in this paper was studied in various medium HCl, NaCl and cooling water. The additional effect of these molecules at different concentrations on the corrosion of metal was studied by measurement of weight loss, potentiodynamic polarization, and electrochemical impedance spectroscopy (EIS); the results obtained from these measurements were summarized in Table 2.

As we can see in Table 2, the corrosion inhibition efficiency of the majority of monosaccharide inhibitors exceeds 75% and can reach 95%, although the different radicals and aggressive solutions have been used. According to the authors' discussions, these compounds can be classified as mixed type inhibitors because the displacement of the potential is less than 85 mV / Ecorr. Also, they reported from the polarization curves that the current densities I_{corr} decrease by the addition of studied inhibitors which mean an increase in the blocked fraction of electrode surface by adsorption and consequently achieve high inhibition efficiencies. In other case, they postulated that The Nyquist plots gotten by EIS measurements are not perfect semicircles and explained this by the dispersion of frequency as a result of the roughness and inhomogeneous of electrode surface. Also, they confirmed that these structures adsorbed into the electrode interface by replacement of water molecules with organic inhibitor molecules. Touir et al. interpreted the adsorption of monoacetoneglucose (P6) and D-galactono-γ-lactone (P4) as the same of those investigated in hydrochloric acid although they tested in different aggressive solutions; They reported that that a layer of corrosion and scale products forms naturally on the ordinary mild steel surface and that the corrosion and scale processes occurring in these systems evolve over time. In addition, it was possible to establish that the corrosion and scale processes are closely related and occur simultaneously at different sites of hemetallic surface. Moreover, it was possible to determine that the modification of the layer formed on the metal surface depends on the molecular structure of the inhibitor applied. It was established that the corrosion inhibitors used for this study, generate a layer with resistive properties that provides good protection against corrosion. Besides, the inhibition efficiency does not depend on the number of hydrogen bonds or the number of free –OH groups but solely on the function of the position of –OH groups on two adjacent carbons. Finally, it was observed a synergistic effect between D-ribonic acid-γ-lactone (P5) and molybdate ions that provides good protection to the ordinary steel against corrosion in cooling media.

Lakhrissi et al. reported the synthesize and characterize the bis-glucobenzimidazolone derivatives by NMR spectroscopy and secondly to evaluate their inhibition effect for mild steel in 200 ppm NaCl using electrochemical techniques. It is noted that the presence of all compounds suppressed the cathodic or anodic reactions exceptionally P7 at 10^{-7} M. These results indicated that these products act as mixed or anodic-type inhibitor and the inhibition efficiency follows the order P8 < P7 < P5 < P6. The cathodic Tafel slopes, bc, are changed with the inhibitor concentrations, indicating a change in the hydrogen evolution reaction and reduction of dissolved oxygen. But in the anodic domain, the I–E characteristics are almost the same in the absence and presence of inhibitors; this can be explained by that these compounds block the reaction sites of iron without affecting the anodic reaction mechanism.

Lakhrissi et al. found their inhibition efficiencies improve with concentration and reached a maximum at 10^{-5} M of each compound and the P6 is better. Electrochemical impedance spectroscopy measure-

Table 1. Monosaccharides compounds used as corrosion inhibitors.

Compound	Chemical Structure	IUPAC Name	Reference
P1	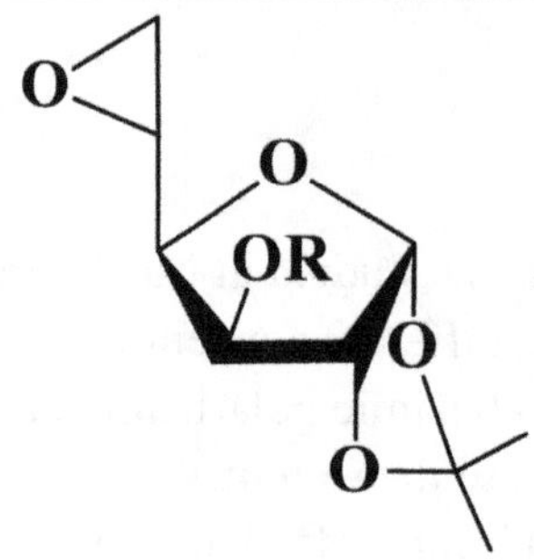 R GC1 (CH_3)	5,6-Anhydro-3-*O*-methyl-1,2-*O*-isopropyli-dene-α-D-glucofuran-ose	(Rbaa et al. 2020 a)
P2	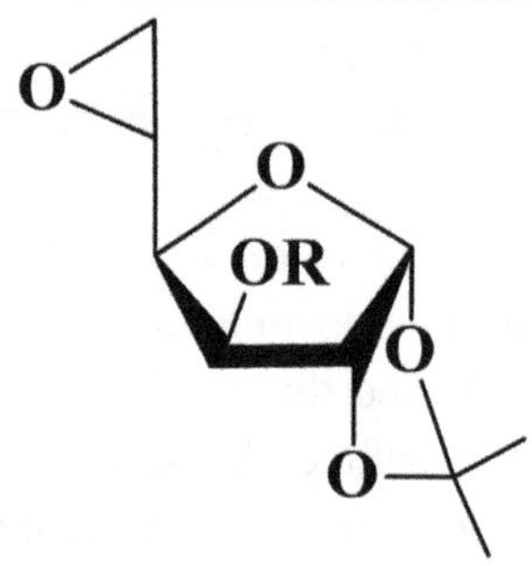 R GC6 (C_6H_{13})	5,6-Anhydro-3-*O*-hexadecyl-1,2-*O*-isopropylidene-α-D-glucofuranose	(Rbaa et al 2020 a)
P3	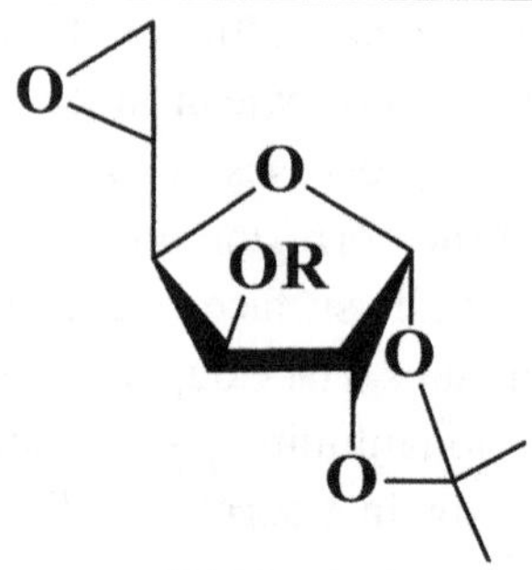 R GC12 ($C_{12}H_{25}$)	5,6-Anhydro-3-*O*-dodecyl-1,2-*O*-isopropylidene-α-D-glucofuranose	(Rbaa et al 2020 b)
P4	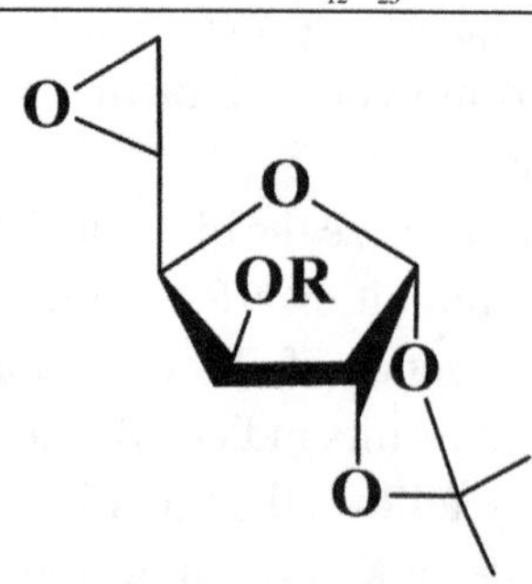 R GC16 ($C_{16}H_{33}$)	5,6-Anhydro-3-*O*-hexadecyl-1,2-*O*-isopropylidene-α-D-glucofuranose	(Rbaa et al. 2020 b)

continued on following page

Table 1. Continued

Compound	Chemical Structure	IUPAC Name	Reference
P5	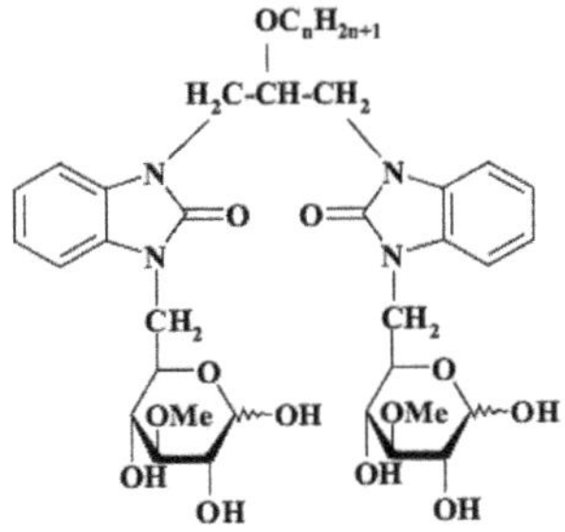 24a (n = 10)	*N,N'*-1,3-bis-[*N*-3-(6-deoxy-3-*O*-methyl-D-glucopyranos-6-yl)-2-oxobenzimidazol-1-yl)]-2-alkyloxypropanes (n = 10, 12, 14, 16).	(Lakhrissi et al., 2017)
P6	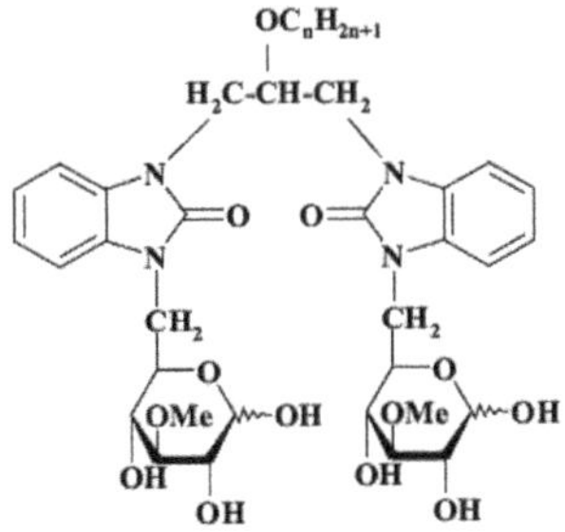 24b (n = 12)	*N,N'*-1,3-bis-[*N*-3-(6-deoxy-3-*O*-methyl-D-glucopyranos-6-yl)-2-oxobenzimidazol-1-yl)]-2-alkyloxypropanes (n = 10, 12, 14, 16).	(Lakhrissi et al., 2017)
P7	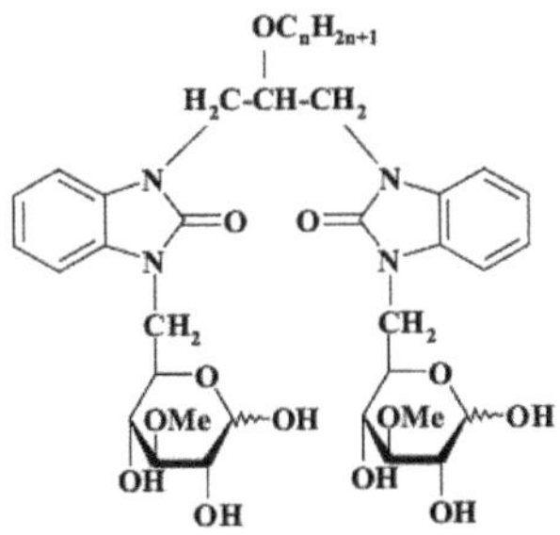 24c (n = 14)	*N,N'*-1,3-bis-[*N*-3-(6-deoxy-3-*O*-methyl-Dglucopyranos-6-yl)-2-oxobenzimidazol-1-yl)]-2-alkyloxypropanes (n = 10, 12, 14, 16).	(Lakhrissi et al., 2017)
P8	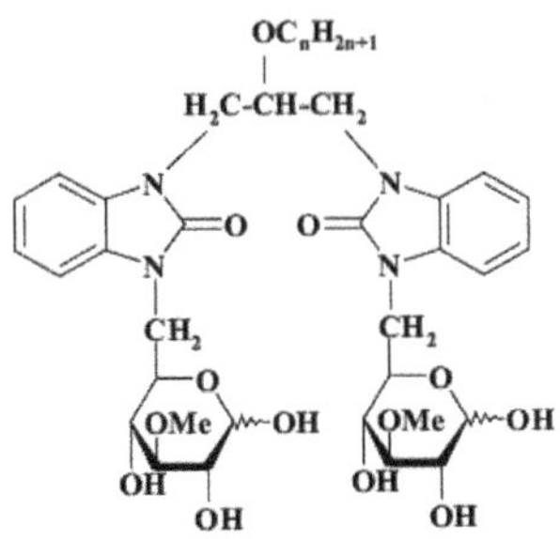 24d (n = 16)	*N,N'*-1,3-bis-[*N*-3-(6-deoxy-3-O-methyl-Dglucopyranos-6-yl)-2-oxobenzimidazol-1-yl)]-2-alkyloxypropanes (n = 10, 12, 14, 16).	(Lakhrissi et al., 2017)

continued on following page

Table 1. Continued

Compound	Chemical Structure	IUPAC Name	Reference
P9	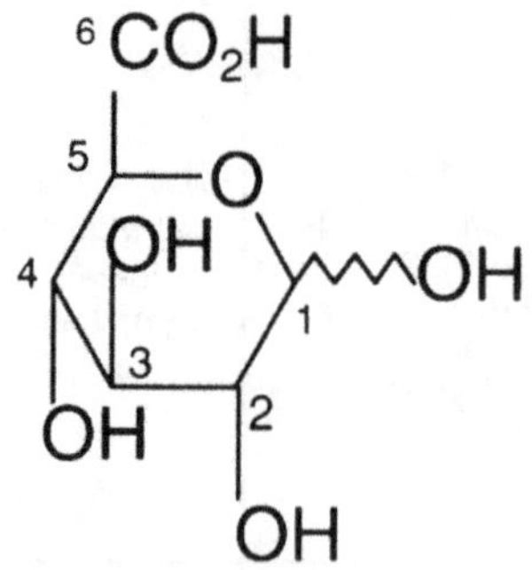	D-glucusonic acid	(Touir et al., 2009)
P10	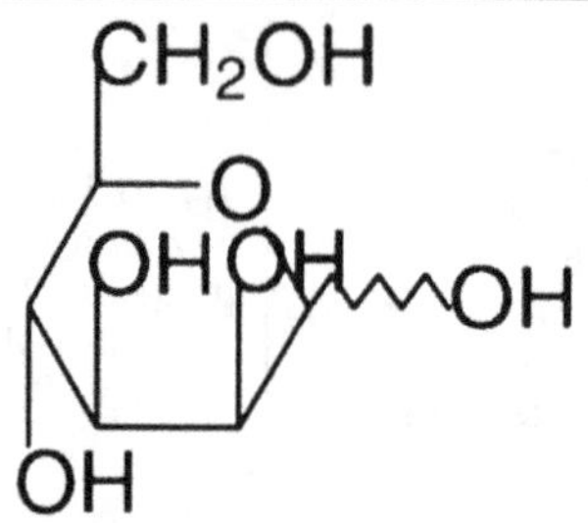	D-Mannose	(Touir et al., 2009; Methal et al., 2015)
P11	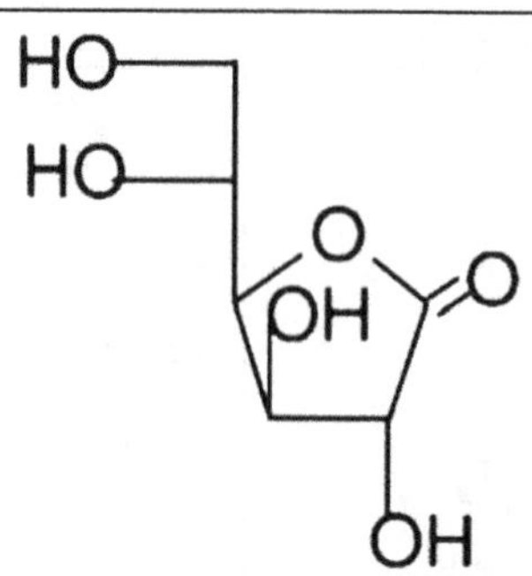	D-Gluconic acid-γ-lactone	(Touir et al., 2009; Methal et al., 2015)
P12	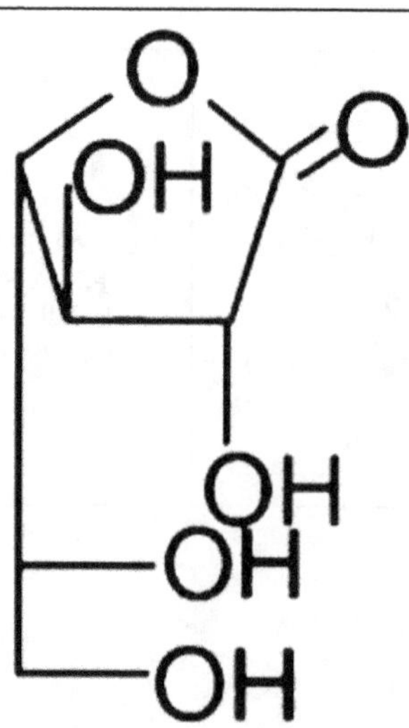	D-Galactono-γ-lactone	(Touir et al., 2009)

continued on following page

Table 1. Continued

Compound	Chemical Structure	IUPAC Name	Reference
P13	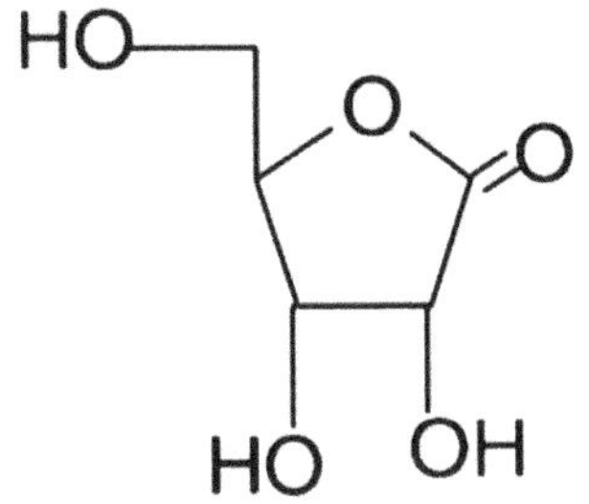	D-Ribonic acid-γ-lactone	(Touir et al., 2009; Methal et al., 2015)
P14	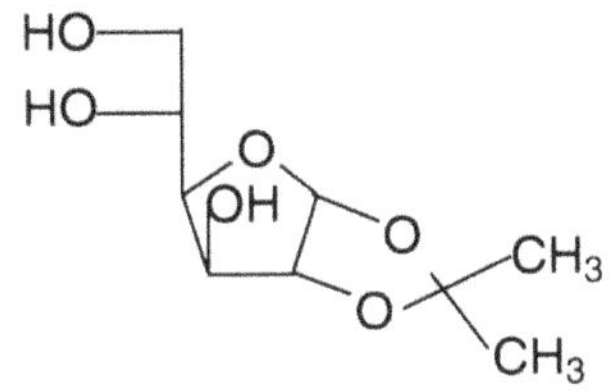	Monoacetoneglucose	(Touir et al., 2009; Methal et al., 2015)
P15	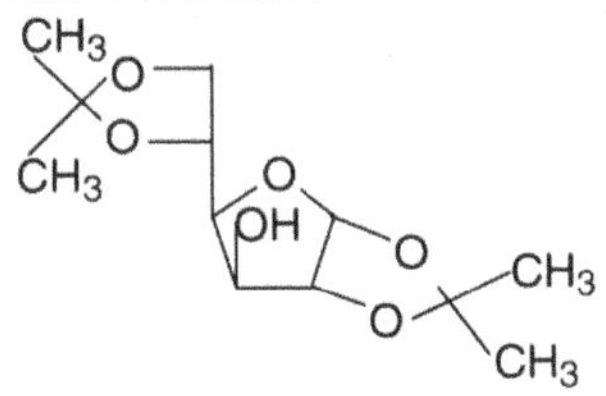	Diacetoneglucose	(Touir et al., 2009; Methal et al., 2015)

ments showed that the diagrams composed of two depressed capacitive loops. This loop can be split into two capacities contributions, although badly separated. The first loop is attributed to the formation of a protective layer and the second is attributed to the charge transfer resistance.

Methal et al. described in their studies the effect of some monosaccharide derivatives, namely: D-mannose (P10), D-gluconic acid-γ-lactone (P11), D-galactono-γ-lactone (P12), monoacetoneglucose (P14), diacetoneglucose (P15), on mild steel corrosion inhibition in an aerated acidic solution of 1.0 M HCl was investigated using weight loss, potentiodynamic polarization curves, and electrochemical impedance spectroscopy measurements. The results indicate that the type of inhibition depends on the type of compounds. Impedance measurements showed that the charge transfer resistance increased and double layer capacitance values decreased with inhibitor's concentration. It is seen that the D-gluconic acid-γ-lactone (P11) performed excellently as a corrosion inhibitor which its efficiency reaches 91% at 10^{-3} M. Adsorption of P11 molecules on mild steel surface was also studied to examine basic information on the interaction between the inhibitors and the metal surface. It is found that the inhibitor adsorb on the metal surface following the Langmuir isotherm model. The temperature effect on the performance of mild steel in 1.0 M HCl in the absence and presence of the best compound was studied in the temperature range from 293 to 323 K. Also, some thermodynamic data for the adsorption and dissolution processes are calculated and discussed.

Furthermore, we focused on previous works the synthesis and presentation of newly glucose derivatives as ecological inhibitors of carbon steel in molar hydrochloric acid (Rbaa et al., 2020b). The study of corrosion inhibiting power is approved using the rheological methods and electrochemical methods (impedance spectroscopy and potentio-dynamic polarization).

Table 2. The impedance data and activation parameters of monosaccharides derivatives used as corrosion inhibitors.

Test Number	Metals/ Alloy	Medium	Abbreviation	IUPAC Name	Concentration	η_{EIS} (%)	E_a kJ mol-1	ΔH_a kJ mol-1	ΔS_a J mol^{-1} K^{-1}	Ref
T1	Carbon steel	1.0 M HCl	P1	5,6-Anhydro-3-*O*-methyl-1,2-*O*-isopropyli-dene-α-D-glucofuran-ose	10^{-3} M	93	63.76	61.19	-61.65	(Rbaa, 2020b)
T2	Carbon steel	1.0 M HCl	P2	5,6-Anhydro-3-O-hexadecyl-1,2-O-isopropylidene-α-D-glucofuranose	10^{-3} M	94.3	72.33	69.73	-18.70	(Rbaa, 2020a)
T3	Carbon steel	1.0 M HCl	P3	5,6-Anhydro-3-O-dodecyl-1,2-O-isopropylidene-α-D-glucofuranose	10^{-3} M	94.4	75.57	72.99	-25.90	(Rbaa, 2020a)
T4	Carbon steel	1.0 M HCl	P4	5,6-Anhydro-3-*O*-hexadecyl-1,2-*O*-isopropylidene-α-D-glucofuranose	10^{-3} M	95	53.50	50.90	-42.84	(Rbaa, 2020b)
T5	Mild steel	200 ppm NaCl	P5	*N,N'*-1,3-bis-[*N*-3-(6-deoxy-3-*O*-methyl-D-glucopyranos-6-yl)-2-oxobenzimidazol-1-yl)]-2-alkyloxypropanes (n=10, 12, 14, 16).	10^{-5} M	63	-	-	-	-
T6	Mild steel	200 ppm NaCl	P6	*N,N'*-1,3-bis-[*N*-3-(6-deoxy-3-*O*-methyl-Dglucopyranos-6-yl)-2-oxobenzimidazol-1-yl)]-2-alkyloxypropanes (n=10, 12, 14, 16).	10^{-5} M	72	-	-	-	
T7	Mild steel	200 ppm NaCl	P7	*N,N'*-1,3-bis-[*N*-3-(6-deoxy-3-*O*-methyl-Dglucopyranos-6-yl)-2-oxobenzimidazol-1-yl)]-2-alkyloxypropanes (n=10, 12, 14, 16).	10^{-5} M	53	-	-	-	
T8	Mild steel	200 ppm NaCl	P8	*N,N'*-1,3-bis-[*N*-3-(6-deoxy-3-*O*-methyl-D-glucopyranos-6-yl)-2-oxobenzimidazol-1-yl)]-2-alkyloxypropanes (n=10, 12, 14, 16).	10^{-5} M	61	-	-	-	
T9	mild steel	Cooling water	P9	D-Glucusonic acid	5×10^{-3} M	24	-	-	-	
T10	mild steel	Cooling water	P10	D-Mannose	10^{-2} M	92	-	-	-	
T11	Mild steel	1.0 M HCl	P10	D-Mannose	10^{-2} M	62	-	-	-	
T12	mild steel	Cooling water	P11	D-Gluconic acid-γ-lactone	5×10^{-3} M	94	-	-	-	
T13	Mild steel	1.0 M HCl	P11		10^{-2} M	90	53.51	50.89	-93.53	
T14	mild steel	Cooling water	P12	D-Galactono-γ-lactone	10^{-3} M	83	-	-	-	
T15	Mild steel	1.0 M HCl	P12	D-Galactono-γ-lactone	10^{-2} M	74	-	-	-	
T16	mild steel	Cooling water	P13	D-Ribonic acid-γ-lactone	10^{-2} M	96	-	-	-	
T17	mild steel	Cooling water	P14	Monoacetoneglucose	10^{-3} M	87	-	-	-	
T18	Mild steel	1.0 M HCl	P14	Monoacetoneglucose	10^{-3} M	75	-	-	-	
T19	mild steel	Cooling water	P15	Diacetoneglucose	10^{-3} M	88	-	-	-	
T20	Mild steel	1.0 M HCl	P15	Diacetoneglucose	10^{-2} M	74	-	-	-	

Impedance Spectroscopy Evaluation

Electrochemical impedance spectroscopy experiments were carried out on a newly polished steel surface in acidic solution without and with the addition of P1, P2, P3 and P4 at open circuit potential at 298 K. Fig. 1 show impedance spectra, as Nyquist and Bode plots, respectively obtained for carbon steel in the absence and presence of different concentrations of P4 as best Monosaccharid inhibitors for the series studies. The Nyquist plots show a single depressed capacitive semicircle for both inhibited and uninhibited

Figure 1. Impedance diagrams obtained with and without P4 inhibitor and electrical equivalent circuit used for modelling the interface carbon steel/1 M HCl solution without and with inhibitor (P4).

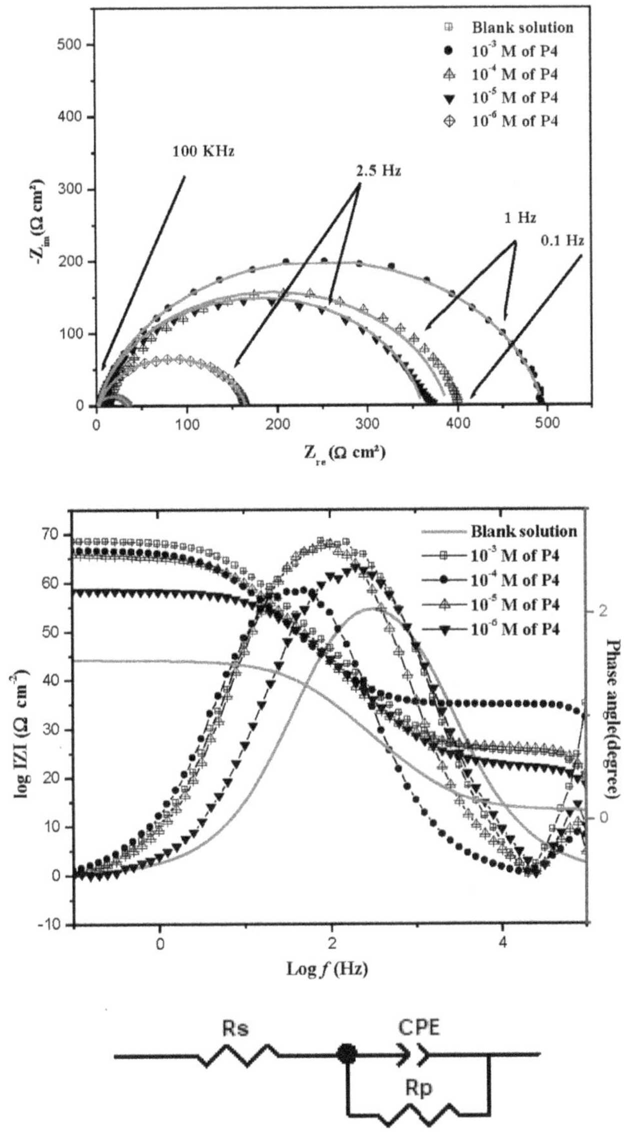

solutions, with one capacitive time constant in the Bode-phase plots. The depressed semicircle in Nyquist diagrams is characteristic of solid electrodes that exhibit frequency dispersion due to the roughness and other inhomogeneities of the surface (Shih, 1989). Also, the slopes of the Bode-modulus plots (log|Z| vs log *f*) are not equal to -1 (Fig. 1), which, and are commonly observed for solid electrodes as a result of frequency dispersion of interfacial impedance. The capacitive loop is usually associated with charge transfer process and electrical double layer. The addition of different concentrations of P1, P2, P3 and P4 does not substantially change the shape of the semicircle indicating that the inhibitor addition does not change the mechanism of corrosion reaction, but inhibits the corrosion by increasing the surface coverage by the adsorbed inhibitor film (Zhihua et al., 2011). As shown in Fig. 1, the diameter of the

Table 3. Impedance data in the absence and in the presence of monosaccharide inhibitors.

Comp	C (M)	R_s (Ω cm^2)	R_p (Ω cm^2)	C_{dl} (μF cm^{-2})	n_{dl}	Q (μF S^{n-1})	η_{EIS} %	θ
Blank solution	00	1.22 ±0.24	34.85±0.43	117.1	0.82±0.05	315.1±2.80	--	--
P1	10^{-6}	2.97±0.22	161.5±0.43	82.1	0.86±0.05	104.3±3.40	78.4	0.784
	10^{-5}	1.34±0.20	358.2±0.43	65.8	0.88±0.04	102.3±1.47	90.4	0.904
	10^{-4}	1.18±0.19	386.4±0.43	54.7	0.86±0.02	129.3±1.61	90.9	0.909
	10^{-3}	1.83±0.20	492.8±0.44	51.3	0.86±0.03	83.85±0.85	92.9	0.929
P2	10^{-6}	1.39±0.21	191.50±0.51	51.81	0.89±0.05	85.0±2.6	81.8	0.818
	10^{-5}	2.24±0.21	263.40±0.51	43.98	0.88±0.05	73.0±1.5	86.8	0.868
	10^{-4}	2.52±0.20	364.00±0.43	41.83	0.88±0.05	68.0±1.0	90.4	0.904
	10^{-3}	2.02±0.20	608.60±0.43	29.72	0.88±0.05	46.3±0.4	94.3	0.943
P3	10^{-6}	1.83±0.21	254.10±0.44	60.03	0.85±0.05	112.0±0.5	86.3	0.863
	10^{-5}	2.50±0.21	351.10±0.44	44.83	0.86±0.05	79.4±1.2	90.1	0.901
	10^{-4}	1.98±0.20	446.90±0.44	30.32	0.83±0.05	63.0±0.5	92.2	0.922
	10^{-3}	2.17±0.20	633.20±0.43	24.67	0.84±0.05	48.0±0.5	94.4	0.944
P4	10^{-6}	0.40±0.21	322.4±0.44	73.3	0.85±0.05	86.14±1.23	89.1	0.891
	10^{-5}	0.97±0.20	394.5±0.45	55.6	0.83±0.04	132.5±1.60	91.1	0.911
	10^{-4}	2.55±0.20	419.4±0.43	47.8	0.87±0.02	89.36±0.80	91.6	0.916
	10^{-3}	3.60±0.20	656.7±0.45	47.8	0.81±0.05	90.62±0.64	94.6	0.946

capacitive loop in the presence of P4 is larger than that in the uninhibited system and increases with increase in concentration of the inhibitor. This implies that P1, P2, P3 and P4 protect the carbon steel surface against acid attack and the protection is enhanced at higher concentration of inhibitor.

Electrochemical impedance kinetic parameters were obtained by fitting the impedance spectra to the equivalent circuit model, given in Fig. 1, in which the solution resistance (R_s) is shorted by a constant phase element (CPE) that in parallel connection to the polarization resistance (R_p), where R_p includes charge transfer resistance (R_{ct}) and the resistance of inhibitor film (R_f) formed on the metal surface in the presence of the inhibitor ($R_p = R_{ct} + R_f$) (Ramezanzadeh et al., 2014). The CPE constant phase element is used instead of a pure capacitor to compensate the non-ideal capacitive response of the interface and to get a more accurate fit of experimental data set. The impedance (Z) of the CPE is defined as (Ehsani et al., 2013):

$$Z_{CPE} = Q^{-1}(iw)^{-n} \quad (1)$$

where Q is the CPE constant (in μF S^{n-1}), i is the imaginary number, ω is the angular frequency (in $rads^{-1}$), and n is CPE exponent, which can be used as a gauge heterogeneity or roughness of the surface. The results of the fitting and simulation analyses of the EIS profiles are listed in Table 2. The data are found to be sufficiently well fitted within the limits of experimental error and reproducibility of data.

In the Table 2 are also given the calculated "double layer capacitance" values (C_{dl}), using the Hsu and Mansfeld formula (Zheludkevich et al., 2005):

$$C_{dl} = (AR_p^{1-n})^{\frac{1}{n}} \quad (2)$$

The inhibition efficiency in the case of impedance study, η_{EIS} (%), was evaluated by R_p values as described previously (Lebrini et al., 2008). The inspection of the results in Table 3 shows that the R_p values increase with increasing concentration of P1, P2, P3 and P4 while the C_{dl} values for the inhibitors are generally lower than that of the acid blank, which suggest that the inhibitors adsorb on the steel surface thereby forming a protective layer on the steel surface and reducing the rate of charge transfer process. Also, the values of Q are lower in the presence of P1, P2, P3 and P4 compared to that of the uninhibited blank system. This can be attributed to the gradual displacement of water molecules by inhibitor molecules at metal/solution interface leading to formation of protective layer on the steel surface (John et al., 2012; Olasunkanmi et al., 2015). The values of the CPE exponent, n are reduced by the addition of the P1, P2, P3 and P4. This behavior can suggest that the surface inhomogeneity of the electrode decreases in the presence of P1, P2, P3 and P4.

As can be seen in Table 3, the polarization resistance value, R_p, increases versus the concentration of P1, P2, P3 and P4 and reaches a maximum value of 492.8 Ω cm² 608.6 Ω cm², 633.20 Ω cm² and 656.7 Ω

Table 4. The polarization data of C-steel in the absence and presence of non-toxic inhibitors.

Medium	Conc.	E_{corr} (mV vs. SCE)	i_{corr} (μA cm^{-2})	β_c (mV dec^{-1})	β_a (mV dec^{-1})	η_{PDP} (%)
Blank solution	0	-464.9	573.17	-138.5	107.9	-------
P1	10^{-6}	-700.4	151.14	-86.6	179.5	47
	10^{-5}	-672.3	141.39	-100.7	368.0	75
	10^{-4}	-669.4	121.62	-98.9	295.2	79
	10^{-3}	-643.8	41.52	-105.0	109.7	93
P2	10^{-6}	-488.3	136.0	-169.4	95.4	75
	10^{-5}	-476.8	127.5	-141.0	62.1	77
	10^{-4}	-448.0	49.3	-94.9	73.2	91
	10^{-3}	-482.7	30.9	-114.1	115.6	94
P3	10^{-6}	-405.8	79.7	-168.8	76.0	86
	10^{-5}	-486.3	77.7	-109.3	95.6	86
	10^{-4}	-485.9	46.4	-133.2	67.6	92
	10^{-3}	-420.3	17.4	-114.5	44.3	94
P4	10^{-6}	-657.8	96.12	-123.5	126.9	83
	10^{-5}	-672.1	58.28	-120.8	141.9	89
	10^{-4}	-667.6	55.44	-99.0	135.7	90
	10^{-3}	-652.4	40.10	-123.3	125.9	95

cm^2, respectively at 10^{-3} M. Similarly, the corrosion inhibition efficiency, η_{EIS} (%), increases continuously with increasing of the P1, P2, P3 and P4 inhibitor concentration and the maximum of 92.9%, 94.3%, 94.4% and 95.0%, respectively at 10^{-3} M (Table 3). Therefore, these results suggest that P4 could serve as an effective corrosion inhibitors for carbon steel in 1.0 M HCl medium.

Potentiodynamic Polarization Evaluation

The potentiodynamic polarization makes it possible to find some basic parameters of the corrosion process during the addition of the inhibitors. As well as the type of inhibitors if it is anodic, cathodic or mixed (Table 4 and Figure 2).

Figure 2. The polarization curves of C-steel in the absence and presence of non-toxic inhibitor (P4).

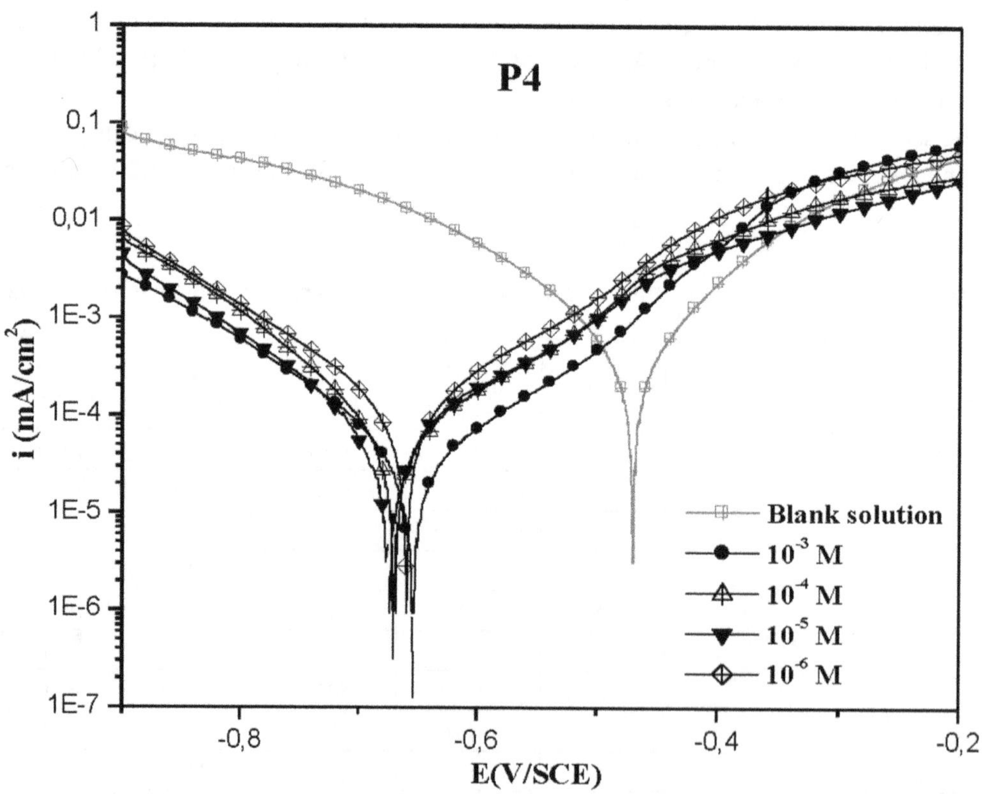

Based on the results in Table 4, we concluded that the addition of these inhibitors causes a decrease in the current density values (i). Consequently, the values of inhibitory efficacy increase with increasing inhibitor concentration. The addition of these inhibitors causes a change in the values of the anodic and cathodic slopes (β_a and β_c) as well as in the values of corrosion potential (E_{corr}). These results show that both inhibitors adsorb to the surface of the steel (Garcia-Ochoa et al., 2016). The shape of the polarization curves shows that these two inhibitors act as cathodic inhibitors (Figure 2) (Rbaa et al., 2020b). From the results obtained by this polarization study these inhibitors are effective for corrosion inhibition of carbon steel in HCl molar. The inhibitory efficiency reached at a maximum value of 95% for the compound P4. Polarization studies are in good agreement with studies of impedance spectroscopy.

7. CONCLUSION

In this work, we set ourselves two objectives, the first one was to review all the monosaccharide used as corrosion inhibitiors for metals surfaces in different medium such as cooling water system, 200 ppm NaCl and also 1.0 M HCl medium. And the second one was to present the application new non-toxic and biodegradable compounds derivatives (D-glucose compounds) as corrosion inhibitors of carbon steel in 1.0 M HCl medium. The electrochemical results show that these compounds are likely to be potentially used against corrosion of carbon steel in 1.0 M HCl medium, and that the inhibitory efficacy reaches an optimal value of 94% for the P4 compound. The results of the potentiodynamic polarization show that these two inhibitors act as cathode-type inhibitors. The thermodynamic study shows that these two non-toxic and biodegradable compounds are adsorbed on the carbon steel surface according to the Langmuir isotherm.

REFERENCES

Cruz-Borbolla, J., Garcia-Ochoa, E., Narayanan, J., Maldonado-Rivas, P., Pandiyan, T., & Vásquez-Pérez, J. M. (2017). Electrochemical and theoretical studies of the interactions of a pyridyl-based corrosion inhibitor with iron clusters (Fe 15, Fe 30, Fe 45, and Fe 60). *Journal of Molecular Modeling*, *23*(12), 342. doi:10.100700894-017-3510-x PMID:29143152

Ehsani, A., Babaei, F., & Nasrollahzadeh, M. (2013). Electrosynthesis and absorbance spectra of TiO2 nanoparticles dispersed in the conductive polymer. *Applied Surface Science*, *283*, 1060–1064. doi:10.1016/j.apsusc.2013.07.067

Fergachi, O., Benhiba, F., Rbaa, M., Glai, M., Touir, R., & Ebn Touhami, M. (2019). Corrosion Inhibition of Ordinary Steel in 5.0 M HCl Medium by Benzimidazole Derivatives: Electrochemical, UV-Visible Spectrometry, and DFT Calculations. *J. Bio-and Tribo-Corros.*, *5*(1), 2–13. doi:10.100740735-018-0215-3

Garcia-Ochoa, E., Guzmán-Jiménez, S. J., Hernández, J. G., Pandiyan, T., Vásquez-Pérez, J. M., & Cruz-Borbolla, J. (2016). Benzimidazole ligands in the corrosion inhibition for carbon steel in acid medium: DFT study of its interaction on Fe30 surface. *Journal of Molecular Structure*, *1119*, 314–324. doi:10.1016/j.molstruc.2016.04.057

Han, X., Zhu, Y., Yang, X., & Li, C. (2010). Electrocatalytic activity of Pt doped TiO2 nanotubes catalysts for glucose determination. *Journal of Alloys and Compounds*, *500*(2), 247–251. doi:10.1016/j.jallcom.2010.04.019

John, S., & Joseph, A. (2012). Effective inhibition of mild steel corrosion in 1 M hydrochloric acid using substituted triazines: An experimental and theoretical study. *RSC Advances*, *2*(26), 9944–9951. doi:10.1039/c2ra20663c

Laabaissi, T., Benhiba, F., Rouifi, Z., Rbaa, M., Oudda, H., Zarrok, H., Lakhrissi, B., Guenbour, A., Warad, I., & Zarrouk, A. (2011). Benzodiazepine Derivatives as Corrosion Inhibitors of Carbon Steel in HCl Media: Electrochemical and Theoretical Studies, T. Zhihua, Z. Shengtao, L. Weihua, H. Baorong, Adsorption and inhibitory mechanism of 1H-1,2,4-triazol-l-yl-methyl-2-(4-chlorophenoxy) acetate on corrosion of mild steel in acidic solution. *Industrial & Engineering Chemistry Research*, *50*(10), 6082–6088. doi:10.1021/ie101793b

Lakhrissi, B., Lakhrissi, L., Massoui, M., Essassi, E. M., Comelles, F., Esquena, J., Solans, C., & Rodriguez-Abreu, C. (2010). Surface and self-aggregation properties of bis-benzimidazolones derivatives of D-glucose. *Journal of Surfactants and Detergents*, *13*(3), 329–338. doi:10.100711743-010-1183-x

Lebrini, M., Bentiss, F., Chihib, N.-E., Jama, C., Hornez, J. P., & Lagrenée, M. (2008). Polyphosphate derivatives of guanidine and urea copolymer: Inhibiting corrosion effect of armco iron in acid solution and antibacterial activity. *Corrosion Science*, *50*(10), 2914–2918. doi:10.1016/j.corsci.2008.07.003

Martin, T. M., & Young, D. M. (2001). Prediction of the acute toxicity (96-h LC50) of organic compounds to the fathead minnow (Pimephalespromelas) using a group contribution method. *Chemical Research in Toxicology*, *14*(10), 1378–1385. doi:10.1021/tx0155045 PMID:11599929

Nam, S., Cho, H. W., Lim, S., Kim, D., Kim, H., & Sung, B. J. (2012). Enhancement of electrical and thermomechanical properties of silver nanowire composites by the introduction of nonconductive nanoparticles: Experiment and simulation. *ACS Nano*, *7*(1), 851–856. doi:10.1021/nn305439t PMID:23237625

Olasunkanmi, L. O., Obot, I. B., Kabanda, M. M., & Ebenso, E. E. (2015). Some quinoxalin-6-yl derivatives as corrosion inhibitors for mild steel in hydrochloric acid: Experimental and theoretical studies. *The Journal of Physical Chemistry C*, *119*(28), 16004–16019. doi:10.1021/acs.jpcc.5b03285

Ramezanzadeh, B., Arman, S. Y., Mehdipour, M., & Markhali, B. P. (2014). Analysis of electrochemical noise (ECN) data in time and frequency domain for comparison corrosion inhibition of some azole compounds on Cu in 1.0 M H2SO4 solution. *Applied Surface Science*, *289*, 129–140. doi:10.1016/j.apsusc.2013.10.119

Rbaa, M., Abousalem, A.S., Ebn Touhami, M., Warad, I., Bentiss, F., Lakhrissi, B., & Zarrouk A. (2019a). Novel Cu (II) and Zn (II) complexes of 8-hydroxyquinoline derivatives as effective corrosion inhibitors for mild steel in 1.0 M HCl solution: Computer modeling supported experimental studies. *J. Mol. Liq., 290.*

Rbaa, M., & Lakhrissi,B. (2019c). Novel oxazole and imidazole based on 8-Hydroxyquinoline as a Corrosion Inhibition of mild steel in HCl Solution: Insights from Experimental and Computational Studies. *S. Interfaces*, (15), 43-59.

Rbaa, M., Benhiba, F., Dohare, P., Lakhrissi, L., Touir, R., Lakhrissi, B., & Lakhrissi, Y. (2020b). Synthesis of new epoxy glucose derivatives as a non-toxic corrosion inhibitors for carbon steel in molar HCl: Experimental, DFT and MD simulation. *Chem. Data Collect.*, *27*, 100394. doi:10.1016/j.cdc.2020.100394

Rbaa, M., Dohar, P., Berish, A., Dagda, O., Lakhrissi, L., Gala, M., Lakhrissi, B., Ebn Touhami, M., Warad, I., & Zarrouk, A. (2020a). New Epoxy sugar based glucose derivatives as eco friendly corrosion inhibitors for the carbon steel in 1.0 M HCl: Experimental and theoretical investigations. *Journal of Alloys and Compounds*, *833*, 154949. doi:10.1016/j.jallcom.2020.154949

Rbaa, M., Galai, M., Abousalem, A. S., Lakhrissi, B., Touhami, M. E., Warad, I., & Zarrouk, A. (2019d). Synthetic, spectroscopic characterization, empirical and theoretical investigations on the corrosion inhibition characteristics of mild steel in molar hydrochloric acid by three novel 8-hydroxyquinoline derivatives. *Ionics*, *13*(8), 1–20. doi:10.100711581-019-03160-9

Rbaa, M., Lakhrissi, B., & Lgaz, H. (2019b). Novel 8-Hydroxyqionoline Derivatives as Effective Corrosion Inhibitors for Mild Steel in HCl Solution: Synthesis, Characterization, Gravimetric, Electrochemical, Surface Morphological and UV-Visible Spectrometry. *Anal. Bioanal. Electrochem.*, *11*(6), 787–811.

Schüürmann, G., Ebert, R. U., & Kühne, R. (2011). R. Ku?hne, Quantitative read-across for predicting the acute fish toxicity of organic compounds. *Environmental Science & Technology*, *45*(10), 461. doi:10.1021/es200361r PMID:21491860

Shih, H., & Mansfeld, F. (1989). A fitting procedure for impedance data of systems with very low corrosion rates. *Corrosion Science*, *29*(10), 1235–1240. doi:10.1016/0010-938X(89)90070-X

Tecuapa-Flores, E. D., Turcio-Ortega, D., Hernandez, J. G., Huerta-Aguilar, C. A., & Thangarasu, P. (2019). The role of keto group in cyclic ligand 1,4,8,11-tetraazacyclotetradecane-5,7-dione as strong corrosion inhibitor for carbon steel surface: Experimental and theoretical studies. *Journal of Molecular Structure*, *1189*, 131–145. doi:10.1016/j.molstruc.2019.04.006

Teissedre, P. L., & Waterhouse, A. L. (2000). Inhibition of oxidation of human low-density lipoproteins by phenolic substances in different essential oils varieties. *Journal of Agricultural and Food Chemistry*, *4*(9), 3801–3805. doi:10.1021/jf990921x PMID:10995274

Zhang, X., Xu, Y., & Ye, B. (2018). An efficient electrochemical glucose sensor based on porous nickel-based metal organic framework/carbon nanotubes composite (Ni-MOF/CNTs). *Journal of Alloys and Compounds*, *767*, 651–656. doi:10.1016/j.jallcom.2018.07.175

Zheludkevich, M. L., Yasakau, K. A., Poznyak, S. K., & Ferreira, M. G. S. (2005). Triazole and thiazole derivatives as corrosion inhibitors for AA2024 aluminium alloy. *Corrosion Science*, *47*(12), 3368–3383. doi:10.1016/j.corsci.2005.05.040

Section 4

Corrosion Inhibitors for Industrial Cleaning, Oil Refinery Equipment, and Pickling Process

Chapter 5
Review of Application of Imidazole and Imidazole Derivatives as Corrosion Inhibitors of Metals

Khaoula Alaoui
Laboratory of Materials Engineering and Environment: Modeling and Application, Faculty of Science, Ibn Tofail University, Morocco

Nadia Dkhireche
Laboratory of Materials Engineering and Environment: Modeling and Application, Faculty of Science, Ibn Tofail University, Morocco

Mohamed Ebn Touhami
Laboratory of Materials Engineering and Environment: Modeling and Application, Faculty of Science, Ibn Tofail University, Morocco

Younes El Kacimi
https://orcid.org/0000-0002-8662-4644
Laboratory of Materials Engineering and Environment: Modeling and Application, Faculty of Science, Ibn Tofail University, Morocco

ABSTRACT

Imidazole is part of the theophylline molecule, found in tea leaves and coffee beans, which stimulates the central nervous system. Imidazole and its derivatives are an important class of heterocyclic compounds that exhibit a wide range of biological and pharmacological activities. Furthermore, synthetic Imidazole are present in many fungicides, analgesic, anti-inflammatory, antibacterial, antitumor activities, antiprotozoal, and antihypertensive medications. Apart of its use for biological and pharmaceutical purpose, it also has varying applications in industries: the imidazole has been used extensively as a corrosion inhibitor on metals such as iron, copper, zinc, carbon steel, and their alloys. Imidazole and its derivatives have been reported as effective corrosion inhibitors due to their excellent excited state intra-molecular proton transfer, which elevates the performances of their adsorption into the metallic surface. In this chapter, the authors have reported a review of some works that investigated the effects of some Imidazole molecule on corrosion inhibition properties.

DOI: 10.4018/978-1-7998-2775-7.ch005

INTRODUCTION

Corrosion prevention is an important issue for industrial application of materials. Among numerous corrosion prevention measures, corrosion inhibition possesses advantages of economy, high-efficiency, and facile-feasibility, and has been widely applied in various fields (El Kacimi et al., 2016; HazwanHussin et al. 2011; Abbouda et al., 2007). These inhibitors can adhere to a metal surface to form a protective barrier against corrosive agents in the environment (James et al., 2007). The effectiveness of an inhibitor to provide corrosion protection depends to a large extent upon the interaction between the inhibitor and the surface (Ebenso et al., 2001). The adsorbed inhibitors can affect the corrosion reaction either by physically blocking the surface or by changing the activation barriers of the anodic and cathodic partial reactions of the corrosion process (Galai et al., 2015). Organic compounds that can donate electrons to unoccupied d-orbitals of a metal surface or to adsorption sites in the surface of an oxide/passive film to form coordinate covalent bonds, or may also accept free electrons from the metal surface by using their antibonding orbitals to form feedback bonds, are generally good corrosion inhibitors.

Recently many researchers have studied the Imidazole and Imidazole derivative as inhibitors of metal corrosion and shown good inhibition efficiency in many different medium.

In this paper, we have reported the work of some Imidazole and Imidazole derivatives which investigated as corrosion inhibitors. These articles were using deferent techniques: potentiodynamic polarization, electrochemical impedance spectroscopy (EIS), and gravimetric measurement. Also, the quantum computations based on DFT methods were performed for many papers in order to find the relationships between the electronic properties and inhibition efficiencies of imidazole derivatives. The chapter summarizes the corrosion inhibition of imidazole and Imidazole derivatives as corrosion inhibitors on different metal surfaces such as mild steel, carbon steel, copper, etc., in different media such as Hydrochloric acid, Sulfuric acid, Phosphoric Acid, Sulfamic acid and Sodium chloride. The information presented in this review is useful for diverse industrial fields and for other applications where corrosion inhibitors for metals are needed.

BACKGROUND

To remove undesirable scales the particular piece of metal is immersed in a suitable acid solution, called an acid pickling bath. The acid solution attacks and dissolves the scale. Once the scale is dissolved, the acid is then free for further attack upon the metal. In order to reduce the metal dissolution and the consumption of acid, corrosion inhibitors are added to the pickling solution. As known, Hydrochloric acid (HCl) is the most important pickling acid. Furthermore, the effect of temperature on a chemical reaction of practical and theoretical important. Like most chemical reactions, the rate of corrosion of iron and steel increases with temperature especially in industrial pickling hydrochloric acid media. Accordingly, acid pickling inhibitors are expected to be chemically stable to provide high protective efficiency under the conditions mentioned above. Hydrochloric acid pickling bath is used to remove the corrosion products from the metal working. Over-pickling can cause a serious damage and physical properties of metals structure can be easily done. Thus, to control the risk and reduce the problem, several organic compounds are used as corrosion inhibitors.

MAIN FOCUS OF THE CHAPTER

imidazole and Imidazole derivatives as corrosion inhibitors on different metal surfaces such as mild steel, carbon steel, copper, etc., in different media such as Hydrochloric acid, Sulfuric acid, Phosphoric Acid, Sulfamic acid and Sodium chloride. In this chapter, we have reported a review of some works that investigated the effects of some Imidazole molecule on corrosion inhibition properties using weight loss, Tafel polarization and electrochemical impedance spectroscopy (EIS).

They reported that the adsorption of some one of this derivatives into the metals surface follows the Langmuir isotherm and making a strong bond sometimes and linked with physical adsorption another time. Other Imidazole derivatives follow the Freundlich and Flory–Huggins isotherm.

Taking into account their excellent performance, the second part of this chapter is dedicated to the corrosion inhibition of mild steel in hydrochloric acid medium using some Triphenylimidazole compounds. These studies were carried out using weight loss, potentiodynamic polarization curves (PP) and electrochemical impedance spectroscopic (EIS). The adsorption and inhibition efficiency of these inhibitors were investigated. The effect of temperature on the corrosion behavior was also studied in the range from 25±2 °C to 55±2 °C. The thermodynamic parameters such as adsorption heat ΔH_a*, entropy of adsorption ΔS_a* and adsorption free energy ΔG_a* were calculated and discussed. Scanning electron microscopy (SEM) was performed and discussed for surface study of uninhibited and inhibited carbon steel samples for the best inhibitor.

CORROSION TECHNIQUE

Recently, several modern techniques were showing up, but the mainly used methods in the investigation of inhibition performance are weight loss method, the Potentiodynamic polarization and Electrochemical impedance spectroscopy (EIS) which uses a cell with three electrodes.

- ***Corrosion weight loss tests:*** Weight loss experiments were done according to ASTM methods. Gravimetric measurements were carried out in an electrolysis cell equipped with a thermostat-cooling condenser. The metals specimens used is immersed in aggressive medium studied. After immersion period, the specimens were cleaned reweighed for determining corrosion rate. Duplicate experiments are performed in each case, and the mean value of the weight loss is reported. Weight loss allows us to calculate the mean corrosion rate as expressed in (mg cm^{-2} h^{-1}). The resulting quantity, corrosion rate (ω_{corr}) is thereby the fundamental measurement in corrosion. This corrosion rate (ω_{corr}) can be determined either by chemical analysis of dissolved metal in solution or by gravimetric method measuring weight of specimen before and after exposure in the aggressive solution applying the following equation 1:

$$\omega_{corr} = \frac{m_i - m_f}{St} \quad (1)$$

where m_i, m_f, S and t denote initial weight, final weight, surface of specimen and immersion time, respectively.

The inhibition efficiency, $\eta_\omega\%$, is determined as follows:

$$\eta_\omega\% = \frac{\omega^0_{corr} - \omega_{corr}}{\omega^0_{corr}} \times 100 \quad (2)$$

where ω^0_{corr} and ω_{corr} are the corrosion rates in the absence and presence of inhibitors, respectively.

- ***Potentiodynamic polarization measurement:*** For electrochemical measurements, the electrolysis cell was a borrosilcate glass (Pyrex®) cylinder closed by a cap with five apertures. Three of them were used for the electrode insertions. The working electrode was pressure-fitted into a polytetrafluoroethylene holder (PTFE) exposing only 1 cm^2 of area to the solution. Platinum and saturated calomel were used as counter and reference electrode (SCE), respectively. All potentials were measured against the last electrode.

The potentiodynamic polarization curves were recorded by changing the electrode potential automatically from negative values to positive values versus E_{corr} using a Potentiostat/Galvanostat type PGZ 100, at a scan rate of 1 mV/s after 1 h of immersion time until reaching steady state. The test solution was thermostatically controlled at 25±2 °C in air atmosphere without bubbling. To evaluate corrosion kinetic parameters, a fitting by Stern-Geary equation was used (Stern & Geary, 1957). To do so, the overall current density values, i, were considered as the sum of two contributions, anodic and cathodic current i_a and i_c, respectively. The corrosion parameters were then evaluated by means of nonlinear least square method by applying equation (3) using Origin software. However, for this calculation, the potential range applied was limited to ± 0.100V around E_{corr}, else a significant systematic divergence was sometimes observed for both anodic and cathodic branches.

The corrosion inhibition efficiency is evaluated from the corrosion current densities values using the relationship (3):

$$\eta_{PP} = \frac{i^0_{corr} - i_{corr}}{i^0_{corr}} \times 100 \quad (3)$$

- ***Electrochemical impedance spectroscopy (EIS):*** The electrochemical impedance spectroscopy measurements were carried out using a transfer function analyzer, with a small amplitude, over a frequency domain from 100 kHz to 100 mHz with five points per decade. The EIS diagrams were done in the Nyquist representation. The results were then analyzed in terms of an equivalent electrical circuit using Bouckamp program (Abdelaal et al., 1983).

The inhibiting efficiency derived from Electrochemical impedance spectroscopy (EIS), η_{EIS} is calculated using the following equation (4):

$$\eta_{EIS} = \frac{R_{ct} - R_{ct}^{0}}{R_{ct}} \times 100 \quad (4)$$

where R_{ct}^{0} and R_{ct} are the charge transfer resistance values in the absence and in the presence of inhibitor, respectively.

In order to ensure reproducibility, all experiments were repeated three times. The evaluated inaccuracy did not exceed 10%.

IMIDAZOLE COMPOUND

Imidazole (1,3-diaza-2,4-cyclopentadiene) is a planar five-member ring system with 3C and 2N atom in 1 and 3 positions with molecular formula $C_3H_4N_2$. The molecular structures of the examined molecule is shown in Figure 1.

Figure 1. Structure of Imidazole (1,3-diaza-2,4-cyclopentadiene)

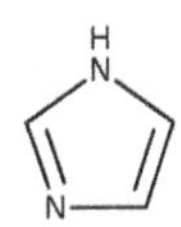

Imidazole is soluble in water and other polar solvents. It exists in two equivalent tautomeric forms because the hydrogen atom can be located on either of the two nitrogen atoms. Imidazole is a highly polar compound, as evidenced by a calculated dipole of 3.61D, and is entirely soluble in water. The compound is classified as aromatic due to the presence of π-electrons, consisting of a pair of electrons from the protonated nitrogen atom and one from each of the remaining four atoms of the ring.

Imidazole is incorporated into many important biological molecules. The most pervasive is the amino acid "histidine", which has an imidazole side chain. Histidine is present in many proteins and enzymes and plays a vital part in the structure and binding functions of hemoglobin. Histidine can be decarboxylated to histamine, which is also a common biological compound. One of the applications of imidazole is in the purification of his tagged proteins in immobilized metal affinity chromatography. Imidazole has become an important part of many pharmaceuticals. Synthetic Imidazoles are present in many fungicides and antifungal, antiprotozoal, and antihypertensive medications. Imidazole is part of the theophylline molecule, found in tea leaves and coffee beans, which stimulates the central nervous system. Apart of its use for pharmaceutical purpose it also have varying applications in industries, the imidazole has been used extensively as a corrosion inhibitor on certain metals, such as copper, mild steel and carbon steel. Preventing metal corrosion is important, especially in aqueous systems, where the conductivity of the metal decreases due to corrosion. Corrosion is a natural and spontaneous phenomenon that results into the conversion of pure metals and their alloys into numerous stable forms such as their oxides, hydroxides, etc. by the chemical and/or electrochemical reactions with the nearby environments (Verma et al., 2018; Kicir et al., 2016; Nazeer & Madkour, 2018). Despite remarkable advancement in the field of corrosion science and technology, the phenomenon of corrosion remains a major obstacle for industries all over the world (Lavanya et al., 2018; Bammou et al., 2014; Daoud et al, 2014). The focal problem of this type of material that is easy to be rusted in the acid pickling process, causing many financial and industrial losses (EL-Hajjaji et al. 2014; Salghi et al. 2015; El Guerraf et al, 2018; Bashir et al, 2018). The acidic solutions especially Hydrochloric and sulphuric acid are applied to remove the undesirable scale and corrosion products from metals. Using inhibitors is one of the most several methods for protecting metals from corrosion and many scientists are directing their research on this theme (Fouda et al., 2014; Ouakki et al. 2018; Karthik et al. 2015; Ghibate et al., 2015). These molecules are added to the aggressive solution in order to reduce the degradation of our metals and isolated from the environment. On the other hand, we can find that the corrosion products such as rust and scale can also act as corrosion inhibitors as they can accumulate on the surface and act as a physical protective barrier (Verma et al., 2017). New inhibitors are discovered every day. Especially, the organic compounds which act as good inhibitors due to the heteroatom exist in there structures such as sulfur, nitrogen, and oxygen besides the aromatic rings (Ghazoui et al., 2014; Bedair et al., 2017; Petersen et al, 2017; Negm et al, 2018). By using these molecules on metal–solution interface, the inhibitors inhibit the deterioration of metal by interacting with the metal surface via adsorption through the donor atoms, p-orbitals, electron density, and the electronic structure of the molecule (Alaoui et al., 2016; Galai et al., 2016; Abd El-Raouf et al., 2018; Verma et al., 2016; Shainy et al. 2016; Jiang et al., 2018). In this item, these compounds make an adsorbed protective film on the metal surfaces, and save metals from the attacking acid solution (Qiao et al., 2014; Ruan et al., 2017; Shaban et al., 2017; He et al. 2014; Ansari et al., 2018). This adsorption can be determined by two ways namely: chemical (chemisorption) or physical (electrostatic) adsorptions. The chemisorption includes a chemical reaction between the inhibitor molecules and the metal surface to create strong electronic bonds. In contrast, the electrostatic interaction is caused by van der Waals force (Arellanes-Lozada et al., 2018; Umoren et al., 2015; Bousskri et al. 2015). The adsorption of these inhibitors is affected by several factors such as nature and magnitude of charge present on metal, nature of electrolyte, electronic structure of inhibitor molecules, nature of substituents, solution temperature, exposure time, etc. (Verma et al.; 2018). Imidazopyridine and its derivatives are important compounds in many fields especially in pharmaceutics (Saddik et al,. 2014; El-Awady et al., 2016; Elaatiaoui et al., 2015; Gao et al., 2016; Li et al., 2016; Ramya et al., 2015; Yuce et al., 2016; Salarvand et al., 2017; Obot et al., 2015; Savas et al., 2016; Kokalj et al. 2012; Topal et al. 2017; Saha et al., 2015; Singh et al. 2018) but recently many researchers have studied this derivative as inhibitors of metal corrosion and shown good inhibition efficiency in many different acid solutions.

In this section, we have reported the work of some Imidazole which investigated as corrosion inhibitors on different metal surfaces such as carbon steel, Iron and copper, in different media such as Hydrochloric acid, Sulfuric acid, Sulfamic acid and Sodium chloride. Comparison between different parameters from the electrochemical impedance measurements, Potentiodynamic polarization and adsorption isotherms for the studied Imidazole compounds were regrouped in Table 1.

The information presented in this review is useful for diverse industrial fields and for other applications where corrosion inhibitors for metals are needed.

Table 1 shown that Imidazole has been used extensively as a corrosion inhibitor on different metals in order to protect them from corrosion in diverse aggressive medium, especially in acidic medium. Imidazole has been reported also as effective corrosion inhibitor for carbon steel, Iron and Copper respectively in 0.5 M of H_2SO_4, 1 M of HCl and in 3% of H_3NO_3S solution, due to their excellent excited state intra-molecular proton transfer which elevates the performances of their adsorption into the metallic

Table 1. Comparison of Names, chemical structures and abbreviations of the studied Imidazole compounds

	Compounds IUPAC Name	Middle/Conc.	Materials	Inhibition Concentration	Inhibition Efficiency		Adsorption Isotherm	Reference
					Imp (%)	IE (%)		
S1	Imidazole	0.5 M H2SO4	Carbon steel	0.01 M	72.4	72.7	Not identified	(Benali et al., 2005)
S2	Imidazole	1M HCl	Iron	14 mM	79.1	-	Langmuir	(Zhang et al., 2009)
S3	Imidazole	0.5 M HCl	Copper	2 mM	37.2		Not identified	(Da-quan et al., 2008)
S4	Imidazole	3% NaCl	Copper	10^{-4}	-	49.79	Freundlich's	(Otmacic & Stupnisek-Lisac, 2003)
S4	Imidazole	0.5 M H2SO4	Copper	0.5 mol x dm^{-3}		44.0	Freundlich	(Wang et al., 2013)
S5	Imidazole	3% H_3NO_3S	Copper	1.0 g/l		95.4	Not identified	(Otmacic & Stupnisek-Lisac, 2002)

surface. However, the inhibition efficiency of this compound is less good for Copper in 0.5 M of HCl, 3% of NaCl and 0.5 M of H_2SO_4 medium. Their inhibition efficiency at the optimum conditions was respectively 37.2%, 49.8% and 44.0%.

IMIDAZOLE SUBSTITUTED COMPOUNDS

Recently many workers have reoriented their attention to the development of new Imidazole derivatives. This section summarizes some Imidazole derivatives used as corrosion inhibitors on different metal surfaces in different media. The information presented in this section is useful for diverse industrial fields and for other applications where corrosion inhibitors for materials are needed. The Imidazole inhibitors cited in this review was studied in acid solution since it is the most applied in industries such (manufacturer fertilizer industries, metallurgical industry, particularly for descaling, pickling of metals and cleaning of industrial plants or metals surface. The additional effect of these molecules at different concentrations on the corrosion of various metals was studied by measurement of weight loss, potentiodynamic polarization, and electrochemical impedance spectroscopy (EIS); the results obtained from these measurements were summarized in Table 2.

It is showed that Imidazole substituted compounds (P1, P2, … to P21) are very good inhibitors for mild steel and carbon steel corrosion in 1.0 M HCl medium which are act as mixed-type inhibitors. EIS showed that all compounds act by the formation of a protective film at the metal surface. The potentiodynamic polarization curves indicated that these compounds act as mixed-type inhibitors and their inhibition efficiencies increased with their concentrations to reach a maximum at 10^{-3} M. It is found

Table 2. Corrosion parameters obtained from electrochemical measurements of various metals in different medium.

	Compounds IUPAC Name	Electrolyte	Metal	Inhibition Concentration	Inhibition Efficiency		Adsorption Isotherm	Reference
					Imp (%)	IE (%)		
P1	2,4,5-triphenyl-4,5-dihydro-1H-imidazole	1M HCl	Mild steel	10^{-3} M	92.5	92.9	Langmuir	(Singh et al., 2018)
P2	2-(4,5-diphenyl-4,5-dihydro-1h-imidazol-2-yl)-5-methoxyphenol	1M HCl	Mild steel	10^{-3} M	91.3	92.9	Langmuir	(Galai et al., 2017)
P3	2-(ethyllthio)-1,4,5-triphenyl-1H-imidazole	1M HCl	Mild steel	3 x 10^{-3} M	84.3	90.4	Langmuir	(Afia et al., 2013)
P4	1,4,5-triphenyl-2-(4-metyoxyphenyl)-1H-imidazole	1M HCl	Mild steel	10^{-3} M	98.9	98.5	-	(Ouakki et al., 2018)
P5	2-(4,5-diphenyl-4,5-dihydro-1H-imidazol-2-yl) phenol	1M HCl	Mild steel	10^{-3} M	96.8	97.5	Langmuir	(Galai et al., 2017)
P6	1,4,5-triphenyl-2-(3,4,5-trimetyoxyphenyl)-1Himidazole	1M HCl	Mild steel	10^{-3} M	96.3	96.6	-	(Ouakki et al., 2018)
P7	2-Chloro-1H-imidazole	1M HCl	Carbon steel	10 mM	57.1	-	...	(Gutiérrez et al., 2010)
P8	2-Methylimidazole	1M HCl	Carbon steel	10 mM	30.3	-	...	(Gutiérrez et al., 2010)
P9	4′-(Imidazol-1-yl)acetophenone	1M HCl	Carbon steel	10 mM	80.9	-	...	(Gutiérrez et al., 2010)
P10	4-(1H-Imidazol-1-yl)aniline	1M HCl	Carbon steel	10 mM	75.3	-	...	(Gutiérrez et al., 2010)
P11	4′-(Imidazol-1-yl) benzaldehyde	1M HCl	Carbon steel	10 mM	79.1	-	...	(Gutiérrez et al., 2010)
P12	1-(4-Bromophenyl)-1H-imidazole	1M HCl	Carbon steel	10 mM	88.1	-	...	(Gutiérrez et al., 2010)
P13	1-(4-Chlorophenyl)imidazole	1M HCl	Carbon steel	10 mM	79.1	-	...	(Gutiérrez et al., 2010)
P14	4-(Imidazol-1-yl)phenol	1M HCl	Carbon steel	10 mM	88.8	-	...	(Gutiérrez et al., 2010)
P15	1-(3-Chlorophenyl)imidazole	1M HCl	Carbon steel	10 mM	70.9	-	...	(Gutiérrez et al., 2010)
P16	1-(2-Chlorophenyl)imidazole	1M HCl	Carbon steel	10 mM	63.0	-	...	(Gutiérrez et al., 2010)
P17	2-(Chloromethyl)benzimidazole	1M HCl	Carbon steel	10 mM	84.8	-	...	(Gutiérrez et al., 2010)
P18	6-Bromo-1H-benzimidazole	1M HCl	Carbon steel	10 mM	89.4	-	Langmuir	(Gutiérrez et al., 2010)
P19	1H-Benzimidazol-5-amine	1M HCl	Carbon steel	10 mM	74.9	-	Langmuir	(Gutiérrez et al., 2010)
P20	2-(Aminomethyl)benzimidazole	1M HCl	Carbon steel	10 mM	69.1	-	...	(Gutiérrez et al., 2010)
P21	1H-Benzimidazol-5-ylmethanol	1M HCl	Carbon steel	10 mM	73.3	-	...	(Gutiérrez et al., 2010)
P22	2-phenyl-2-imidazoline	1M HCl	Iron	6.5 mM	84.3	-	Langmuir	(Zhang et al. 2009)
P23	1-n-undecyl-imidazole	0.5 M HCl	Copper	2 mM	72.4			(Da-quan et al., 2008)
P24	3-butyl-2,4,5-triphenylimidazole	1% NaCl	Carbon steel	60 ppm	-	33.5	Langmuir	(Zunita et al., 2012)
P25	3-butyl-2-(2-butoxyphenyl)-4,5-diphenylimidazole	1% NaCl	Carbon steel	60 ppm	-	53.8	Langmuir	(Zunita et al., 2012)
P26	4-methylimidazole	3% NaCl	Copper	10^{-4} M	-	61.31	Freundlich's	(Otmacic et al., 2003)
P27	4-methyl-5-imidazole carbaldehyd	3% NaCl	Copper	3 x 10^{-4} M	-	71.24	Freundlich's	(Otmacic et al., 2003)
P28	Ethyl-4-methyl-5-imidazole carboxylate	3% NaCl	Copper	10^{-3} M	-	80.50	Freundlich's	(Otmacic et al., 2003)
P29	1-phenyl-4-methyl imidazole	3% NaCl	Copper	5 x 10^{-4} M	-	94.31	Freundlich's	(Otmacic et al., 2003)
P30	1-p-tolyl)-4-methyl imidazole	3% NaCl	Copper	7 x 10^{-4} M	-	93.03	Freundlich's	(Otmacic et al., 2003)
P31	4-methyl-1(4-methoxyphenyl) imidazole	3% NaCl	Copper	2 x 10^{-4} M	-	83.95	Freundlich's	(Otmacic et al., 2003)
P32	4-methyl-1-phenyl imidazole	0.5 M NaCl	Copper	0.005 M		93		(Otmacic Curkovic et al., 2010)

continued on following page

Table 2. Continued

	Compounds IUPAC Name	Electrolyte	Metal	Inhibition Concentration	Inhibition Efficiency		Adsorption Isotherm	Reference
					Imp (%)	IE (%)		
P33	4-methyl-1-(p-tolyl) imidazole	0.5 M NaCl	Copper	0.001 M		92		(Otmacic Curkovic et al., 2010)
P34	Benzimidazole	1M H3PO4	Mild steel	5 x 10^{-2} M	62.1	62.0	Flory–Huggins	(Ghanbaria et al., 2010)
P35	2-methyl benzimidazole	1M H3PO4	Mild steel	5 x 10^{-2} M			Flory–Huggins	(Ghanbaria et al., 2010)
P36	2-aminobemzimidazole	1M H3PO4	Mild steel	5 x 10^{-2} M	69.5	69.3	Flory–Huggins	(Ghanbaria et al., 2010)
P37	1-benzyl-1H-imidazole	0.5 M H2SO4	Carbon steel	0.01 M	79.5	-	...	(Benali et al. 2005)
P38	(E)-3-(4-((1H-imidazol-1-yl)methyl)phenyl)acrylic acid	0.5 M H2SO4	Carbon steel	0.01 M	80.3	-	...	(Benali et al. 2005)
P39	1-buty1-1H-imidazole	0.5 M H2SO4	Carbon steel	0.01 M	76.5	-	...	(Benali et al. 2005)
P40	1-tosyl-1H-imidazole	0.5 M H2SO4	Carbon steel	0.01 M	80.0	-	...	(Benali et al. 2005)
P41	(E)-methyl 3-(4-((1H-imidazol-1-yl)methyl)phenyl) acrylate	0.5 M H2SO4	Carbon steel	0.01 M	80.5	-	...	(Benali et al. 2005)
P42	4-methyl-5-hydrocymethylimidazole	0.5 M H2SO4	Copper	0.3 mol x dm^{-3}	-	65.4	Freundlich	(Otmacic et al., 2002)
P43	1-phenyl-4-methylimidazole	0.5 M H2SO4	Copper	0.5 mol x dm^{-3}	-	94.3	Freundlich	(Otmacic et al., 2002)
P44	1-(p-tolyl)-4-methylimidazole	0.5 M H2SO4	Copper	0.05 mol x dm^{-3}	-	86.0	Freundlich	(Otmacic et al., 2002)

also that the mechanism of the hydrogen reduction and the iron dissolution reactions changed by the inhibitor addition.

They act by adsorption mechanism at the mild steel surface and their inhibition depended on their molecular structures and the 1,4,5-triphenyl-2-(4-metyoxyphenyl)-1H-imidazole (P4) compound is the best inhibitor compared to others in same medium.

It has shown also that these compounds act as excellent inhibitors for mild steel compared to carbon steel and their inhibition efficiencies follow the order: P4 > P5 ≈ P6 > P1 > P2 > P3 and depend on the type of the substituent.

On the other hand, for the second series of Imidazole substituted corrosion inhibitors studied, we can conclude that, the inhibition efficiency for carbon steel in hydrochloric acid was increased in the order P18 ≈ P14 ≈ P12 > P17 > P10 > P19 > P21 > P9 ≈ P11 ≈ P13 ≈ P15 > P20 > P16 > P7 > P8, which depends on their Imidazole substitution. EIS measurements are not perfect semicircles and explained this by the dispersion of frequency as a result of the roughness and inhomogeneous of electrode surface. Also, they confirmed that these structures adsorbed into the electrode interface by replacement of water molecules with organic inhibitor molecules. Electrochemical Impedance Spectroscopy (EIS) results showed that the best corrosion inhibition performance is for 6-Bromo-1H-benzimidazole (P18), 4-(Imidazol-1-yl)phenol (P14) and 1-(4-Bromophenyl)-1H-imidazole (P12).

TRIPHENYLIMIDAZOLE COMPOUNDS

To remove undesirable scales the particular piece of metal is immersed in a suitable acid solution, called an acid pickling bath. The acid solution attacks and dissolves the scale. Once the scale is dissolved, the acid is then free for further attack upon the metal. In order to reduce the metal dissolution and the consumption of acid, corrosion inhibitors are added to the pickling solution. As known, Hydrochloric acid (HCl) is the most important pickling acid. Furthermore, the effect of temperature on a chemical reaction of practical and theoretical important. Like most chemical reactions, the rate of corrosion of iron and steel increases with temperature especially in industrial pickling hydrochloric acid media. Accordingly, acid pickling inhibitors are expected to be chemically stable to provide high protective efficiency under the conditions mentioned above. Hydrochloric acid pickling bath is used to remove the corrosion products from the metal working. Over-pickling can cause a serious damage and physical properties of metals structure can be easily done. Thus, to control the risk and reduce the problem, several organic compounds are used as corrosion inhibitors.

Taking into account their excellent performance, the third part of this chapter is dedicated to the corrosion inhibition of mild steel in hydrochloric acid medium using the best Imidazole molecules listed above. These studies were carried out using weight loss and electrochemical methods. The effect of temperature on the corrosion behavior and the thermodynamic parameters was also studied and discussed.

Gravimetric Measurements

The effect of different concentration of substituted triphenylimidazole compounds on carbon steel corrosion in 1.0 M HCl was studied using gravimetric method due to its simplicity and good reliability. The observed weight-loss values of triplicate measurements are highly reproducible giving standard deviations. The inhibition efficiency (η_w%) and other parameters such as corrosion rate (ω_{corr}) and surface coverage (θ) at various concentration of the inhibitors are given in Table 3.

Table 3. Weight loss data of carbon steel in 1.0 M HCl without and with different concentrations of substituted triphenylimidazole compounds at 25±2 °C after 6 h of immersion.

Compounds	Conc. (M)	ω_{corr} / mg cm^{-2} h^{-1}	η_ω %	θ
Blank solution	0	31.1	-	-
P1	10^{-6}	13.3	57.2	0.572
	10^{-5}	9.0	71.0	0.710
	10^{-4}	4.2	86.5	0.865
	10^{-3}	2.4	92.3	0.923
P2	10^{-6}	4.3	86.2	0.862
	10^{-5}	2.1	93.2	0.932
	10^{-4}	1.3	95.8	0.958
	10^{-3}	0.9	97.1	0.971
P3	10^{-6}	5.6	82.0	0.820
	10^{-5}	4.5	85.5	0.855
	10^{-4}	3.9	87.5	0.875
	10^{-3}	2.2	93.1	0.931

Careful examination of the obtained results showed that the inhibition efficiencies of the studied inhibitors increase with increasing concentrations where the P2 is the best inhibitor. So, the maximum values of the inhibition efficiency are 92.3% for P1, 97.1% for P2 and 93.1% for P3 obtained at 10^{-3} M.

The inhibitors are expected to get adsorbed through the lone pairs of electrons on N atoms of amino group and imidazole ring as well as π-electron density on the phenyl and imidazole ring by their coordination with metal surface. In addition, the participation of phenyl ring to N atom during the adsorption process may be confirmed by changing the π-electron density on phenyl ring by substituting electron donating (-OH) group. Generally, electron donating groups increase the inhibition efficiency of the inhibitors. The inhibitors P1, P2 and P3 have nearly same size and number of active centers but P2 shows higher inhibition efficiency (η_w%) than P1 and P3 due to higher delocalized π-electron density at benzene ring.

Potentiodynamic Polarization Curves

Potentiodynamic polarization curves of carbon steel in 1.0 M HCl without and with different concentrations of each substituted triphenylimidazole compounds are presented in Figure 2, 3 and 4. Their extrapolation parameters and inhibition efficiency are plotted in Table 4. It can be shown that the inhibitors addition hindered the acid attack on carbon steel surface. In addition, an increase in their concentration gives a decrease in anodic and cathodic current densities indicating that the substituted compounds acted as mixed-type inhibitors. However, the inhibitor addition does not change the hydrogen evolution reaction mechanism such as indicated by the slight changes in the cathodic Tafel slopes (β_c) values. This indicates that hydrogen evolution is activation controlled (Peters, 1970; Szauer, 1981).

Figure 2. Potentiodynamic polarization curves for carbon steel in 1.0 M HCl in the absence and presence of various concentrations of P1.

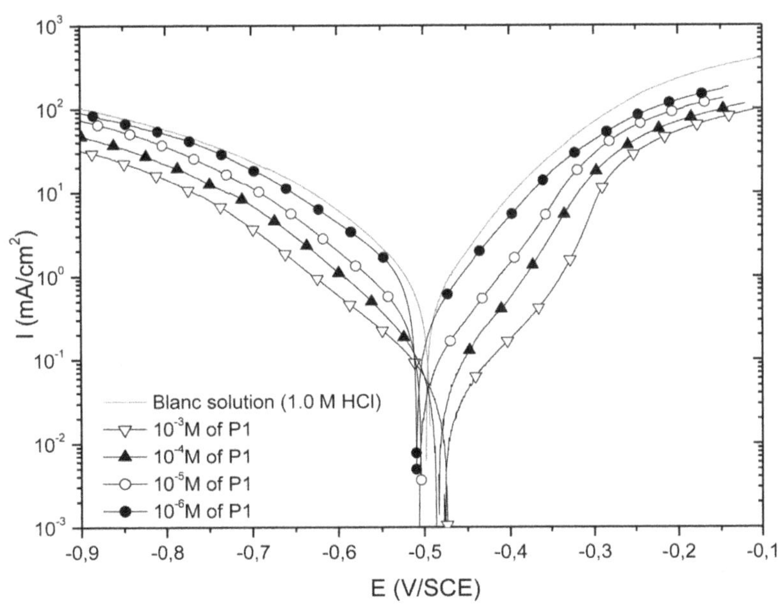

It is apparent also that the η_{PP} followed the order: P2 > P1 ≈ P3 such as found by the weight loss measurements. It is noted also, that all inhibitors shifted slightly the corrosion potential (E_{corr}) toward the anodic direction.

Figure 3. Potentiodynamic polarization curves for carbon steel in 1.0 M HCl in the absence and presence of various concentrations of P2.

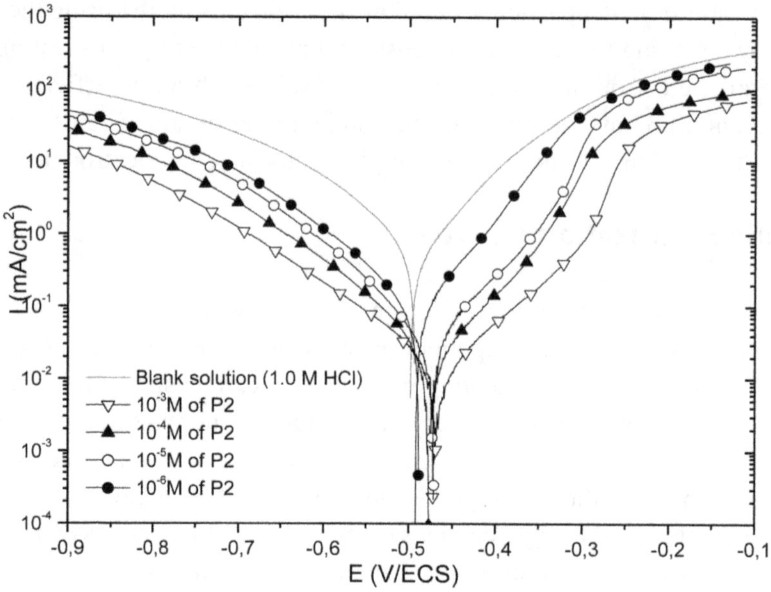

Figure 4. Potentiodynamic polarization curves for carbon steel in 1.0 M HCl in the absence and presence of various concentrations of P3.

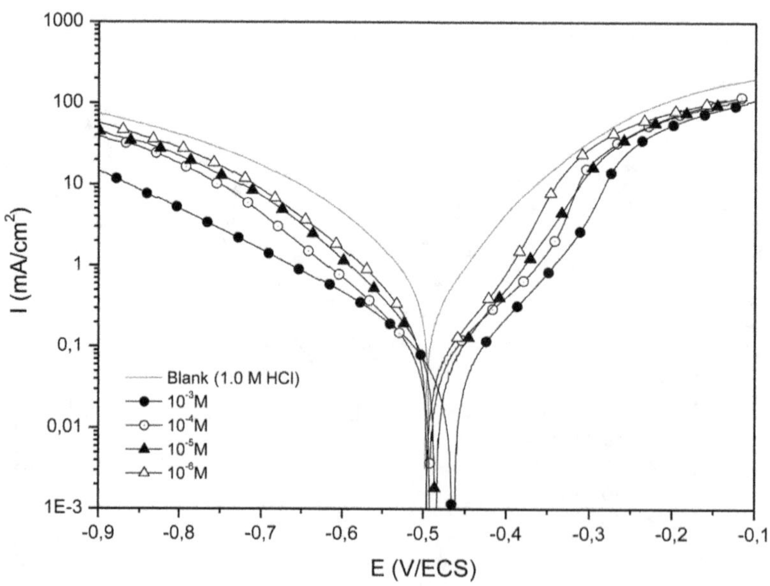

Electrochemical Impedance Spectroscopy (EIS)

Figures 5, 6, 7 and 8 show the Nyquist plots obtained for carbon steel in 1.0 M HCl without and with different concentrations of substituted Triphenylimidazole compounds at the open circuit potential.

Table 4. Electrochemical parameters for carbon steel in 1.0 M HCl at various concentrations of Triphenilimidazole at 25±2°C.

Compounds	Concentration	E_{corr}	i_{corr}	β_c	β_a	η_{Tafel}
Blank	00	-498	983	-92	104	-
P1	10^{-6}	-505	430	-125	55	56.3
	10^{-5}	-499	300	-129	60	69.5
	10^{-4}	-480	130	-120	72	86.7
	10^{-3}	-465	70	-108	79	92.9
P2	10^{-6}	-485	145	-78	62	85.2
	10^{-5}	-465	50	-85	68	94.9
	10^{-4}	-473	40	-94	74	95.9
	10^{-3}	-462	25	-119	82	97.5
P3	10^{-6}	-492	160	-96	87	83.7
	10^{-5}	-479	137	-100	82	86.0
	10^{-4}	-482	100	-106	80	89.8
	10^{-3}	-458	70	-116	79	92.9

Note: Concentration in M, E_{corr} in mV (vs. SCE), i_{corr} in μA/cm², β_a and β_c are respectively slope of anode and cathode polarization curve, in mV dec.

Figure 5. Nyquist plots for carbon steel in 1.0 M HCl solution without inhibitors at 25±2°C.

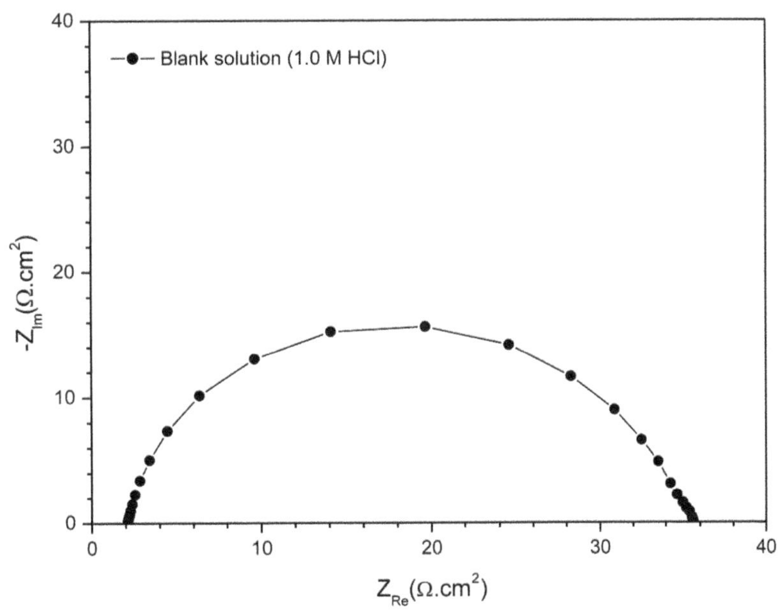

Figure 6. Nyquist plots for carbon steel in 1.0 M HCl solution in the absence and presence of various concentrations of P1 at 25±2°C.

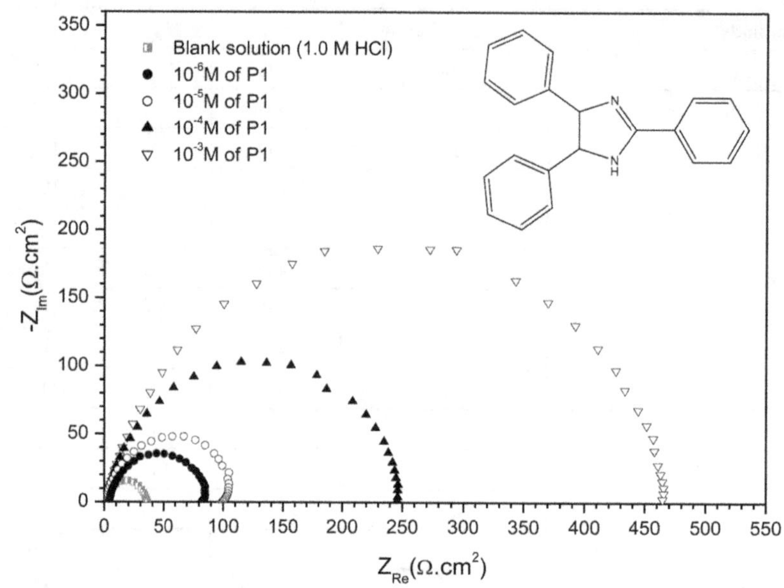

Figure 7. Nyquist plots for carbon steel in 1.0 M HCl solution in the absence and presence of various concentrations of P2 at 25±2°C.

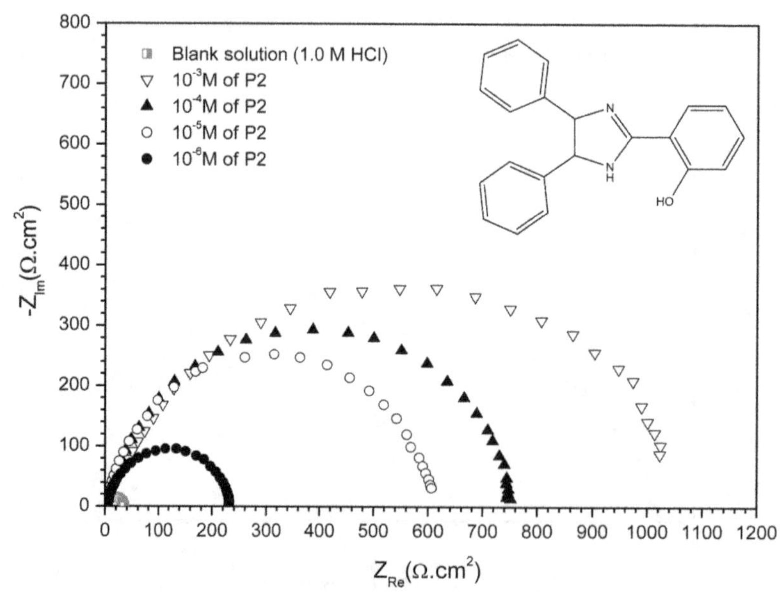

Their corresponding parameters which were extracted using the equivalent circuit presented in Figures 9, are shown in Table 5. Thus circuit has been used previously to model the iron/acid solution interface (Obot, 2010). It is apparent that the obtained spectra were composed of one capacitive loop which its diameter was significantly increased after inhibitors addition to reach a maximum at 10^{-3} M of all com-

Figure 8. Nyquist plots for carbon steel in 1.0 M HCl solution in the absence and presence of various concentrations of P3 at Ecorr (T=298±2 K).

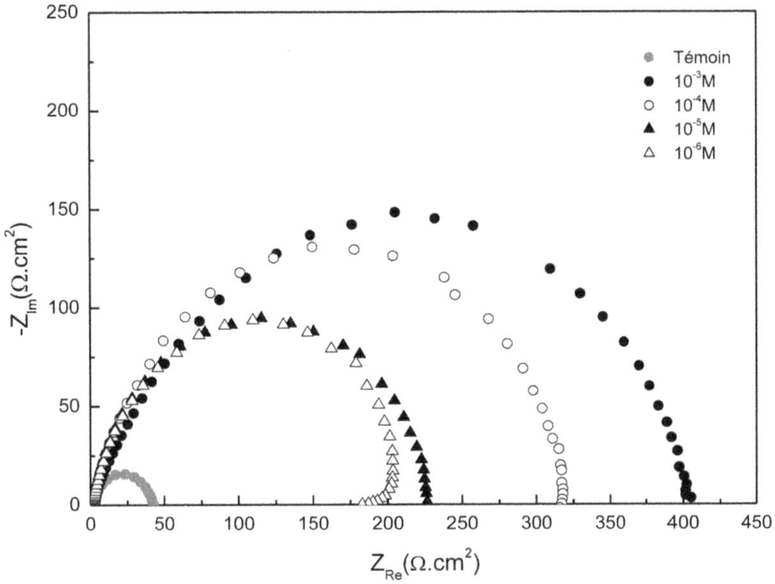

Table 5. Electrochemical impedance parameters and inhibition efficiency for carbon steel in 1.0 M HCl solution without and with different concentration of Triphenylimidazole at 25±2°C.

Compounds	Conc. / M	R_{ct}/ Ω cm²	C_{ct}/μF cm⁻²	n_{ct}	η_{EIS}/ %
Blank	0	35	298	0.79	-
P1	10^{-6}	80	90	0.87	56.2
	10^{-5}	110	72	0.90	68.2
	10^{-4}	245	65	0.94	85.7
	10^{-3}	467	54	0.95	92.5
P2	10^{-6}	230	90	0.76	84.8
	10^{-5}	611	74	0.86	94.3
	10^{-4}	756	65	0.88	95.4
	10^{-3}	1089	58	0.89	96.8
P3	10^{-6}	205	70	0.78	82.9
	10^{-5}	235	62	0.75	85.0
	10^{-4}	318	55	0.90	89.0
	10^{-3}	404	50	0.94	91.3

pounds. In addition, these diagrams are not perfect semicircles which has been attributed to frequency dispersion (Watson, 2001). However, the inhibitors addition is found to enhance R_{ct} values and bring down C_{dl} values. These finding can be explained by the fact that the carbon steel corrosion in 1.0 M HCl was controlled by a charge transfer process and the corrosion inhibition occurs by the adsorption of the

Figure 9. Equivalent circuit model for system mild steel/1.0 M HCl/Triphenylimidazole Substituted compounds.

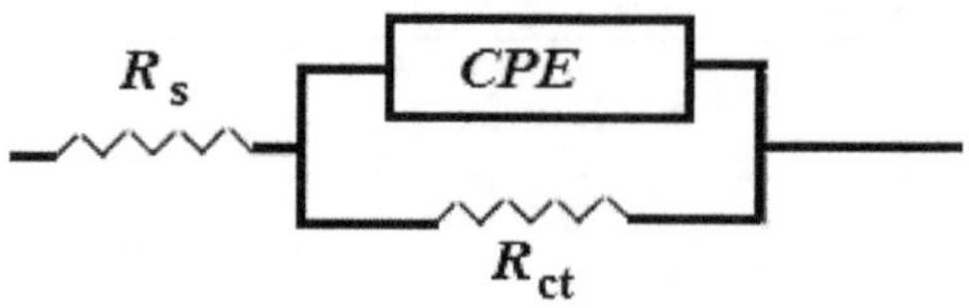

substituted Triphenylimidazole compounds on carbon steel surface. However, the decrease in the C_{dl} values, with can result from a decrease in local dielectric constant and/or an increase in courant density, the thickness of the electrical double layer, suggested by adsorption of the Triphenylimidazole molecules at the metal/solution interface (Atwell, 1989; Xia, 1998). It is noted also that the inhibition efficiency follow the order: P2 > P1 ≈ P3, which is in a good agreement with results obtained from weight loss and potentiodynamic polarization measurements.

In the other hand, the effectiveness of organic compounds mainly depends on their size and their active centers [84]. The best performance of compound P2 as corrosion inhibitors over compound P1 may be attributed to the presence of –OH group in compound P2. Indeed, the protection efficiency increases with increasing of inhibitor concentration, the maximum η_{EIS} (%) of 96.8% for P2 was achieved at 10^{-3}M.

Figure 10 (a) and (b) shows Nyquist plots and the representative Bode diagrams for carbon steel in 1.0 M HCl in the presence of various concentrations of 10^{-3} M of P2 to the aggressive solution leads to a change of the impedance diagrams in both shape and size, in which a depressed semicircle at the high frequency part of the spectrum was observed. The increase in size of the semicircle with inhibitor concentration means that the inhibitor effect increases as well.

Values of the charge transfer resistance R_{ct} were obtained from these plots by determining the difference in the values of impedance at low and high frequencies [84]. The effective capacity C_{dl} can be estimated using the following mathematical formulas from the CPE:

$$C = Q^{1/n_{ct}} * R^{(1-n_{ct})/n_{ct}} \quad (5)$$

With n_{ct} is the degree of heterogeneity.

The equivalent circuit model employed for these systems is presented in figure 6.

The results described below can be interpreted in terms of the equivalent circuit of the electrical double layer shown in fig. 6, which has been used previously to model the iron-acid interface (Ouali, 2010). In this equivalent circuit, R_s is the solution resistance, R_{ct} is the charge transfer resistance and CPE is a constant phase element. Excellent fit with this model was obtained for all experimental data. As an example, the Nyquist and Bode plots for 1.0 M HCl solution at 10^{-3} M with P2 are presented in Fig. 7.

Temperature Effect

Temperature can modify the interaction between the steel electrode and the acidic media without and with substituted quinolone compounds inhibitor's. Polarisation curves for carbon steel in 1.0 M HCl in the absence and presence of 10^{-3}M of Triphenylimidazole inhibitor's in the temperature range 25±2°C

Figure 10. EIS Nyquist and Bode diagrams for carbon steel/1.0 M HCl with 10^{-3} M of P2 interface: (···) experimental; (—) fitted data using structural model in Fig. 6.

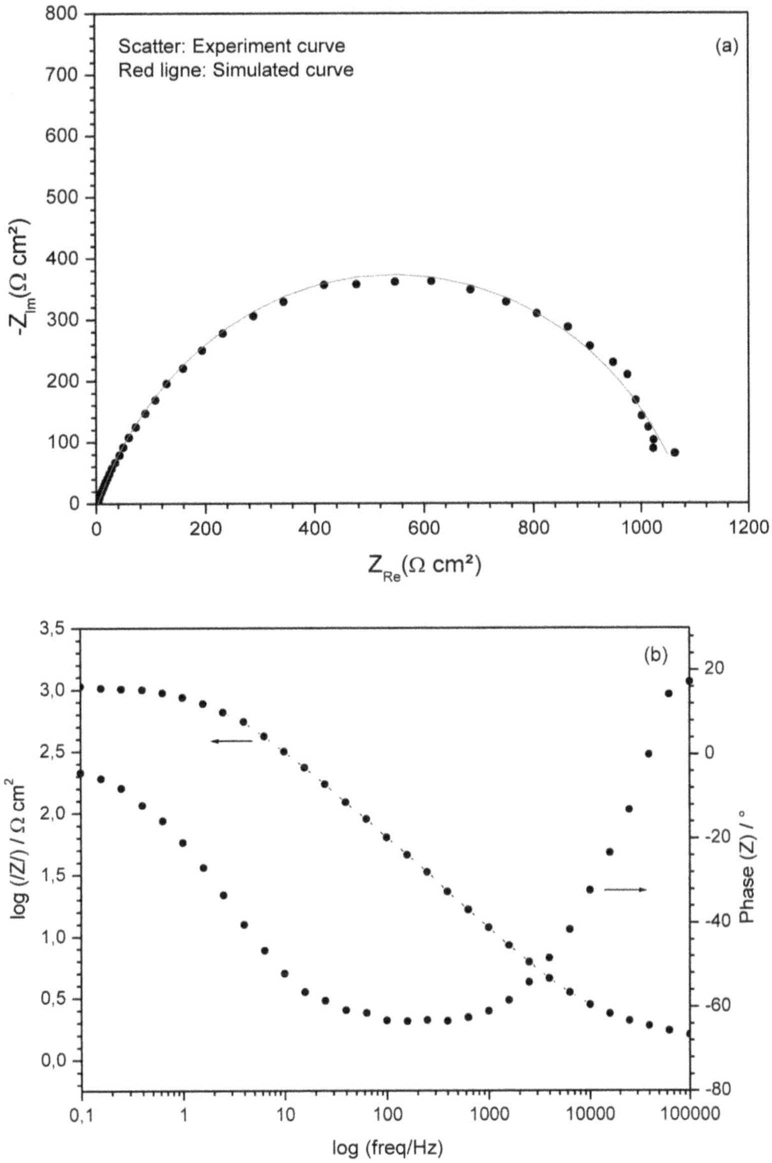

to 55±2°C are shown in Figures 11 to 14 presented the obtained potentiodynamic polarization curves and their corresponding data are presented in Table 6.

It is clear that all curves exhibit Tafel behaviour and show a little different effect in the anodic and cathodic branches. It is seen also that the inhibition efficiency decreased slightly with temperature.

Figure 11. Potentiodynamic polarization curves for carbon steel in 1.0 M HCl in the absence of inhibitors at different temperatures between 25±2 and 55±2 °C.

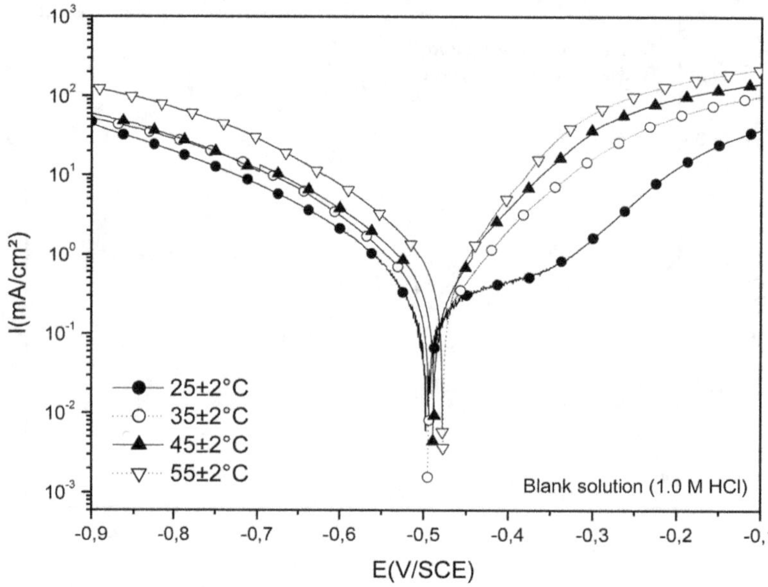

Figure 12. Potentiodynamic polarization curves for carbon steel in 1.0 M HCl in the presence of 10^{-3} M of P1 at different temperatures between 25±2°C and 55±2°C.

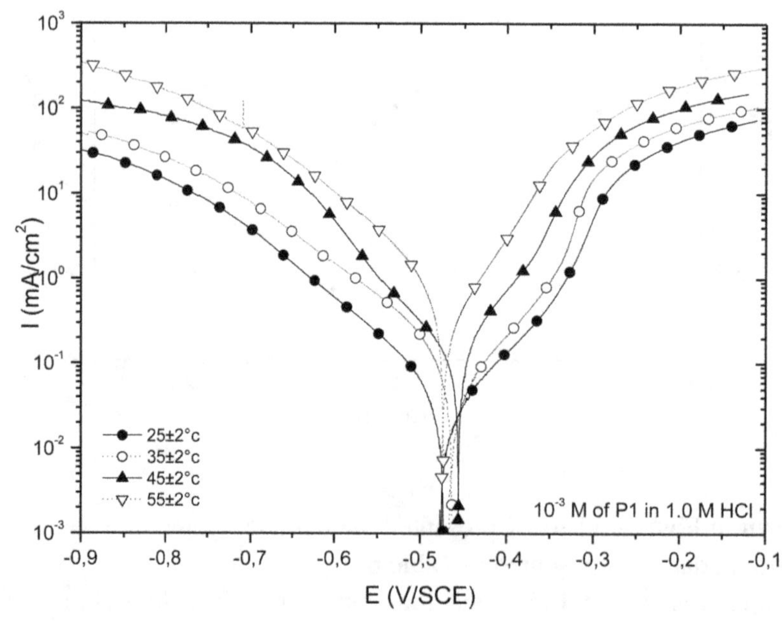

Thermodynamic Parameters of Triphenylimidazole on Carbon Steel Surface

Thermodynamic parameters are important to study the inhibitive mechanism. The thermodynamic functions for dissolution of mild steel in the absence and in the presence of various concentrations of Triphenylimidazole substituted compounds were obtained by applying the Arrhenius equation and the transition state equation (Gunasekaran, 2004; Singh, 1979; Brug, 1984).

Figure 13. Potentiodynamic polarization curves for carbon steel in 1.0 M HCl in the presence of 10^{-3}M of P2 at different temperatures between 25±2°C and 55±2°C.

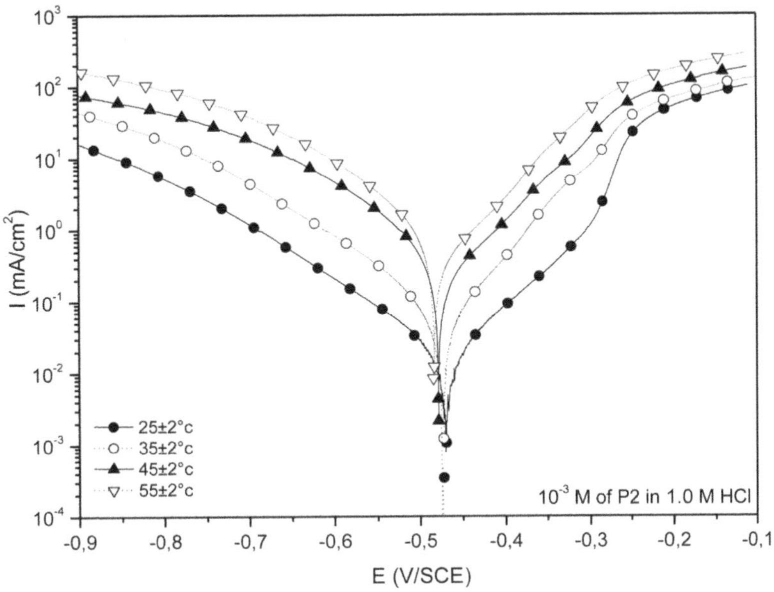

Figure 14. Potentiodynamic polarization curves for carbon steel in 1.0 M HCl in the presence of 10^{-3} M of P3 at different temperature.

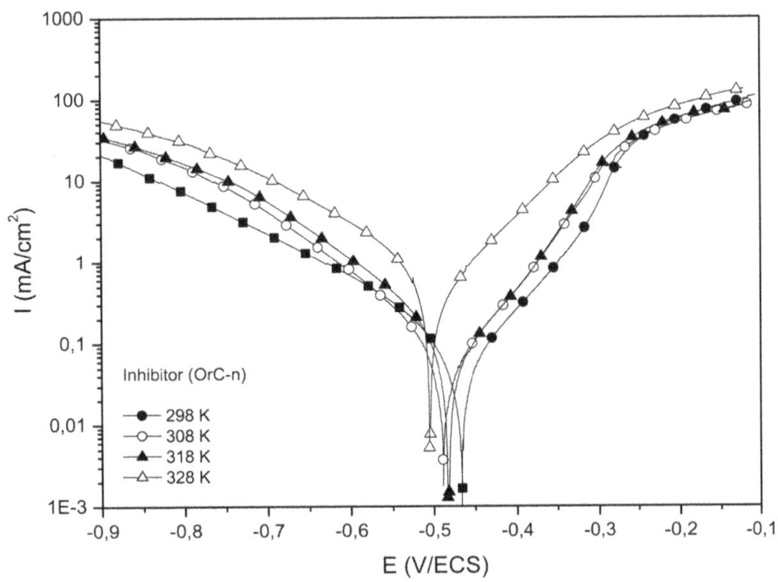

Table 6. The influence of temperature on the electrochemical parameters for carbon steel in 1.0 M HCl with 10^{-3} M of substituted Triphenylimidazole compounds

Inhibitors	Temperature	E_{corr}	I_{corr}	β_c	β_a	η_{ct}
Blank	25±2	-498	983	-92	104	-
	35±2	-491	1200	-184	112	-
	45±2	-475	1450	-171	124	-
	55±2	-465	2200	-161	118	-
P1	25±2	-465	70	-108	79	92.9
	35±2	-459	100	-137	74	91.6
	45±2	-453	150	-93	68	89.6
	55±2	-466	260	-87	60	88.2
P2	25±2	-462	25	-119	82	97.5
	35±2	-467	45	-100	72	96.2
	45±2	-474	80	-94	67	94.5
	55±2	-477	160	-86	60	92.7
P3	25±2	-458	70	-116	79	92.9
	35±2	-480	100	-103	73	91.7
	45±2	-477	140	-100	70	90.3
	55±2	-498	230	-125	67	89.5

Note: Temperature in °C, E_{corr} in mV (vs. SCE), i_{corr} in mA/cm², β_a and β_c are respectively slope of anode and cathode polarization curve, in mV dec.

Figure 15. Arrhenius plots of carbon steel in 1.0 M HCl without (a) and with (b) 10^{-3} M of P1, P2 and P3.

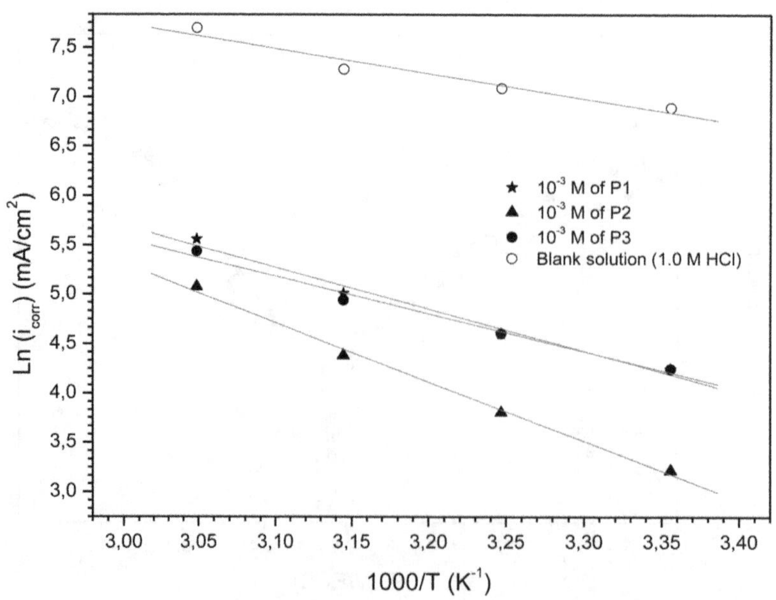

However, the logarithm of corrosion rates (Ln i_{corr}) versus reciprocal of absolute temperature (1/T) for 1.0 M HCl without and with substitutedquinolonewas examined (Figure 15) using the Arrhenius equation:

$$Lni_{corr} = -\frac{E_a}{RT} + LnA \quad (6)$$

where A is the Arrhenius pre-exponential constant, R is the universal gas constant, E_a is the apparent activation energy and T is the absolute temperature. The values obtained from the slope of the linear plots are shown in Table 7.

Table 7. The values of activation parameters E_a, ΔH_a and ΔS_a* for carbon steel in 1.0 M HCl and added of 10^{-3} M of the inhibitor's.*

Compounds	E_a (KJ×mol^{-1})	ΔH_a* (KJ×mol^{-1})	ΔS_a* (J×mol^{-1}×K^{-1})
Blank solution	21	18.5	-126
10^{-3} M of P1	35.2	32.6	-100.7
10^{-3} M of P2	49.8	47.2	-59.9
10^{-3} M of P3	31.7	29.0	-112.5

It is found that all the linear regression coefficients are close to 1, indicated that the corrosion of carbon steel in hydrochloric acid can be explained using the kinetic model. As observed from the Table 7, the E_a increased with Triphenylimidazole addition compared to the uninhibited solution (Blank solution).

Inspection of these data reveals that the thermodynamic parameters ΔH_{ads} and ΔS_{ads} of dissolution reaction of carbon steel in 1.0 M HCl in the presence of Triphenylimidazole are higher than in the absence of inhibitor. The positive sign of enthalpies reflect the endothermic nature of steel dissolution process meaning that dissolution of steel is difficult (Moradi et al., 2013). The increase in E_a in the presence of Triphenylimidazole may be interpreted as physical adsorption. Indeed, a higher energy barrier for the corrosion process in the presence of inibitor's was associated with physical adsorption or weak chemical bonding between the inhibitors species and the carbon steel surface (Xia et al, 1998; Finley et al., 1960). (Finley et al. 1960) have explained that the increase in E_a can be attributed to decrease in the inhibitor adsorption at metallic surface with the rise of temperature. In this context, Singh et al. have considered that the increase in temperature caused an increase in the electron density at the adsorption centers, which improved the inhibition efficiency (Singh et al., 1979).

The other kinetic parameters such as enthalpy of adsorption (ΔH_a) and entropy of adsorption (ΔS_a) were obtained from transition state equation:

$$Ln\frac{i_{corr}}{T} = \ln\left(\frac{R}{Nh}\right) + \left(\frac{\Delta S_a}{R}\right) - \frac{\Delta H_a}{RT} \quad (7)$$

where i_{corr} is the corrosion rate, h the Plank's constant and N is Avogrado's number, ΔH_a the enthalpy of activation and ΔS_a the entropy of activation.

Figure 16. Transition-state plots for mild steel corrosion rates $\ln j_{corr}$ *versus 1/T in 1.0 M HCl without (a) and with (b) 10^{-3} M of P1, P2 and P3.*

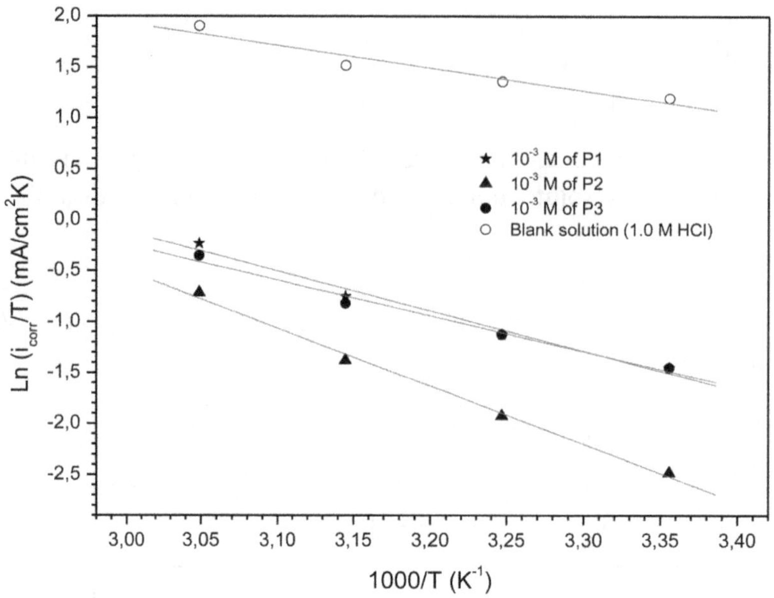

Figure 16 shows the variation of Ln (i_{corr}/T) function (1/T) as a straight line with a slope of ($-\Delta H_a$/ R) and the intersection with they-axis is[Ln(R/Nh) + (ΔS_a /R)]. From these relationships, valuesof ΔS_a and ΔH_a can be determined. The activation parameters (ΔH_a and ΔS_a) which determined from the slopes of Arrhenius lines without and with inhibitors, are summarized in Table 5. It is seen that the ΔH_a value for dissolution reaction of mild steel in 1.0 M HCl in the presence of P2 is higher than that in the presence P1 and the free solution. In addition, the ΔH_a values in the presence P1, P2 and P3 are lower than that in their absence.

However, the positive signs of ΔH_a values reveal the endothermic nature of the mild steel dissolution process suggesting that is difficult (Moradi, 2013) with inhibitors. The same remarks were observed for the E_a values indicating that the corrosion process must involved a gaseous reaction, simply the hydrogen evolution reaction, associated with a decrease in the total reaction volume (Alaoui, 2016).

Additionally, Table 7 shows that the ΔS_a values increase with the presence of P2 compared to blank solution, which mean an increase in disorder during the transition from reactant to the activated complex during corrosion process. Also the ΔS_a values tend to more negative values as the P1 addition showing more ordered behaviour leading to increase inhibition efficiency.

Adsorption Isotherm

It is generally assumed that the inhibitor adsorption onto the metal/solution interface is the first step in the mechanism of inhibition in aggressive media. Four types of adsorption may take place by heterocyclic molecules at the metal/solution interface: (1) electrostatic attraction between the charged metal and the charged molecules, (2) interaction of uncharged electrons pair in the molecule with the metal, (3) interaction of π electrons with the metal and (4) combination of (1) and (3). Chemical adsorption

Figure 17. Plot of the Langmuir adsorption isotherm of Triphenylimidazole substituted compounds on the carbon steel surface at 25±2°C

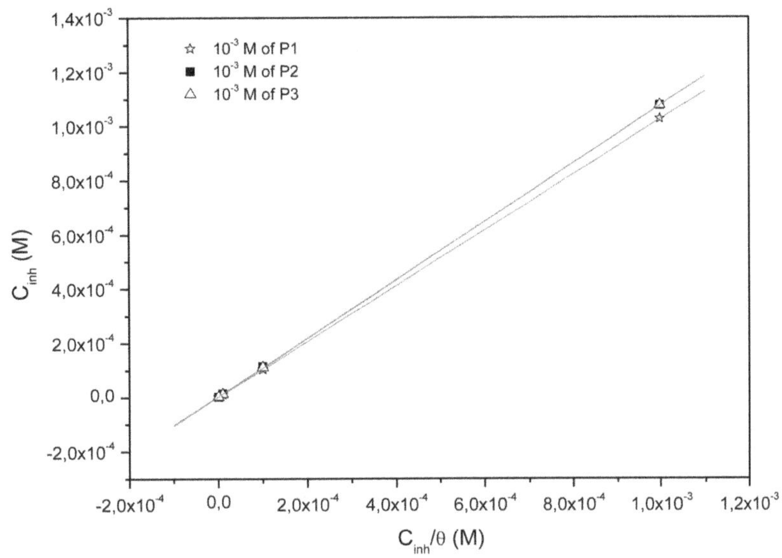

involves the share or charge transfer from the molecules onto the surface to form a coordinate type bond. Electron transfer is typical for transition metals having vacant low energy electron orbital. As for inhibitors, the electron transfer can be expected with compounds having relatively loosely bound electrons. A correlation between θ and inhibitor concentration C in the aqueous solution can be represented by the Langmuir adsorption isotherm (Brug et al., 1984; Moradi et al., 2013):

$$\theta = \frac{KC}{1+KC} \tag{8}$$

Rearranging this equation, it becomes:

$$\frac{C}{\theta} = \frac{1}{K} + C \tag{9}$$

Where K represents the constant of adsorption reaction.

Table 8. Regression coefficient, Equilibrium constant and free enthalpy of adsorption of different Triphenylimidazole substituted compounds on carbon steel in 1.0 M HCl at 25±2 °C.

Compounds	R^2	K_{ads} (mol^{-1})	ΔG_{ads} ($KJ.mol^{-1}$)
P1	0.99998	249.5	-17.7
P2	1	1425.5	-19.6
P3	1	649.0	-18.7

Figure 17 exemplifies the relation between C/θ and C at 25±2°C. It yields a straight line with slope close to unity. The strong correlation ($r^2 \approx 1$) for the Langmuir adsorption isotherm plot confirms the validity of this approach. Thus, we obtained the value of K for the inhibitors used is 249,5 mol^{-1}, 1425,2 mol^{-1} and 649,0 mol^{-1} respectively for P1, P2 and P3. The adsorptive equilibrium constant K is in relation with the standard free energy of adsorption (Brug et al., 1984; Tang et al., 2013).

$$K = \frac{1}{55.55}\exp\left(-\frac{\Delta G_{ads}}{RT}\right) \tag{10}$$

Where R is the universal gas constant, T is the thermodynamic temperature and value of 55.55 is the water concentration in the solution ($mol.L^{-1}$).

We found that all Triphenylimidazole substituted compounds show a good linear fit proving that the adsorption of these compounds from 1.0 M HCl solution on the carbon steel surface obeys the Langmuir adsorption isotherm. The equilibrium constant K_{ads} (mol^{-1}) is calculated from the intercept from which the free enthalpy of adsorption ΔG (kJ. mol^{-1}) is obtained. The calculated values are tabulated in Table 8. The values obtained are negative. The negative values ensure the spontaneity of the adsorption process and the stability of adsorbed layer on the carbon steel surface (Szauer et al., 1981; Alaoui et al. 2016; Watson et al., 2001). The values of free enthalpy of adsorption obtained in this study is around to 20 KJ/mol. This is consistent with electrostatic interactions between the charged metals and the charged molecules, which signifies physical adsorption.

FUTURE RESEARCH DIRECTIONS

On the other hand, it could be interesting also to study the action mode of these compounds using Quantum and Molecular Modeling simulation. It could provide information about the nature of adsorption of the Triphenylimidazole substituted compounds on the metallic surface. It could be interesting also to study the effect of these compounds on the corrosion of another metal in another acidic medium such as H_2SO_4 and H_3PO_4. This study confirmed the performance and the other application of the Triphenylimidazole substituted compounds.

CONCLUSION

Although a number of publications are witnessing that the Imidazole compounds and Imidazole derivatives molecules are excellent inhibitors under a variety of corrosive environments for most of the metals. The non-toxicity and biodegradability are the major advantages for these inhibitors. All the examined Imidazole substituted compounds are effective corrosion inhibitors for mild steel compared to carbon steel in 1.0 M HCl solution. They act by adsorption mechanism where their inhibition depends on the concentration and the type of substitution their structures and the molecules. In addition, the triphenylimidazole inhbitors marks the best corrosion inhibition performance compared to others Imidazole inhibitors studies. We can note also, that their inhibition efficiency decreases slightly with temperature. So, it is found also that the order of this inhibition was confirmed by all techniques measurements. The

P2 give the highest inhibition efficiency. The adsorption of all compounds follows the Langmuir isotherm with negative values of free enthalpy, suggesting a spontaneous adsorption process.

REFERENCES

Abbouda, Y., Abourriche, A., Saffaj, T., Berrada, M., Charrouf, M., Bennamara, A., Al Himidi, N., & Hannache, H. (2007). 2,3-Quinoxalinedione as a novel corrosion inhibitor for mild steel in 1 M HCl. *Materials Chemistry and Physics*, *105*(1), 1–5. doi:10.1016/j.matchemphys.2007.03.037

Abd El-Raouf, M., Khamis, E. A., Abou Kana, M. T. H., & Negm, N. A. (2018). Electrochemical and quantum chemical evaluation of new bis (coumarins) derivatives as corrosion inhibitors for carbon steel corrosion in 0.5 M H2SO4. *Journal of Molecular Liquids*, *255*, 341–353. doi:10.1016/j.molliq.2018.01.148

Abdelaal, M. S., Radwan, S., & El Saied, A. (1983). Phenothiazines as corrosion inhibitors for zinc in NHAC Solution. *British Corrosion Journal*, *18*(2), 102–106. doi:10.1179/bcj.1983.18.2.102

Alaoui,, K., El Kacimi,, Y., Galai,, M., Dahmani,, K., Touir,, R., & El Harfi,, A., & EbnTouhami, M. (2016). Poly(1-phenylethene): As a novel corrosion inhibitor for carbon steel/hydrochloric acid interface. *Anal Bioanal Electrochem.*, *8*, 830.

Alaoui, K., El Kacimi, Y., Galai, M., Touir, R., Dahmani, K., Harfi, A., & Ebn Touhami, M. (2016). Anti-corrosive properties of Polyvinyl-Alcohol for carbon steel in Hydrochloric acid media: Electrochemical and Thermodynamic investigation. *J. Mater. Environ. Sci.*, *7*, 2389–2403.

Ansari, F. A., Verma, C., Siddiqui, Y. S., Ebenso, E. E., & Quraishi, M. A. (2018). Volatile corrosion inhibitors for ferrous and non-ferrous metals and alloys: A review. *Int J Corros Scale Inhib.*, *7*, 126.

Arellanes-Lozada, P., Olivares-Xometl, O., Likhanova, N. V., Lijanova, I. V., Vargas-García, J. R., & Hernández-Ramírez, R. E. (2018). Adsorption and performance of ammonium-based ionic liquids as corrosion inhibitors of steel. *Journal of Molecular Liquids*, *265*, 151–163. doi:10.1016/j.molliq.2018.04.153

Atwell, G. J., Baguley, B. C., & Denny, W. A. (1989). Potential antitumor agents. 57. 2-Phenylquinoline-8-carboxamides as minimal DNA-intercalating antitumor agents with in vivo solid tumor activity. *Journal of Medicinal Chemistry*, *32*(2), 396–401. doi:10.1021/jm00122a018 PMID:2913299

Bammou, L., Belkhaouda, M., Salghi, R., Benali, O., Zarrouk, A., Zarrok, H., & Hammouti, B. (2014). Corrosion inhibition of steel in sulfuric acidic solution by the chenopodium ambrosioides extracts. *J Assoc Arab Univ Basic Appl Sci.*, *16*(1), 83–90. doi:10.1016/j.jaubas.2013.11.001

Bashir, S., Sharma, V., Lgaz, H., Chung, M., Singh, A., & Kumar, A. (2018). The inhibition action of analgin on the corrosion of mild steel in acidic medium: A combined theoretical and experimental approach. *Journal of Molecular Liquids*, *263*, 454–462. doi:10.1016/j.molliq.2018.04.143

Bedair, M. A., El-Sabbah, M. M. B., Fouda, A. S., & Elaryian, H. M. (2017). Synthesis,electrochemical and quantum chemical studies of some prepared surfactants based on azodye and Schiff base as corrosion inhibitors for steel in acid medium. *Corrosion Science*, *128*, 54–72. doi:10.1016/j.corsci.2017.09.016

Benali, O., Larabi, L., Tabti, B., & Harek, Y. (2005). Influence of 1-methyl 2-mercapto imidazole on corrosion inhibition of carbon steel in 0.5 M H2SO4. *Anti-Corrosion Methods and Materials*, *52*(5), 280–285. doi:10.1108/00035590510615776

Bousskri, A., Anejjar, A., Messali, M., Salghi, R., Benali, O., Karzazi, Y., Jodeh, S., Zougagh, M., Ebenso, E. E., & Hammouti, B. (2015). Corrosion inhibition of carbon steel in aggressive acidic media with 1-(2-(4-chlorophenyl)-2-oxoethyl)pyridazinium bromide. *Journal of Molecular Liquids*, *211*, 1000–1008. doi:10.1016/j.molliq.2015.08.038

Brug, G. J., Van DenEeden, A. L. G., Sluyters-Rehbach, M., & Sluyters, J. H. (1984). The analysis of electrode impedances complicated by the presence of a constant phase element. *J. Electroanal. Chem.*, *176*(1-2), 275–295. doi:10.1016/S0022-0728(84)80324-1

Daoud, D., Douadi, T., Issaadi, S., & Chafaa, S. (2014). Adsorption and corrosion inhibition of new synthesized thiophene Schiff base on mild steel X52 in HCl and H2SO4 solutions. *Corrosion Science*, *79*, 50–58. doi:10.1016/j.corsci.2013.10.025

Ebenso, E. E. (2001). Inhibition of corrosion of mild steel hydrochloric acid by some azo dyes. *Niger J Chem Res*, *6*, 8–12.

El-Awady, A. R., Semreen, M. H., Saber, M. M., Cyprian, F., Menon, V., & Al-Tel, T. H. (2016). Modulation of DNA damage response and induction of apoptosis mediates synergism between doxorubicin and a new imidazopyridine derivative in breast and lung cancer cells. *DNA Repair*, *37*, 1–11. doi:10.1016/j.dnarep.2015.10.004 PMID:26590797

El Guerraf, A., Titi, A., Cherrak, K., Mechbal, N., El Azzouzi, M., Touzani, R., Hammouti, B., & Lgaz, H. (2018). The synergistic effect of chloride ion and 1,5-diaminonaphthalene on the corrosion inhibition of mild steel in 0.5 M sulfuric acid: Experimental and theoretical insights. *Surfaces and Interfaces*, *13*, 168–177. doi:10.1016/j.surfin.2018.09.004

El-Hajjaji,, F., Belkhmima,, R. A., Zerga,, B., Sfaira,, M., & Taleb,, M., EbnTouhami, M., Hammouti, B., Al-Deyab, S. S., & Ebenso, E. (2014). Temperature performance of a thionequinoxaline compound as mild steel corrosion inhibitor in hydrochloric acid medium. *International Journal of Electrochemical Science*, *9*, 4721.

El Kacimi, Y., Touir, R., Galai, M., Belakhmima, R. A., Zarrouk, A., Alaoui, K., Harcharras, M., El Kafssaoui, H., & Ebn Touhami, M. (2016). Effect of silicon and phosphorus contents in steel on its corrosion inhibition in 5 M HCl solution in the presence of Cetyltrimethylammonium/KI. *J. Mater. Environ. Sci*, *7*, 371–381.

El Ouali, I., Hammouti, B., Aouniti, A., Ramli, Y., Azougagh, M., Essassi, E. M., & Bouachrine, M. (2010). Thermodynamic characterisation of steel corrosion in HCl in the presence of 2-phenylthieno (3, 2-b) quinoxaline. *J. Mater. Environ. Sci.*, *1*, 1–8.

Elaatiaoui, A., Rokni, Y., Mohammed, K., Asehraou, A., Chelfi, T., Saddik, R., Oussaid, A., Villalgordo, J. M., Abouricha, S., El Mahi, B., Oussaid, A., Zarrouk, A., & Benchat, N. (2015). Synthesis, spectroscopicand antimicrobial activityof some new 7-methyl-2-phenylimidazo[1,2-a]pyridin-3-amine derivatives. *J Mater Environ Sci.*, *6*, 2083.

Finley, H. F., & Hackerman, N. (1960). Effect of adsorption of polar organic compounds on the reactivity of steel. *Journal of the Electrochemical Society*, *107*(4), 259–263. doi:10.1149/1.2427675

Fouda, A. S., Shalabi, K., & Elmogazy, H. (2014). Corrosion inhibition of α-brass in HNO3 by indole and 2-oxyindole. *J Mater Environ Sci.*, *5*, 1691.

Galai, M., El Gouri, M., Dagdag, O., El Kacimi, Y., El Harfi, A., & Ebn Touhami, M. (2015). The effect of temperature on the inhibition of acid hydrochloric corrosion of carbon steel by hexa propylene glycol cyclotriphosphazene. *Journal of Chemical and Pharmaceutical Research*, *7*, 712–719.

Galai, M., Rbaa, M., El Kacimi, Y., Ouakki, M., Dkhirech, N., Touir, R., Lakhrissi, B., & Ebn Touhami, M. (2017). Anti-corrosion Properties of some Triphenylimidazole Substituted Compounds in Corrosion Inhibition of Carbon Steel in 1.0 M Hydrochloric Acid Solution. *Anal. Bioanal. Electrochem.*, *9*, 80–101.

Gao, M., Wang, M., & Zheng, Q. H. (2016). Synthesis of carbon-11-labeled imidazopyridine- and purinethioacetamide derivatives as new potential PET tracers for imaging of nucleotide pyrophosphatase/phosphodiesterase. *Bioorganic & Medicinal Chemistry Letters*, *26*(5), 1371–1375. doi:10.1016/j.bmcl.2016.01.081 PMID:26856922

Ghanbaria, A., Attara, M. M., & Mahdavianb, M. (2010). Corrosion inhibition performance of three imidazole derivatives on mild steel in 1M phosphoric acid. *Materials Chemistry and Physics*, *124*(2-3), 1205–1209. doi:10.1016/j.matchemphys.2010.08.058

Ghazoui, A., Zarrouk, A., Bencaht, N., Salghi, R., Assouag, M., El Hezzat, M., Guenbour, A., & Hammouti, B. (2014). New possibility of mild steel corrosion inhibition by organic heterocyclic compound. *Journal of Chemical and Pharmaceutical Research*, *6*, 704.

Ghibate,, R., Sabry,, F., & Kharbach,, Y., KandriRodi, Y., Skalli, M. K., Haoudi, A., Senhaji, O., Touzani, M., Taouil, R., Aouniti, A., & Hammouti, B. (2015). Valuation of surfactant phosphonates synthesized in the protection of metal surfaces against corrosion of mild steel in 0.5M H2SO4 media. *International Journal of Engineering Research and Applications*, *5*, 22.

Gunasekaran, G., & Chauhan, L. R. (2004). Eco-Friendly Inhibitor for Corrosion Inhibition of Mild Steel in Phosphoric Acid. *Electrochimica Acta*, *49*(25), 4387–4395. doi:10.1016/j.electacta.2004.04.030

Gutiérrez, E., Rodríguez, J. A., Cruz-Borbolla, J., & Alvarado-Rodríguez, J. G. (2016). Pandiyan Thangarasu, Development of a predictive model for corrosion inhibition of carbonsteel by imidazole and benzimidazole derivatives. *Corrosion Science*, *108*, 23–35. doi:10.1016/j.corsci.2016.02.036

HazwanHussin, M., Jain Kassim, M., Razali, N. N., Dahon, N. H., & Nasshorudin, D. (2011). The effect of tinospora crispa extracts as a natural mild steel corrosion inhibitor in 1m hcl solution. *Arabian Journal of Chemistry*, S616–S624.

He, X., Jiang, Y., Li, C., Wang, W., Hou, B., & Wu, L. (2014). Inhibition properties and adsorption behavior of imidazole and 2-phenyl-2-imidazoline on AA5052 in 1.0 M HCl solution. *Corrosion Science*, *83*, 124–136. doi:10.1016/j.corsci.2014.02.004

James, A. O., Oforka, N. C., & Abiola, K. (2007). Inhibition of Acid Corrosion of Mild Steel by Pyridoxal and Pyridoxol Hydrochlorides. *International Journal of Electrochemical Science*, *2*, 278–284.

Jiang, L., Qiang, Y., Lei, Z., Wang, J., Qin, Z., & Xiang, B. (2018). Excellent corrosion inhibition performance of novel quinoline derivatives on mild steel in HCl media: Experimental and computational investigations. *Journal of Molecular Liquids*, *255*, 53–63. doi:10.1016/j.molliq.2018.01.133

Karthik, R., Muthukrishan, P., Chen, S. M., Jeyaprabha, B., & Prakash, P. (2015). Anti-corrosion inhibition of mild steel in 1M hydrochloric acid solution by using tiliacoraaccuminata leaves extract. *International Journal of Electrochemical Science*, *10*, 3707.

Kıcır, N., Tansuğ, G., Erbil, M., & Tüken, T. (2016). Investigation of ammonium (2, 4-dimethylphenyl)-dithiocarbamate as a new, effective corrosion inhibitor for mild steel. *Corrosion Science*, *105*, 88–99. doi:10.1016/j.corsci.2016.01.006

Kokalj, A. (2012). On the HSAB based estimate of charge transfer between adsorbates and metal surfaces. *Chemical Physics*, *393*(1), 1–12. doi:10.1016/j.chemphys.2011.10.021

Lavanya, K., Saranya, J., & Chitra, S. (2018). Recent reviews on quinolone derivative as corrosion inhibitors. *Corrosion Reviews*, *36*(4), 365–371. doi:10.1515/corrrev-2017-0129

Li, C., Chen, L., Steinhuebel, D., & Goodman, A. (2016). Rapid construction of imidazopyridines from ortho-haloaminopyridines. *Tetrahedron Letters*, *57*(25), 2708–2712. doi:10.1016/j.tetlet.2016.04.118

Mai, W., Soghrati, S., & Buchheit, R. G. (2016). A phase field model for simulating the pitting corrosion. *Corrosion Science*, *110*, 157–166. doi:10.1016/j.corsci.2016.04.001

Moradi, M., Duan, J., & Du, X. (2013). Investigation of the effect of of 4, 5-dichloro-2-n-octyl-4-isothiazolin-3-one inhibition on the corrosion of carbon steel in Bacillus sp. inoculated artificial seawater. *Corrosion Science*, *69*, 338–345. doi:10.1016/j.corsci.2012.12.017

Murulana, L. C., Kabanda, M. M., & Ebenso, E. E. (2015). Experimental and theoretical studies on the corrosion inhibition of mild steel by some sulphonamides in aqueous HCl. *RSC Advances*, *5*(36), 28743–28761. doi:10.1039/C4RA11414K

Nazeer, A. A., & Madkour, M. (2018). Potential use of smart coatings for corrosion protection of metals and alloys: A review. *Journal of Molecular Liquids*, *253*, 11–22. doi:10.1016/j.molliq.2018.01.027

Negm, N. A., El Hashash, M. A., Abd-Elaal, A., Tawfik, S. M., & Gharieb, A. (2018). Amide type nonionic surfactants: Synthesis and corrosion inhibition evaluation against carbon steel corrosion in acidic medium. *Journal of Molecular Liquids*, *256*, 574–580. doi:10.1016/j.molliq.2018.02.078

Obot, I. B., Macdonald, D. D., & Gasem, Z. M. (2015). Density functional theory (DFT) as a powerful tool for designing new organic corrosion inhibitors. part 1: An overview. *Corrosion Science*, *99*, 1–30. doi:10.1016/j.corsci.2015.01.037

Obot, I. B., & Obi-Egbedi, N. O. (2010). Indeno-1-one [2, 3-b] quinoxaline as an effective inhibitor for the corrosion of mild steel in 0.5 M H2SO4 solution. *Materials Chemistry and Physics*, *122*(2-3), 325–328. doi:10.1016/j.matchemphys.2010.03.037

Otmacic, H., & Stupnisek-Lisac, E. (2002). Copper corrosion inhibitors in near neutral media. *Electrochimica Acta*, *47*, 4189–4194.

Otmacic, H., & Stupnisek-Lisac, E. (2010). Copper corrosion inhibitors in near neutral media. *Electrochimica Acta, 48*(2003), 985-991.

Otmacic Curkovic, H., Stupnisek-Lisac, E., & Takenouti, H. (2010). The influence of pH value on the efficiency of imidazole based corrosion inhibitors of copper. *Corrosion Science*, *52*(2), 398–405. doi:10.1016/j.corsci.2009.09.026

Ouakki, M., Rbaa, M., Galai, M., Lakhrissi, B., Rifi, E. H., & Cherkaoui, M. (2018). Experimental and Quantum Chemical Investigation of Imidazole Derivatives as Corrosion Inhibitors on Mild Steel in 1.0 M Hydrochloric Acid. *Journal of Bio- and Tribo-Corrosion, 4*, 35.

Peters,W. (1970)*Chemotherapy and Drug Resistance in Malaria*. Academic Press.

Petersen, A., Rodrigues, S. R., Dalmoro, V., Falcade, T., & Tamborim, S. M. (2017). Anthocyanins as a corrosion inhibitor for 2024-T3 aluminum alloys: A study of electrochemical behavior. *Int J Corros Scale Inhib.*, *6*, 29.

Qiao, L., Wang, Y., Wang, W., Mohedano, M., Gong, C., & Gao, J. (2014). The preparation and corrosion performance of self-assembled monolayers of stearic acid and MgO layer on pure magnesium. *Materials Transactions*, *55*(8), 1337–1343. doi:10.2320/matertrans.M2014117

Ramya, K., Mohan, R., Anupama, K. K., & Joseph, A. (2015). Electrochemical and theoretical studies on the synergistic interaction and corrosion inhibition of alkyl benzimidazoles and thiosemicarbazide pair on mild steel in hydrochloric acid. *Materials Chemistry and Physics*, *149*, 632–647. doi:10.1016/j.matchemphys.2014.11.020

Ruan, L., Zhang, Z., Huang, X., Lyu, Y., Wen, Y., Shang, W., & Wu, L. (2017). Evaluation of corrosion inhibition of two schiff bases selfassembled films on carbon steel in 0.5 M HCl. *International Journal of Electrochemical Science*, *12*, 103–115. doi:10.20964/2017.01.38

Saddik, R., Gaadaoui, A., Hamal, A., Zarrouk, A., Touzani, R., & Benchat, N. (2014). Synthesis, antibacterial and antifungal activity of some new imidazo[1,2-a]pyridine derivatives. *Der Pharm Lett.*, *6*, 343.

Saha, S. K. R., Ghosh, P., Hens, A., Murmu, N. C., & Banerjee, P. (2015). Density functional theory and molecular dynamics simulation study on corrosion inhibition performance of mild steel by mercapto-quinoline Schiff base corrosion inhibitor. *Physica E, Low-Dimensional Systems and Nanostructures*, *66*, 332–341. doi:10.1016/j.physe.2014.10.035

Salarvand, Z., Amirnasr, M., Talebian, M., Raeissi, K., & Meghdadi, S. (2017). Enhanced corrosion resistance of mild steel in 1M HCl solution by trace amount of 2-phenyl-benzothiazole derivatives: Experimental, quantum chemical calculations and molecular dynamics (MD) simulation studies. *Corrosion Science*, *114*, 133–145. doi:10.1016/j.corsci.2016.11.002

Salghi, R., Ben Hmamou, D., Benali, O., Jodeh, S., Warad, I., Hamed, O., Ebenso, E. E., Oukacha, A., Tahrouch, S., & Hammouti, B. (2015). Study of the corrosion inhibition effect of pistachio essential oils in 0.5 M H2SO4. *International Journal of Electrochemical Science*, *10*, 8403.

Savas, K., Tüzün, B., Kaya, C., & Obot, I. B. (2016). Determination of corrosion inhibition effects of amino acids: Quantum chemical and molecular dynamic simulation study. *Journal of the Taiwan Institute of Chemical Engineers*, *58*, 528–535. doi:10.1016/j.jtice.2015.06.009

Shaban, A., Felhosi, I., & Telegdi, J. (2017). Laboratory assessment of inhibition efficiency and mechanism of inhibitor blend (P22SU) on mild steel corrosion in high chloride containing water. *Int J Corros Scale Inhib.*, *6*, 262.

Shainy, K. M., Ammal, P. R., Unni, K. N., Benjamin, S., & Joseph, A. (2016). Surface interaction and corrosion inhibition of mild steel in hydrochloric acid using pyoverdine, an eco-friendly biomolecule. *J Bio-Tribo Corros.*, *2*(3), 20. doi:10.100740735-016-0050-3

Singh, A., Ansari, K. R., Haque, J., Dohare, P., Lgaz, H., Salghi, R., & Quraishi, M. A. (2018). Effect of electron donating functional groups on corrosion inhibition of mild steel in hydrochloric acid: Experimental and quantum chemical study. *Journal of the Taiwan Institute of Chemical Engineers*, *82*, 233–251. doi:10.1016/j.jtice.2017.09.021

Singh, D., Chaudhary, R. S., Prakash, B., & Agrawal, C. V. (1979). Inhibitive Efficiency of some Substituted Thioureas for the Corrosion of Aluminium in Nitric Acid. *British Corrosion Journal*, *14*(4), 235–239. doi:10.1179/000705979798358571

Stern, M., & Geary, A. L. (1957). Electrochemical Polarization: I. A Theoretical Analysis of the Shape of Polarization Curves. *J. Electrochem.*, *104*(1), 56–63. doi:10.1149/1.2428496

Szauer, T., & Brand, A. (1981). Adsorption of oleates of various amines on iron in acidic solution. *Electrochimica Acta*, *26*(9), 1253–1256. doi:10.1016/0013-4686(81)85107-9

Tang, Y., Zhang, F., Huc, S., Cao, Z., Wu, Z., & Jing, W. (2013). Novel benzimidazole derivatives as corrosion inhibitors of mild steel in the acidic media. Part I: Gravimetric, electrochemical, SEM and XPS studies. *Corrosion Science*, *74*, 271–282. doi:10.1016/j.corsci.2013.04.053

Topal, E., & Gece, G. (2017). Untangling the inhibition effects of aliphatic amines on silver corrosion: A computational study. *Chem J Moldova.*, *12*(2), 64–70. doi:10.19261/cjm.2017.411

Umoren, S. A., & Solomon, M. M. (2015). Effect of halide ions on the corrosion inhibition efficiency of different 4 organic species—A review. *Journal of Industrial and Engineering Chemistry*, *21*, 81–100. doi:10.1016/j.jiec.2014.09.033

Verma, C., Ebenso, E. E., Bahadur, I., & Quraishi, M. A. (2018). An overview on plant extracts as environmental sustainable and green corrosion inhibitors for metals and alloys in aggressive corrosive media. *Journal of Molecular Liquids*, *266*, 577–590. doi:10.1016/j.molliq.2018.06.110

Verma, C., Ebenso, E. E., & Quraishi, M. A. (2017). Ionic liquids as green and sustainable corrosion inhibitors for metals and alloys: An overview. *Journal of Molecular Liquids*, *233*, 403–414. doi:10.1016/j.molliq.2017.02.111

Verma, C., Olasunkanmi, L. O., Ebenso, E. E., & Quraishi, M. A. (2018). Adsorption characteristics of green 5-arylaminomethylene pyrimidine2,4,6-triones on mild steel surface in acidic medium: Experimental and computational approach. *Results Phys.*, *8*, 657–670. doi:10.1016/j.rinp.2018.01.008

Verma, C., Olasunkanmi, L. O., Ebenso, E. E., Quraishi, M. A., & Obot, I. B. (2016). Adsorption behavior of glucosamine-based, pyrimidine-fused heterocycles as green corrosion inhibitors for mild steel: Experimental and theoretical studies. *The Journal of Physical Chemistry C, 120*(21), 11598–11611. doi:10.1021/acs.jpcc.6b04429

Wang, H. P., Wu, Q., Li, C. M., & Gu, N. (2013). Copper corrosion inhibition by polyaspartic acid and imidazole. *Materials and Corrosion, 64*, 347-352.

Watson, A. A., Fleet, G. W. J., Asano, N., Molyneux, R. J., & Nugh, R. J. (2001). Polyhydroxylated alkaloids— Natural occurrence and therapeutic applications. *Phytochemistry*, *56*(3), 265–295. doi:10.1016/S0031-9422(00)00451-9 PMID:11243453

Xia, Y., Yang, Z. Y., Xia, P., Bastow, K. F., Tachibani, Y., Kuo, S. C., Hamel, E., Hackl, T., & Lee, K. H. (1998). Antitumor Agents. 181. Synthesis and Biological Evaluation of 6,7,2′,3′,4′-substituted-1,2,3,4-tetrahydro-2-phenyl-4-quinolones as a New Class of Antimitotic Antitumor Agents. *Journal of Medicinal Chemistry*, *41*, 1155. doi:10.1021/jm9707479 PMID:9544215

Yüce, A. O., Telli, E., Mert, B. D., Kardaş, G., & Yazıcı, B. (2016). Experimental and quantum chemical studies on corrosion inhibition effect of 5,5 diphenyl 2-thiohydantoin on mild steel in HCl solution. *Journal of Molecular Liquids*, *218*, 384–392. doi:10.1016/j.molliq.2016.02.087

Zhang, D., Gao, L., Zhou, G., & Lee, K. Y. (2008). Undecyl substitution in imidazole and its action on corrosion inhibition of copper in aerated acidic chloride media. *Journal of Applied Electrochemistry*, *38*(1), 71–76. doi:10.100710800-007-9401-6

Zhang, Z., Chen, S., Li, Y., Li, S., & Wang, L. (2009). A study of the inhibition of iron corrosion by imidazole and its derivativesself-assembled films. *Corrosion Science*, *51*(2), 291–300. doi:10.1016/j.corsci.2008.10.040

Zunita, M., Wahyuningrum, D., & Bundjali, B. (2012). Investigation of Corrosion Inhibition Activity of 3-butyl-2,4,5-triphenylimidazole and 3-butyl-2-(2-butoxyphenyl)-4,5-diphenylimidazole toward Carbon Steel in 1% NaCl Solution. *International Journal of Electrochemical Science*, *7*, 3274–3288.

Chapter 6
Novel Compounds of Imidazole Derivatives as Effective Corrosion Inhibitors for Mild Steel in Hydrochloric Acid Solution:
Electrochemical and Spectroscopic Characterization

Moussa Ouakki
https://orcid.org/0000-0002-6265-4734
Laboratory of Materials, Electrochemistry, and Environment, Ibn Tofail University, Morocco

M. Galai
Laboratory of Materials Engineering and Environment: Modeling and Application, Ibn Tofail University, Morocco

Z. Aribou
Laboratory of Materials Engineering and Environment: Modeling and Application, Ibn Tofail University, Morocco

M. Rbaa
Laboratory of Agro-Resources, Polymers, and Process Engineering, Ibn Tofail University, Morocco

B. Lakhrissi
Laboratory of Agro-Resources, Polymers, and Process Engineering, Ibn Tofail University, Morocco

M. Cherkaoui
Laboratory of Materials, Electrochemistry, and Environment, Ibn Tofail University, Morocco

ABSTRACT

The inhibition performance of two imidazole derivatives, IM-Cl and IM-CH3, on the corrosion behavior of MS in 1M HCl acid solution was studied through weight loss method and electrochemical tests. The results obtained from the electrochemical methods show that inhibition efficiency increases with the increase of inhibitors concentration. The adsorption of inhibitors on the mild steel surface obeys Langmuir adsorption isotherm. The corrosion protection was once also investigated with the aid of

DOI: 10.4018/978-1-7998-2775-7.ch006

UV–Vis spectrophotometry. SEM-EDX was performed and discussed for surface study of uninhibited and inhibited mild steel samples.

1. INTRODUCTION

Metal corrosion is an unavoidable difficulty for industries. Mild steel is the most extensively employed metal in industrial applications because of its excellent mechanical characteristics and inexpensiveness, (Verma, 2015a). A Large quantity of the mild steel, as well as other iron alloys, is destroyed by corrosion in aggressive media, especially in acid environments which cost billions of dollars every year (Verma, 2015b). This situation has grown to be a more important task for corrosion engineers and the scientific neighborhood. From an economic point of view, stopping or slowing down this deterioration has become obligatory. The only solution to this undesirable problem is to use corrosion safety methods. Corrosion inhibitors are commonly added to the aggressive acids to prevent the severe metal dissolution during acid treatment (Tawfik, 2015). Therefore, specific amounts of various synthetic corrosion inhibitors are added to the HCl solution for mild steel corrosion control. Recently, imidazole derivatives compounds have gained significant attention in corrosion science (Ouakki, 2019, 2018; Rbaa, 2017a; Galai, 2016). Literature reviews reveal that heterocyclic compounds based on imidazole are used as efficient corrosion inhibitors for metals in an acid medium (Yadav, 2016; Zhang, 2015)

Most organic compounds inhibit corrosion via adsorption on the metal surface. This adsorption mainly depends on some physico-chemical properties of the molecule, related to its functional groups (e.g. nitrogen, oxygen and sulfur atoms), aromaticity, the possible steric effects and electronic density. The aim of this work is to study the inhibition properties of two imidazole derivatives on mild steel corrosion in 1M HCl using gravimetric measurements, potentiodynamic polarization curves, and electrochemical impedance spectroscopy. It also aims to predict the thermodynamic feasibility of these compounds on the steel surface. In addition, detailed investigation of temperature were also carried and discussed to improve a better understanding of the adsorption mechanism of the studied inhibitors. The surface of the mild steel upstream and downstream of corrosion was characterized by SEM-EDAX, and the gravimetric solution was analyzed by UV–visible spectrometry. The corrosion inhibition performance of the compound IM-Cl is significant compared with that of the compound IM-CH_3 and this is explained by the existence of a mesomeric electron donor group (-Cl) carried by IM-Cl improving thus its adsorption capacity on the metal (Rbaa, 2017b, 2017c).

2. MATERIALS AND METHODS

2.1. Materials and Sample Preparation

The aggressive solution 1M HCl was prepared by dilution of analytical grade HCl (37%) with distilled water. The concentrations of imidazole derivatives inhibitors in the work were set as 10^{-6} to10^{-3} M and the blank solution was also prepared for comparison. Mild steel samples having composition 0.17 wt.% C, 0.37 wt.% Mn, 0.20 wt.% Si, 0.03 wt.% S, 0.01 wt.% P and the remainder iron, were cut into 2.00 cm * 1.00 cm * 0.20 cm for weight loss tests. For electrochemical experiments, the steel samples were used with an exposed surface of 1.0 cm^2 to the acid solution. Prior to use, the steel samples were polished successively with fine grade Emery papers from 180 to 1200 grades. The specimens were washed thoroughly with distilled water and finally degreased with ethanol and dried at room temperature. The chemical structures of the studied inhibitors are shown in Table 1 and Fig 1.

Table 1. Names, chemical structures and abbreviations of the studied Imidazole compounds.

Inhibitor Code	Chemical Names	Mol Formulas (F.wt)
IM-Cl	2-(4-chlorophenyl)-1,4,5-triphenyl-1H-imidazole	$C_{27}H_{19}N_2Cl$ (406,90 g/ mol)
IM-CH_3	1,4,5-triphenyl-2-(p-tolyl)-1H-imidazole	$C_{28}H_{22}N_2$ (386,48 g/ mol)

Figure 1. Chemical structures of IM-Cl (A) and IM-CH_3(B).

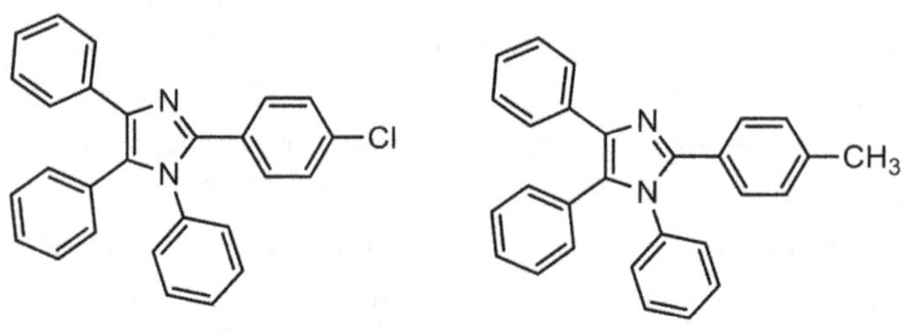

A B

2.2. Weight Loss Experiments

For the weight loss measurements, the experiments were achieved using the standard method described earlier (Gupta, 2016a, 2016b). The corrosion rate of mild steel was determined using the following Eq. (1):

$$CR = \frac{w}{S * t} \tag{1}$$

Where, W is the weight loss of mild steel sample, S is the area of mild steel sample and t is the immersion time (6 h).

The inhibition efficiency, $\eta_{CR\%}$, is determined according to the- following Eq. (2)

$$\eta_{CR\%} = \frac{CR - CR_{inh}}{CR} * 100 \tag{2}$$

where CR and CR_{inh} are the corrosion rates in the absence and presence of the inhibitors, respectively.

2.3. Electrochemical Measurements

Electrochemical measurements were carried by the method described previously (Gupta, 2016a, 2016b). The electrochemical impedance measurements (EIS) were performed on mild steel samples at constant potential (OCP) in the frequency range from 100 kHz to 100 mHz under potentiostatic conditions using an AC signals with amplitude of 10 mV peak to peak. The charge transfer resistance was calculated from Nyquist plot from which corrosion inhibition efficiency was calculated using the following Eq. (3) (Hermas, 2008):

$$\eta_{imp\%} = \frac{R_p - \overset{\circ}{R_p}}{R_p} * 100 \tag{3}$$

$$\theta = \frac{R_p - \overset{\circ}{R_p}}{R_p} \tag{4}$$

Where $\overset{\circ}{R_p}$ and R_p are the charge transfer resistance values in the absence and presence of the inhibitors, respectively and θ the surface coverage, given by Eq. (4).

The Potentiodynamic polarization studies were performed on mild steel specimens by automatically changing the electrode potential from 900 to -100 mV/Ag/AgCl versus OCP at a scan rate of 1 mVs^{-1}. The corrosion current density (i_{corr}) was calculated by extrapolating the linear Tafel slopes of anodic and cathodic curves with reference to the corrosion potential. The inhibition efficiency was determined by using the following Eq. (5) (Liu, 2009):

$$\eta_{pp\%} = \frac{\overset{\circ}{i_{corr}} - i_{corr}}{\overset{\circ}{i_{corr}}} * 100 \tag{5}$$

where $\overset{\circ}{i}_{corr}$ and i_{corr} are the corrosion current densities values in the absence and presence of the inhibitor, respectively.

2.4. Surface Characterization by SEM/EDX

The surface morphological observation for the steel samples in the medium 1M HCl without and with inhibitors, after an exposure period of 6 h was carried out by Scanning Electron Microscopy (Quanta FEG 450) coupled with EDX analyses. These analyses were performed in foundation MAScIR.

2.5. UV–Visible Spectra

To discover more data on the bonding mechanism of mild steel surfaces/ compounds/aggressive environment, corrosion protection was also investigated by means of ultraviolet-visible spectrophotometry

(UV-Vis), before and after immersion of the mild steel sample for 48 h. The Jenway ultraviolet-visible spectrophotometer (series 67) was once utilized for this analysis.

3. RESULTS AND DISCUSSION

3.1. Gravimetric Measurement

The corrosion rate and inhibition efficiencies obtained from the gravimetric experiments in the presence of different concentrations of inhibitors in 1M HCl are presented in Table 2. The observed weight- loss values of triplicate measurements are highly reproducible giving standard deviations. The results show that the mild steel corroded severely in the uninhibited 1M HCl solution, whereas the presence of inhibitors reduced the dissolution rate considerably. Shahin et al. pointed out in their work (Shahin, 2002) that the relative inhibition efficiencies of the studied compounds can be rationalized qualitatively under the assumptions that:

- The corrosion inhibition is essentially based on the coverage of the metal surface by the inhibitor molecules, thus making the contact of the corrosive medium difficult;
- Attachment of the molecules on the metal surface is facilitated through the coordination of π-electron system to the metal atom;
- The stability of the complex formed at the metal/solution interface is somewhat related to the molecule's being planar.

Both IM-Cl and IM- CH_3 exhibited substantial protection efficiency for mild steel in 1M HCl. As the concentration of the inhibitor is increased $\%\eta_{CR}$ also increases. At higher concentration, Maximum inhibition efficiencies of 96.2% and 95.9% were observed for IM-Cl and IM-CH_3 respectively at 10^{-3}M. The $\%\eta_{CR}$ is almost leveled off and only slight variation is observed. This can be explained by the adsorption of the IM-Cl and IM-CH_3 onto mild steel surface resulting in the blocking of the reaction sites.

Table 2. Corrosion parameters obtained from weight loss for mild steel in 1M HCl solution in the absence and presence of the different concentrations of IM-Cl and IM-CH_3 at 298 K.

Medium	Concentration (M)	CR ($mg/cm^2.h$)	η_{CR} (%)
1M HCl	--	6.8930	--
IM-Cl	10^{-6}	0.6232	90.9
	10^{-5}	0.3441	95.0
	10^{-4}	0.3301	95.2
	10^{-3}	0.2619	96.2
IM-CH_3	10^{-6}	0.6832	90.0
	10^{-5}	0.5441	92.1
	10^{-4}	0.3221	95.3
	10^{-3}	0.2819	95.9

The increase in concentration of the inhibitor leads to the gradual formation of multilayers that further reduce the rate of corrosion from the attack of the corrosion active ions in the acid medium beyond what can be achieved with mono layer coverage below that concentration (Amin, 2009).

3.2. Electrochemical Studies

3.2.1. Polarization Studies

In order to know the kinetics of anodic and cathodic reactions, polarization experiments were carried out potentiodynamically in unagitated 1M HCl solutions in the absence and presence of different concentrations of inhibitors and the obtained polarization curves are shown in Fig. 2. The electrochemical parameters derived from Tafel curves are given in Table 3.

In acidic solutions, the anodic reaction is the movement of metal ions from the working electrode into the corrosive solution, and the cathodic reaction is the discharge of the ions in the electrolyte such as the reduction of hydrogen ions to form hydrogen gas. The inhibitors may affect either the anodic, cathodic, or both reactions. Since the anodic Tafel slope (βa) and cathodic Tafel slope (βc) of imidazole were found to change with inhibitor concentration, the inhibitors affected both these reactions and referred to as mixed-type inhibitors. The adsorption of inhibitors can affect the corrosion rate in two ways:

- by decreasing the available reaction area, the so-called geometric blocking effect and
- by modifying the activation energy of the cathodic and/or anodic reactions occurring in the inhibitor-free metal in the course of the inhibited corrosion process.

It is a difficult task to determine which aspect of the inhibiting effect is connected to the geometric blocking action and which are connected to the energy effect. Theoretically, no shifts in E_{corr} should be observed after addition of the corrosion inhibitors if the geometric blocking effect is stronger than the energy effect (De Souza, 2009). It is observed from Table 3 that the corrosion current density (i_{corr}) decreases with increasing inhibitor concentration and it can be seen also that the corrosion potentials (E_{corr}) do not shift, which suggests that the both inhibitors act as very good organic inhibitors for mild steel in 1M HCl solution.

The Tafel curves in solutions containing inhibitors do not shift, which suggests that the inhibitors could also suppress the anodic and cathodic reactions. An inhibitor can be classified as an anodic or cathodic type inhibitor when the displacement in E_{corr} value is >85 mV (Kowsari, 2014). In the present study, it is found that the displacement exhibited by the studied inhibitors was <85 mV, which indicates that IM-Cl and IM-CH_3 act as mixed type inhibitors. Otherwise, the anodic and cathodic tafel slope values of inhibited solution have changed with respect to uninhibited solution, which also states that the tested inhibitors are of mixed type. The higher values of βa compared to those of βc at each concentration of the inhibitor further confirm that these inhibitors are regarded as mixed type inhibitors with anodic predominance. The obtained inhibition efficiencies values are in good agreement with those obtained by gravimetric measurements.

Figure 2. Polarization curves for mild steel recorded in 1M HCl for selected concentrations of the inhibitors IM-Cl (a) and IM-CH_3 (b) at 298 K.

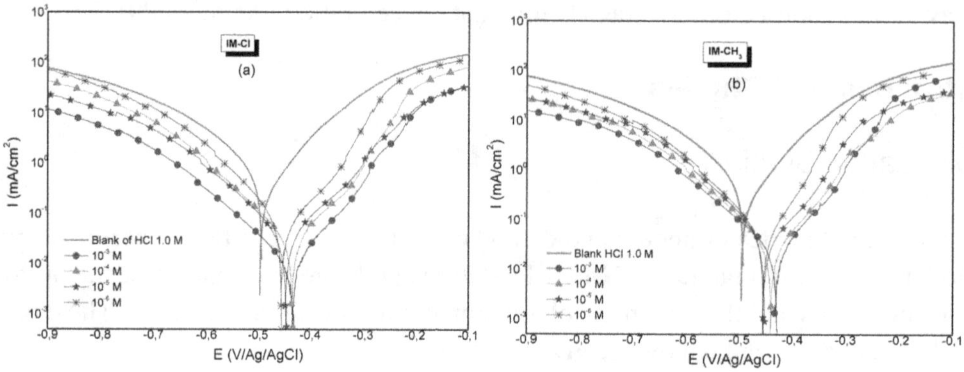

Table 3. Polarization parameters for mild steel in 1M HCl in the absence and presence of different concentrations of IM-Cl and IM-CH_3 at 298 K.

Medium	Conc. *M*	$-E_{corr}$ mV/Ag/AgCl	i_{corr} $\mu A\ cm^{-2}$	$-\beta_c$ $mV\ dec^{-1}$	β_a $mV\ dec^{-1}$	η_{PP} %
1M HCl	--	498	983	140	150	-
IM-Cl	10^{-6}	451	83	105	130	91.6
	10^{-5}	441	47	112	134	95.2
	10^{-4}	429	44	104	126	95.5
	10^{-3}	428	32	114	138	96.7
IM-CH_3	10^{-6}	452	96	105	132	90.2
	10^{-5}	450	74	104	128	92.5
	10^{-4}	438	42	108	130	95.7
	10^{-3}	427	32	114	126	96.7

3.2.2. EIS Studies

Electrochemical impedance spectroscopy (EIS) technique is popularly used in corrosion inhibition to study the electrical properties of the metal/solution interface. The mechanistic information of the system under subject is provided by the nature and shape of the impedance curves (Asan, 2006). This technique is extensively used in corrosion inhibition investigations. The electrochemical characteristics of mild steel in 1M HCl solution at 298 K in the absence and presence of different concentrations of IM-Cl and IM-CH_3 (corrosion inhibitors) are presented as EIS Nyquist and Bode's plots in Figs. 3 and 4. In the Nyquist plot (Fig. 3) a depressed charge transfer semicircle was appeared which is related to a single time constant shown in Bode's plot (El Mehdi, 2003; Wang, 2003). It can be observed from Fig. 3 that the impedance response of IM-Cl and IM-CH_3 has been considerably increased with increasing concentration of inhibitors to the electrolyte. The results obtained from EIS were of the similar order as those obtained from polarization and gravimetric measurements.

Figure 3. Nyquist plots of mild steel in 1M HCl solution in the absence and presence of various concentrations of IM-Cl (a) and IM-CH_3 (b) at 298K.

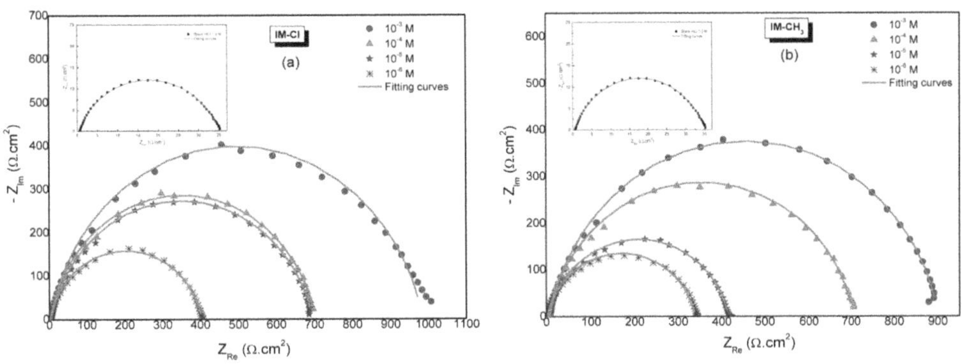

The capacitive behavior of mild steel in the presence of IM-Cl and IM-CH_3 could be visualized from Bode's plot (Fig. 4). A destructive outbreak of sulfuric acid corroded the mild steel and made the metal surface highly asymmetrical, which could be explained out by small phase angle. On the contrary, adsorption of the inhibitors on mild steel surface effectively lowered surface irregularities; additionally, we address the evolution of phase angle by measuring their values at arbitrarily chosen frequency (100 Hz) (Fig. 4). Clearly, with increasing concentration of both inhibitors the phase angle values dramatically increase as well.

Figure 4. Bode's plots and Phase angle values at 100 Hz at 298 K for mild steel in 1M HCl (c) solution in the absence and presence of various concentrations of IM-Cl (a) and IM-CH_3

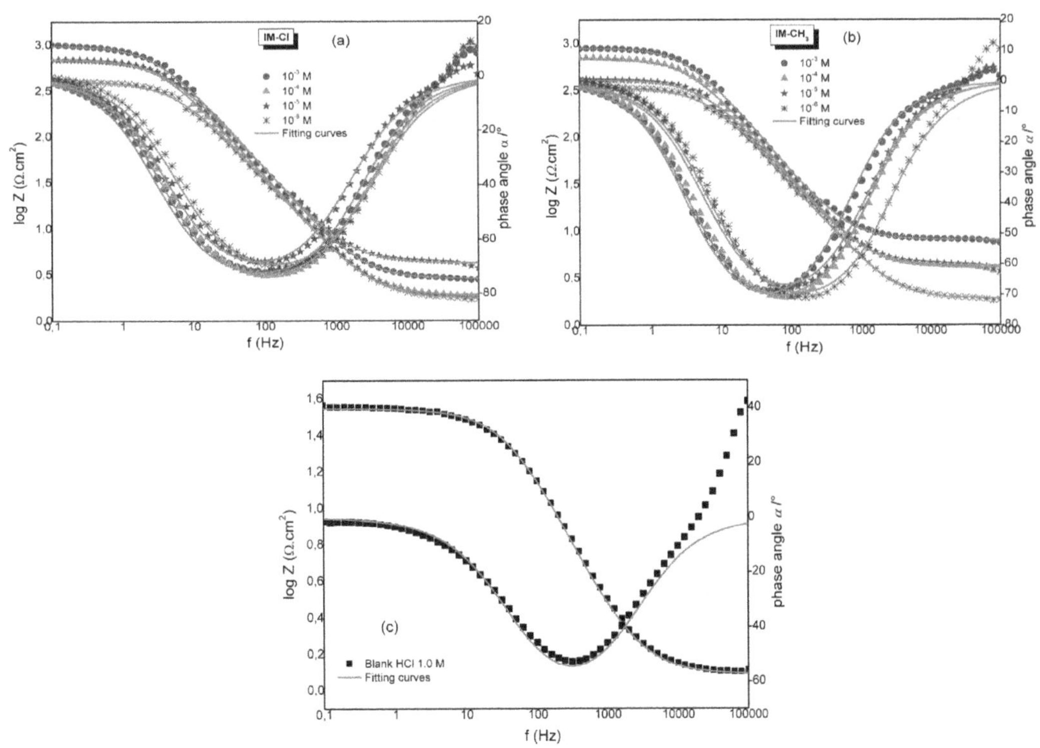

Figure 5. Electrochemical equivalent circuit used to fit the impedance and bode's spectra.

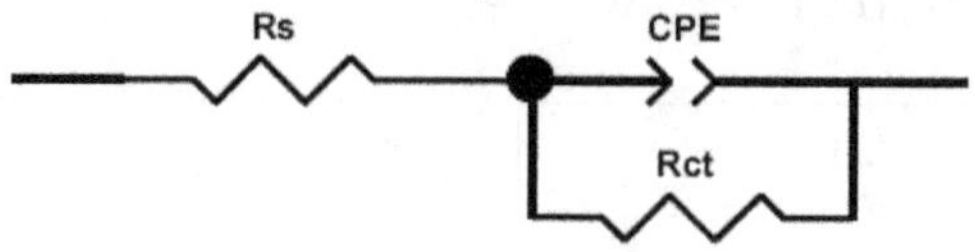

Furthermore, whatever the concentration used, the phase angle values are higher than that of the blank but lower than – 90° which also describes the non-ideal capacitor (Lgaz, 2020, 2019). This indicates an increased capacitance of the interference due to the presence of adsorbed inhibitors molecules at the interface (Oguzie, 2007). EIS and bode's spectra were analyzed using the equivalent circuits in Fig. 5. The equivalent circuits used to model mild steel/solution interface for these systems are presented in Fig. 5. The Table 4 results described below can be interpreted in terms of the equivalent circuit of the electrical double. The results show that the R_p values increases and C_{dl} decreases with increasing inhibitors concentration, which is attributed to the increase in surface coverage of mild steel by inhibitor molecules leading to a decrease in the corrosion rate (El Ouali, 2010; Singh, 2010; Ech-chihbi, 2016, 2017). The decreased value of C_{dl} attributed towards decrease in local dielectric constant and/or an increase in the thickness of the protective layer, suggesting that IM-Cl and IM-CH_3 act by adsorption mechanism at mild steel/acid interface (Gerengi, 2012). The capacitance of the double layer (C_{dl}) is linked with the thickness of the protective layer (d) by the following relation (6):

Table 4. Impedance parameters for mild steel in 1M HCl at 298 K in the absence and presence of various concentrations of IM-Cl and IM-CH_3.

Medium	Conc (M)	R_s (Ω cm^2)	R_{ct} (Ω cm^2)	C_{dl} (μF cm^{-2})	n_{dl}	Q ($\Omega^{-1} cm^2 s^1$)	θ	η_{imp} %
1M HCl	--	1.12±0.29	34.7±0.59	121.0	0.773±0.02	419.0±0.028	-	-
IM-Cl	10^{-6}	1.72±0.2	403.5±0.44	69.84	0.848±0.009	119.8±0.14	0.914	91.4
	10^{-5}	4.18±0.19	691.6±0.42	56.5	0.852±0.006	91.26±0.6	0.949	94.9
	10^{-4}	1.88±0.19	700.4±0.40	64.7	0.873±0.008	95.73±0.5	0.950	95.0
	10^{-3}	2.84±0.19	981.9±0.37	55.05	0.868±0.007	80.75±0.35	0.964	96.5
IM-CH_3	10^{-6}	1.89±0.2	337.5±0.44	71.45	0.862±0.009	119.2±0.17	0.897	89.7
	10^{-5}	4.26±0.2	411.2±0.43	67.35	0.864±0.007	109.5±0.1	0.915	91.5
	10^{-4}	4.07±0.19	700.7±0.4	64.97	0.874±0.006	95.62±0.6	0.950	95.0
	10^{-3}	8.09±0.19	893.2±0.39	55.00	0.889±0.006	76.75±0.3	0.961	96.1

$$C_{dl} = \frac{\varepsilon\varepsilon_0}{e}. \qquad (6)$$

Where ε and ε_o are dielectric constant of the medium and vacuum permittivity, e is the thickness of the protective layer and S is the surface area of the electrode.

With increase in the concentration of IM-Cl and IM-CH_3 the C_{dl} values are decreasing (Table 4) which depicts that the thickness of the protective layer (e) is increasing. The values of n parameter are known as "phase shifts" which are often related to the degree of surface heterogeneity according to the following Eq.(7) (Ouakki, 2018):

$$C_{dl} = Q\left(\omega_{max}\right)^{\alpha-1} \tag{7}$$

Where, $\omega_{max} = 2\pi f_{max}$ is the angular frequency at the maximum value of the imaginary part of the impedance spectrum, Q is the admittance of the constant phase element (CFE) in Ω^{-1} cm^2 s^1.

The values of the n parameter for inhibited systems are close to unity at higher concentrations implying that the interface behaves nearly capacitive (Singh, 2009; Quartarone, 2009). The inhibition efficiencies of both products at 10^{-3}M reached 96.5% of IM-Cl and 96.1% of IM-CH_3, indicating that the synthesized compounds act as an effective corrosion inhibitor for mild steel in HCl medium.

3.3. Adsorption Isotherm and Effect of Temperature

3.3.1. Adsorption Parameters

The interaction between the both inhibitor molecules and steel surface is best explained by adsorption isotherms. The efficiency of inhibitors is dependent mainly on their adsorption ability on the metal surface. The adsorption process involves replacement of water molecules at a metal solution interface according to the following process (Gabrielli, 1981; Kalman., 1994; El-Taib, 2017):

$$Org_{(sol)} + nH_2O_{(ads)} \text{ ® } Org_{(ads)} + nH_2O_{(sol)}$$

where $Org_{(sol)}$ and $Org_{(ads)}$ are the organic molecules in the solution and adsorbed on the metal surface, respectively, and n is the number of water molecules replaced by the organic molecules. In order to acquire more information regarding the interaction of these compounds and the surface of mild steel, it is important to determine the mode of adsorption and the type of adsorption isotherm as there are ten different adsorption isotherms (Langmuir, Temkin, Freundlich, Frumkin, Modified, Langmuir, Henry, Viral, Damaskin, Volmer, and Flory–Huggins) (Oguzie, 2004; Kadiri, 2018; Obot, 2010) which were tested for their fit to the experimental data for the current investigation.

The best description of the adsorption behavior of studied inhibitors was explained by Langmuir adsorption isotherm (Fig. 9) (Kalman, 1994), as the average linear regression coefficient values (R^2) obtained for IM-Cl and IM-CH_3 and slope are very close to unity. All the values are listed in Table 5. The isotherm is given by the following Eq. (8).

$$\frac{C_{inh}}{\theta} = \frac{1}{K_{ads}} + C_{inh} \tag{8}$$

where C_{inh} is the concentration of inhibitor, θ is the surface coverage degree, and K_{ads} is the adsorptive equilibrium constant. The linear relationships of C_{inh}/θ versus C_{inh} are depicted in Fig. 6.

Figure 6. Langmuir adsorption isotherm for mild steel in 1M HCl solution containing different concentrations of IM-Cl and IM-CH_3 at 298 K.

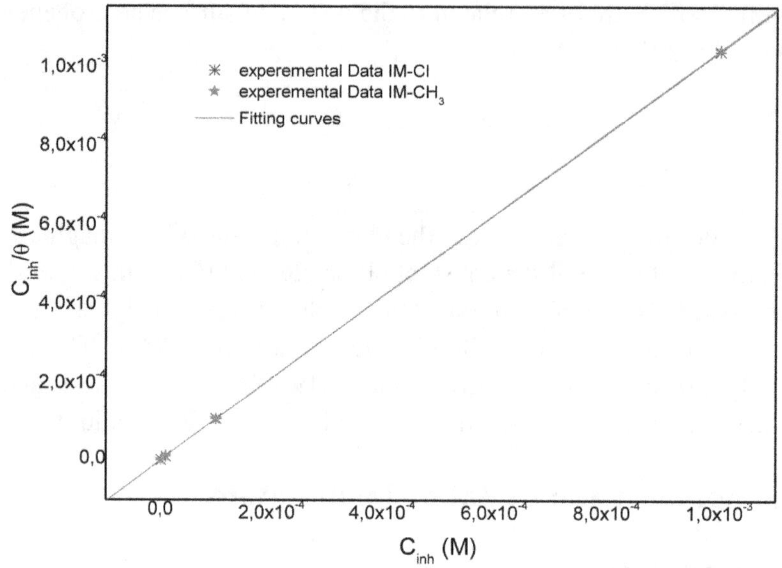

The value of K_{ads} was found to be large, namely, 43.4 10^4 (l/mol) of IM-Cl and 37.4 10^4 (l/mol) of IM-CH_3, implying efficient adsorption and based on which the free energy of adsorption (ΔG_{ads}) was calculated using the following Eq. (9):

Table 5. Thermodynamic parameters for IM-Cl and IM-CH_3 adsorption on mild steel surface in 1M HCl at 298 K.

Medium	K_{ads} (L/mol)	ΔG_{ads} (Kj/mol)	R^2	Slopes
1M HCl + IM-Cl	17.58 10^5	-45.57	1	1.03687
1M HCl + IM-CH_3	16.69 10^5	-45.45	1	1.04005

$$\Delta G_{ads} = RT \ln\left(55.5K_{ads}\right) \qquad (9)$$

Where, R is universal gas constant and the absolute temperature is denoted by T. The molar concentration of water is expressed in mol/L and in solution its value is 55.5 (Flis, 1996).

Generally, values of ΔG_{ads} around or less than 20 kJ mol^{-1} suggested that the adsorption process is related with the electrostatic interaction between charged inhibitor molecules and the charged surface of the metal, which is termed as "physisorption" (Be Bellman, 2008; Tazouti, 2016); while values around or higher than 40 kJ mol^{-1} are associated with sharing of charge or transfer from the inhibitor molecules to the metal surface to form a coordinate type of metal bond termed as "chemisorption" (Mu, 2006; Issaadi, 2011).

In the present case of the adsorption of imidazole derivatives, ΔG_{ads} was determined to be -45.57 kJ/mol of IM-Cl and -45.45 kJ/mol of IM-CH_3, suggesting the chemical adsorption of the inhibitor molecules into the corroding metal surface.

3.3.2. Temperature Kinetics

The influence of the reaction temperature on the potentiodynamic polarization curves in the absence and presence of 10^{-3}M of IM-Cl and IM-CH_3 inhibitor's for mild steel in 1M HCl in the temperature range 298±2 K to 328±2 K are represented in Fig. 7.

There is a clear acceleration of both the cathodic and anodic reactions with an increase in temperature. As it is observed from the polarization parameters (Table 6), for mild steel, the corrosion current densities increased with the increase of temperature in 1M HCl solution in the absence and presence of 10^{-3}M of the studied inhibitors. These results could be attributed to the acceleration of chemical process as the temperature is increased (Nesic, 2007). It is noted from Table 6 that, the inhibition efficiency reduced with temperature rise, indicating chemical adsorption of the imidazole molecules on the mild steel surface (Okafor, 2008).

Figure 7. Potentiodynamic polarization curves for mild steel immersed in 1M HCl in the present of 10^{-3} M of IM-Cl (a), IM-CH_3 (b) and Blank (c) at various temperatures.

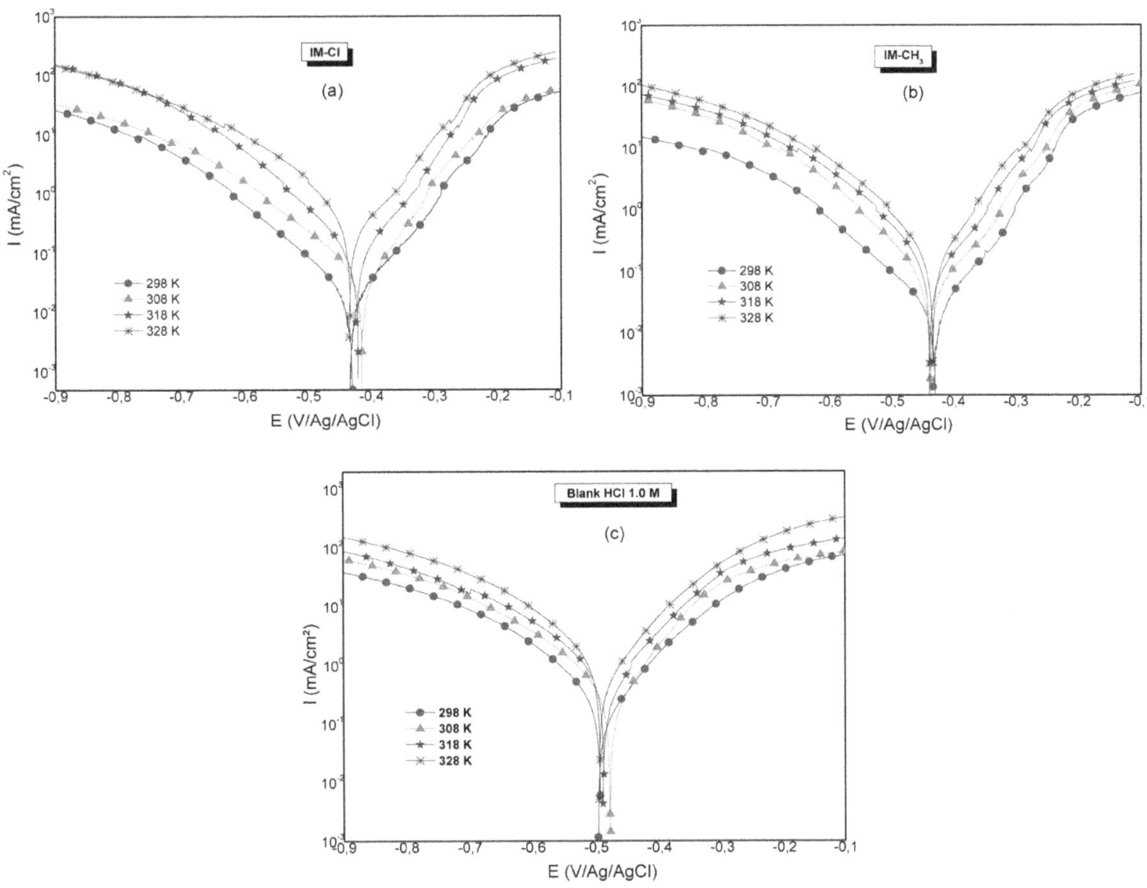

Table 6. Polarization parameters for mild steel in 1M HCl with 10^{-3} M of IM-Cl and IM-CH_3 at various temperatures.

Medium	Temperature K	$-E_{corr}$ mV/Ag/AgCl	i_{corr} μA cm^{-2}	$-\beta_c$ mV dec^{-1}	β_a mV dec^{-1}	η_{PP} %
Blank	298	498	983	140	150	-
	308	491	1200	184	112	-
	318	475	1450	171	124	-
	328	465	2200	161	118	-
IM-Cl	298	428	32	114	138	96.7
	308	411	50	119	138	95.7
	318	419	72	105	126	95.0
	328	430	125	129	135	94.3
IM-CH_3	298	427	32	114	126	96.7
	308	436	54	112	136	95.5
	318	437	75	129	138	94.7
	328	431	135	133	142	93.8

3.3.5. Activation Energy

The thermodynamic and kinetic parameters including the apparent activation energy (E_a), the enthalpy of activation (ΔH_a) and the entropy of activation (ΔS_a) for mild steel dissolution in 1M HCl solutions in the absence and presence of 10^{-3} M of IM-Cl and IM-CH at 298–328 K were calculated from the Arrhenius Eq. (10) (El Hezzat, 2015; Ekanem, 2010; Ahamad, 2010):

$$i_{corr} = Ae^{\frac{-Ea}{RT}} \tag{10}$$

Taking the logarithm of both sides of the Arrhenius Eq. (11) could be obtained

$$\ln i_{corr} = \ln A - \frac{E_a}{RT} \tag{11}$$

The change of enthalpy {ΔH_a} and entropy {ΔS_a} for the formation of activated complex in the transition state was obtained from the transition state Eq. (12) (Ekanem, 2010; Ahamad, 2010; Bochris &. Reddy, 1977).

$$\ln\left(\frac{i_{corr}}{T}\right) = \left[\ln\left(\frac{R}{hN_a}\right) + \left(\frac{\Delta S_a}{RT}\right)\right] - \frac{\Delta H_a}{RT} \tag{12}$$

Figure 8. Arrhenius plots for mild steel corrosion in 1M HCl in the absence and in presence of 10^{-3} M of IM-Cl and IM-CH_3/(a) of ln (i_{corr}) vs 1000/T and (b) ln (i_{corr}/T) vs 1000/T.

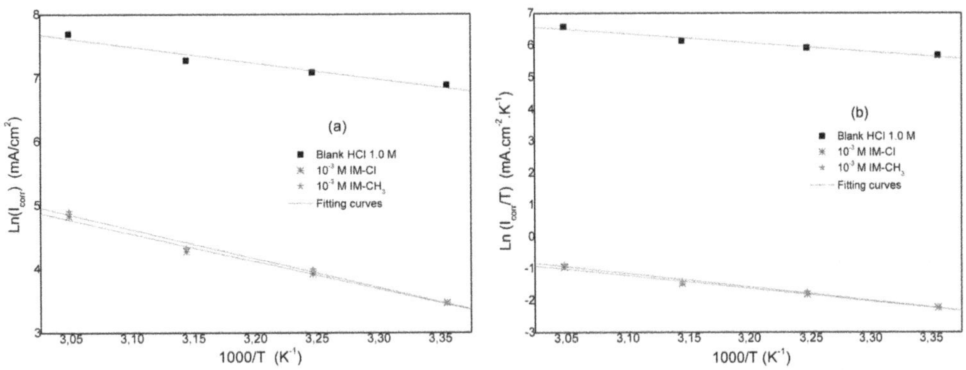

where i_{corr} is the corrosion current density, A is the pre-exponential factor, h is the Planck's constant, N_a is the Avogadro's number, E_a is the apparent activation energy, R is the gas constant (R=8.314 J mol^{-1}. K^{-1}) and T is the absolute temperature.

A plot of ln (i_{corr}) vs 1000/T presented in Fig. 8a show straight line with slope of the line ($-E_a/R$). The values of E_a were calculated from the slope and given in Table 7. From the Table 7 it is clear that the E_a values of solution containing 10^{-3} M of IM-Cl and IM-CH_3 is higher than that in case of uninhibited solution which may be attributed to the formation of compact barrier film on the mild steel surface (Shaban, 2016). The higher energy barrier for corrosion process in case of inhibited solutions suggest that adsorbed inhibitor's film prevents the charge/mass transfer reaction occurring on the surface (Murulana, 2016) and thus protecting metal from dissolution. The higher value of E_a for IM-Cl than that of IM-CH_3 can be attributed to the fact that IM-Cl is substituted by the donor group corresponding to the mesomeric effect (Cl) whereas IM-CH_3 is substituted by an inductive effect group (CH_3).

A plot of ln (i_{corr}/T) vs 1000/T (Fig. 8b) gave a straight line with slope of the line ($-\Delta H_a/R$) and an intercept (ln (R/hN_a)+($\Delta S_a/R$) from which the values of ΔH_a and ΔS_a were calculated and summarized in Table 7. From these data, it was found that the thermodynamic parameters, ΔH_a and ΔS_a of dissolution reaction of mild steel in 1M HCl in the presence of 10^{-3} M of IM-Cl and IM-CH_3 were higher than uninhibited solution. The positive signs of the enthalpies ΔH_a reflected the endothermic nature of the steel dissolution process. The higher value of the entropy of activation, ΔS_a in the presence of 10^{-3} M of IM-Cl and IM-CH_3 in solution compared to the case of the blank acid solution might be due to the enhanced randomness of the molecules (Martinez, 2016).

Table 7. Activation parameters for mild steel dissolution in 1M HCl solution in the absence and presence of optimum concentration of IM-Cl and IM-CH_3.

Medium	Conc (10^{-3} M)	E_a (KJ/mol)	ΔH_a (KJ/mol)	ΔS_a (J/mol.K)	E_a- ΔH
1M HCl	Blank	21.0	18.5	-126.0	2.5
	IM-Cl	36.0	33.4	-104.0	2.6
	IM-CH_3	37.9	35.3	-97.5	2.6

We note also that the variation of the activation energy Ea and the enthalpy of activation ΔHa vary in the same way with the concentration of inhibitor (Ghazoui, 2013):

$$E_a - \Delta H_a = RT \tag{13}$$

3.4. Surface Analysis

Scanning electron micrographs (Fig. 9) of the mild steel surface in 1M HCl with and without addition of IM-Cl and IM-CH_3 were taken in order to establish whether inhibition is due to the formation of an organic film on the metal surface. The resulting of the high-resolution SEM micrograph shows that the steel surface was strongly damaged in the absence of the two inhibitors. However, there are less pits and cracks observed in the micrographs in the presence of IM-Cl and IM-CH_3 (Fig. 9) which suggests a formation of protective film on steel surface which was responsible for the corrosion inhibition. Indeed, IM-Cl and IM-CH_3 has a strong tendency to adhere to the steel surface and can be regarded as good inhibitors for steel corrosion in hydrochloric medium. The high inhibitive performance of these imidazole derivatives suggests a strong bonding of the IM-Cl and IM-CH_3 on the metal surface due to

Figure 9. Surface morphology of mild steel after immersion for 6 h in 1M HCl with 10^{-3} M of IM-Cl and IM-CH_3

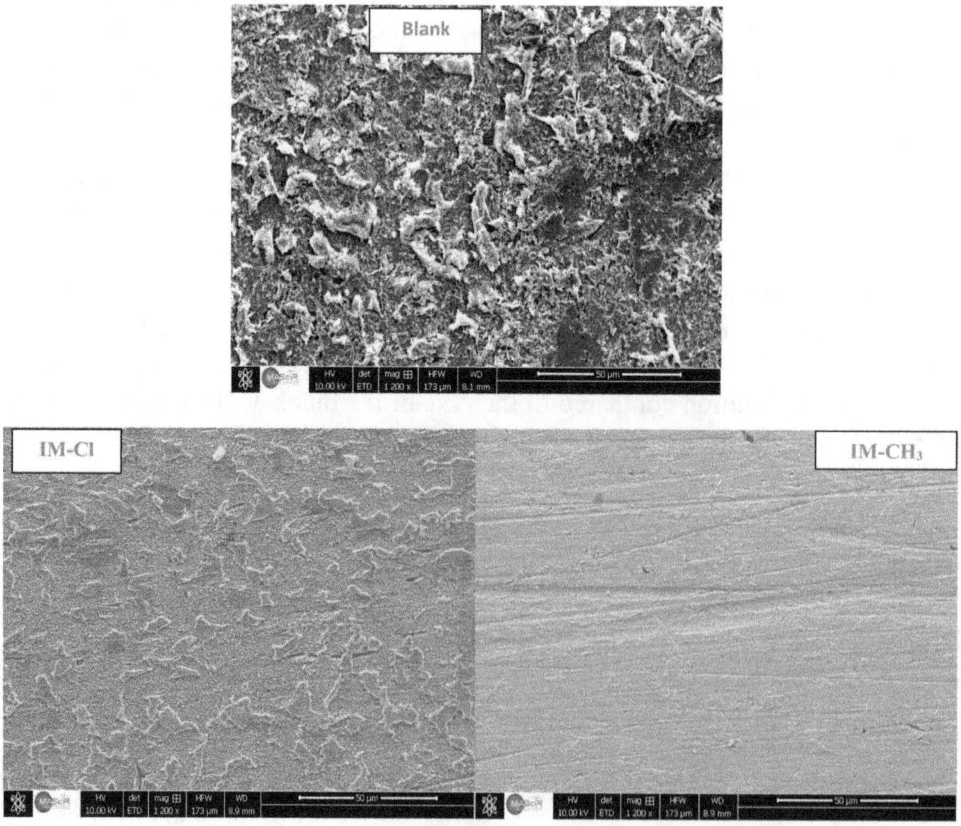

presence of lone pairs from heteroatom (nitrogen) and π-orbitals, blocking the active sites and therefore decreasing the corrosion rate.

EDX analyses (Figure 10) shows a decrease in the peaks of O atoms and apparition of peaks of Nitrogen N, indicating that the molecules of the products IM-Cl and IM-CH_3 are strongly adsorbed on the metal surface and justifies the inhibitory role of this product (Table 8) (Suzuki, 1996; Shetty, 2017; Rbaa, 2019a) . As a consequence, these results confirm those obtained with electrochemical measurements.

Figure 10. Qualitative and quantitative EDAX analysis of the corroded surface after 6h of immersion in the blank solution containing 1M HCl and solution with 10^{-3} M of inhibitors: IM-Cl and IM-CH_3 at 298 K

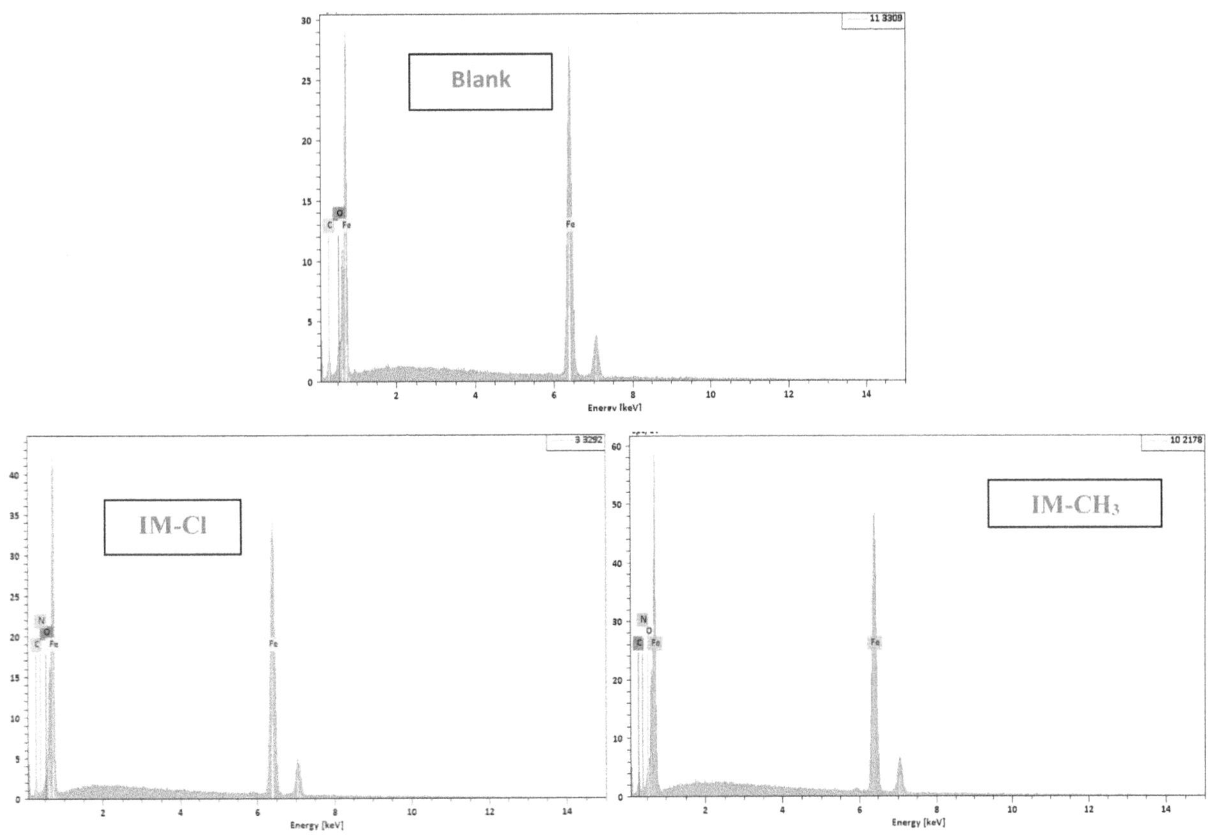

Table 8. %mass of the different elements resulting from the EDX analysis of the surface steel in 1M HCl in the absence and presence of the inhibitors: IM-Cl and IM-CH_3 at 298 K

Elements	% Mass of Steel in Blank Solution	% Mass of Steel in Blank Solution With IM-Cl	% Mass of Steel in Blank Solution With IM-CH_3
C	11.5	3.35	3.56
O	19.3	1.39	1.74
N	--	0.36	0.40
Fe	69.2	94.9	96.0

3.5. UV-visible Spectrometry

In order to confirm the possibility of complex formation (inh-Fe), the UV-visible absorption spectra obtained from a 1M HCl solution containing 10^{-3}M of IM-Cl and IM-CH_3 before and after 48 h of steel immersion at 298 K are shown in Fig 11 As described in the literature (Rbaa, 2019a; Goel, 2010; Wang, 2019; Fergachi, 2019) reported that the shift in wavelength with variation in absorbance indicates the formation of a complex between the two species in solution.

The absorption spectra of imidazole IM-Cl and IM-CH_3 before immersion of steel have absorption bands between 200-400 nm and are almost identical and can be attributed to the electronic transitions π-π*of aromatic cycle of the compound tested (Ðlhan, 2002; Rbaa, 2019b).. After 48 h of immersion of the steel in the 1M HCl solution containing 10-3 M of IM-Cl and IM-CH_3, the UV-Visible spectra obtained are presented in Fig. 11. that there is a shift in the absorbance of these bands without there being a significant difference in the shape of the spectra before and after the immersion of the two inhibitors IM-Cl and IM-CH_3 and of steel. These experimental findings clearly show the possibility of the formation of a complex between the cation Fe^{2+} and the inhibitors studied in 1M HCl (Alaa, 2006; Anacona, 2005; Song, 2007).

Figure 11. UV–visible spectra of 1M HCl solution containing 10^{-3} M of IM-Cl and IM-CH_3 before (black) and after (red) After 48 h of mild steel immersion.

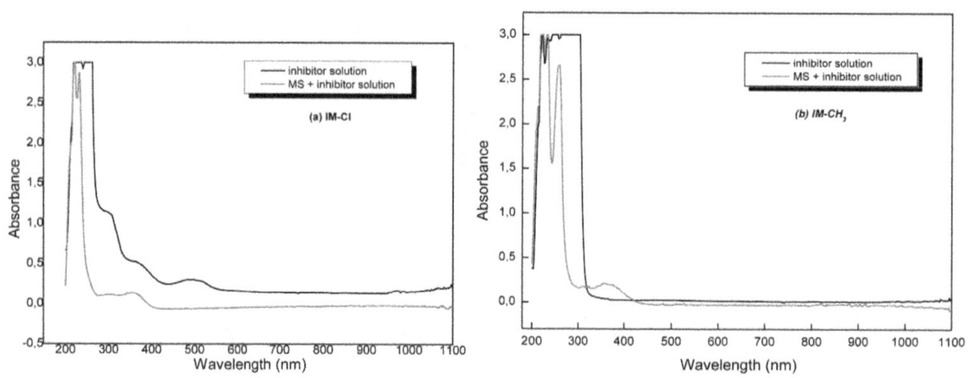

4. CONCLUSION

Compound IM-Cl exhibits the best corrosion inhibition performance that compound IM-CH_3 which is confirmed by electrochemical studies and weight loss. From the Polarization results, it could be confirmed that both the compounds are mixed type inhibitors with anodic predominance and their adsorption obey the Langmuir Adsorption isotherm. The UV–Visible studies clearly reveal the formation of a complex that may also be responsible for the observed inhibition. SEM images also confirm the inhibitor adsorption process and reduction in the surface damage of mild steel due to corrosion.

BACKGROUND

Corrosion of materials is a natural phenomenon that is a cause of concern as it has incurred a total damage of billions of dollars to many industries. Hence, various corrosion protection technologies have been employed, among them; the use of inhibitors is one of the practical methods for preventing corrosion of mild steel especially in acid media. Organic compounds containing hetero atoms in the functional groups, pi-electron in triple or conjugated double bonds and presence of aromatic rings in their structure are the major adsorption centres and are usually good inhibitors. Imidazole derivatives are better known, and more widely studied by researchers because of their many important chemical and biological properties, they have been commonly used in medicinal chemistry. In addition to these biological activities, this family of organic compounds has nitrogen heteroatom in their molecular structure as well as imidazole ring with benzene moieties which can create the inhibition properties in acidic medium and can be strongly adsorbed onto the surface.

MAIN FOCUS OF THE CHAPTER

The main focus of this chapter is to evaluate the inhibitory properties of two compounds of imidazole derivatives (IM-Cl and IM-CH_3), which show a high protective action for mild steel in 1 M HCl solution. Effect of concentration and temperature on the inhibitory efficiency of imidazole has been studied. Moreover, In order to predict the inhibition mechanism, some thermodynamic and kinetic parameters were calculated. Furthermore, the adsorption of inhibitor on the steel surface was discussed by using the electrochemical measurements. The purpose of the present chapter is also to visualize the surface morphology of steel samples without and with these compounds in the electrolytic solution using SEM/EDX to confirm the presence of the film or the layer formed.

REFERENCES

Ahamad, I., Prasad, R., & Quraishi, M. A. (2010). Inhibition of mild steel corrosion in acid solution by Pheniramine drug: Experimental and theoretical study. *Corrosion Science*, *52*(9), 3033–3041. doi:10.1016/j.corsci.2010.05.022

Alaa, S. A., & Ayman, A. (2006). El-Fetouh. Gouda., Ragaa, El-Sheikh., Faten, Zahran. *Spectrochimica Acta. Part A: Molecular and Biomolecular Spectroscopy*, *14*, 171.

Amin, M. A., Abd-El-Rehim, S. S., El-Sherbini, E. E. F., & Hazzazi, O. A. (2009). Polyacrylic acid as a corrosion inhibitor for aluminium in weakly alkaline solutions. Part I: Weight loss, polarization, impedance EFM and EDX studies. *Corrosion Science*, *51*(3), 658–667. doi:10.1016/j.corsci.2008.12.008

Anacona, J. R., Martell, T., & Sanchez, I. (2005). Metal complexes of a new ligand derived from 2, 3-quinoxalinedithiol and 2, 6-bis (bromomethyl) pyridine. *Journal of the Chilean Chemical Society*, *50*(1), 375–378. doi:10.4067/S0717-97072005000100001

Asan, A., Soylu, S., Kıyak, T., Yıldırım, F., Oztas, S. G., Ancın, N., & Kabasakaloğlu, M. (2006). Investigation on some Schiff bases as corrosion inhibitors for mild steel. *Corrosion Science*, *48*(12), 3933–3944. doi:10.1016/j.corsci.2006.04.011

Be Bellman, C., & Stamm, M. (2008). *Polymer Surfaces and Interfaces*. Springer.

Bochris, J. O'M., & Reddy, A.K.N. (1977). Modern Electrochemistry (vol. 2). Plenum Press.

De Souza, F. S., & Spinelli, A. (2009). Caffeic acid as a green corrosion inhibitor for mild steel. *Corrosion Science*, *51*(3), 642–649. doi:10.1016/j.corsci.2008.12.013

Ðlhan, S., Temel, H., & Sekerci, M. (2002). Synthesis and characterization of new N, N¢-bis (cinnemaldeydene)-1, 2-bis (p-aminophenoxy) ethane and its transition metal complexes. Synt. React. Inorg. Met-Org. *Chem*, *32*, 1625–1634.

Ech-chihbi, E., Belghiti, M. E., Salim, R., Oudda, H., Taleb, M., Benchat, N., Hammouti, B., & El-Hajjaji, F. (2017). Experimental and computational studies on the inhibition performance of the organic compound "2-phenylimidazo [1,2-a]pyrimidine-3-carbaldehyde" against the corrosion of carbon steel in 1.0M HCl solution. *J. Surf. Interface*, *9*, 206–217. doi:10.1016/j.surfin.2017.09.012

Ech-chihbi, E., Nahlé, A., Salim, R., Oudda, H., El Hajjaji, F., El Kalai, F., El Aatiaoui, A., & Taleb, M. (2019). An Investigation into Quantum Chemistry and Experimental Evaluation of Imidazopyridine Derivatives as Corrosion Inhibitors for C-Steel in Acidic Media. *Journal of Bio- and Tribo-Corrosion*, *5*(24).

Ekanem, U. F., Umoren, S. A., Udousoro, I. I., & Udoh, A. P. (2010). Inhibition of mild steel corrosion in HCl using pineapple leaves (Ananas comosus L.) extract. *Journal of Materials Science*, *45*(20), 5558–5566. doi:10.100710853-010-4617-y

El Hezzat, M., Assouag, M., Zarrok, H., Benzekri, Z., El Assyry, A., Boukhris, S., Souizi, A., Galai, M., Touir, R., Ebn Touhami, M., Oudda, H., & Zarrouk, A. (2015). Correlated DFT and electrochemical study on inhibition behavior of ethyl 6-amino-5-cyano-2-methyl-4-(p-tolyl)-4H-pyran-3-carboxylate for the corrosion of mild steel in HCl. *Der Pharma Chemica*, *7*(10), 77–88.

El Mehdi, B., Mernari, B., Traisnel, M., Bentiss, F., & Lagrenee, M. (2003). Synthesis and comparative study of the inhibitive effect of some new triazole derivatives towards corrosion of mild steel in hydrochloric acid solution. *Materials Chemistry and Physics*, *77*(2), 489–496. doi:10.1016/S0254-0584(02)00085-8

El Ouali, I., Hammouti, B., Aouniti, A., Ramli, Y., Azougagh, M., & Essassi, E.M., Bouachrine. (2010). Thermodynamic characterisation of steel corrosion in HCl in the presence of 2-phenylthieno (3, 2-b) quinoxaline. *J. Mater. Envir. Sci*, *1*, 1–8.

El-Taib Heakal, F., Deyab, M.A., Osman, M.M., Nessim, M.I., & Elkholy, A.E. (2017). Synthesis and assessment of new cationic gemini surfactants as inhibitors for carbon steel corrosion in oilfield water. *RSC Adv*, 47335–47352.

Fergachi, O., Benhiba, F., Rbaa, M., Ouakki, M., Galai, M., Touir, R., Lakhrissi, B., Oudda, H., & Ebn Touhami, M. (2019). Corrosion inhibition of ordinary steel in 5.0 M HCl medium by benzimidazole derivatives: electrochemical, UV–visible spectrometry, and DFT calculations. *Journal of Bio- and Tribo-Corrosion, 5*, 21.

Flis, J., & Zakroczymski, T. (1996). Impedance study of reinforcing steel in simulated pore solution with tannin. *Journal of the Electrochemical Society, 143*(8), 2458. doi:10.1149/1.1837031

Gabrielli, C. (1980). *Identification of Electrochemical Processes by Frequency Response Analysis*. Solartron.

Galai, M., Rbaa, M., El Kacimi, Y., Ouakki, M., Dkhirech, N., Touir, R., & Lakhrissi, B., EbnTouhami, M. (2016). Anti-corrosion properties of some triphenylimidazole substituted compounds in corrosion inhibition of carbon steel in 1. 0 M hydrochloric acid solution. *Anal. Bioanal. Electrochem, 9*, 80.

Gerengi, H., & Sahin, H. I. (2012). Schinopsis lorentzii Extract As a Green Corrosion Inhibitor for Low Carbon Steel in 1 M HCl Solution. *Industrial & Engineering Chemistry Research, 51*(2), 780–787. doi:10.1021/ie201776q

Ghazoui, A., Bencaht, N., Al-Deyab, S. S., Zarrouk, A., Hammouti, B., Ramdani, M., & Guenbour, M. (2013). An Investigation of two novel pyridazine derivatives as corrosion inhibitor for C38 steel in 1.0 M HCl. *International Journal of Electrochemical Science, 8*, 2272–2292.

Goel, R., Siddiqi, W. A., Ahmed, B., Khan, M. S., & Chaubey, V. M. (2010). Synthesis characterization and corrosion inhibition efficiency of N-C2 {(2E)-2-[4-(dimethylamino) benzylidene] hydrazinyl} 2-oxo ethyl benzamide on mild steel. *Desalination, 263*(1-3), 45–57. doi:10.1016/j.desal.2010.06.033

Gupta, N. K., Quraishi, M. A., Verma, C., & Mukherjee, A. K. (2016a). Spiro [indoline-3,4¢-pyrano[2,3-c] pyrazole] Derivatives as Novel Class of Green Corrosion Inhibitors for Mild Steel in Hydrochloric Acid Medium: Theoretical and Experimental Approach. *RSC Advances, 6*, 102076–102087. doi:10.1039/C6RA22116E

Gupta, N. K., Quraishi, M. A., Verma, C., & Mukherjee, A. K. (2016b). Schiff's bases derived from l-lysine and aromatic aldehydes as green corrosion inhibitors for mild steel: Experimental and theoretical studies. *Journal of Molecular Liquids, 215*, 47–57. doi:10.1016/j.molliq.2015.12.027

Hermas, A. A., & Morad, M. S. (2008). A comparative study on the corrosion behaviour of 304 austenitic stainless steel in sulfamic and sulfuric acid solutions. *Corrosion Science, 50*(9), 2710–2717. doi:10.1016/j.corsci.2008.06.029

Issaadi, S., Douadi, T., Zouaoui, A., Chafaa, S., Khan, M. A., & Bouet, G. (2011). Novel thiophene symmetrical Schiff base compounds as corrosion inhibitor for mild steel in acidic media. *Corrosion Science, 53*(4), 1484–1488. doi:10.1016/j.corsci.2011.01.022

Kadiri, L., Galai, M., Ouakki, M., Essaadaoui, Y., Ouass, A., Cherkaoui, M., Rifi, E. H., & Lebkiri, A. (2018). Coriandrum Sativum. L Seeds Extract as a Novel Green Corrosion Inhibitor for Mild Steel in 1.0 M Hydrochloric and 0.5 M Sulfuric Solutions. *Anal. Bioanal. Electrochem, 10*(2), 249–268.

Kalman, E., Varhegi, B., Bako, I., Felhosi, I., Karman, F. H., & Shaban, A. (1994). Corrosion Inhibition by 1-hydroxy-ethane-1, 1-diphosphonic acid: An electrochemical impedance spectroscopy study. *Journal of the Electrochemical Society*, *141*(12), 3357–3360. doi:10.1149/1.2059339

Kowsari, E., Payami, M., Amini, R., Ramezanzadeh, B., & Javanbakht, M. (2014). Task-specific ionic liquid as a new green inhibitor of mild steel corrosion. *Applied Surface Science*, *289*, 478–486. doi:10.1016/j.apsusc.2013.11.017

Lgaz, H., Chung, I. M., Albayati, M. R., Chaouiki, A., Salghi, R., & Mohamed, S. K. (2020). Improved corrosion resistance of mild steel in acidic solution by hydrazone derivatives: An experimental and computational study. *Arabian Journal of Chemistry*, *13*(1), 2934–2954. doi:10.1016/j.arabjc.2018.08.004

Lgaz, H., Chung, I. M., Salghi, R., Ali, I. H., Chaouiki, A., El Aoufir, Y., & Khan, M. I. (2019). On the understanding of the adsorption of Fenugreek gum on mild steel in an acidic medium: Insights from experimental and computational studies. *Applied Surface Science*, *463*, 647–658. doi:10.1016/j.apsusc.2018.09.001

Liu, F. G., Du, M., Zhang, J., & Qiu, M. (2009). Electrochemical behavior of Q235 steel in saltwater saturated with carbon dioxide based on new imidazoline derivative inhibitor. *Corrosion Science*, *51*(1), 102–109. doi:10.1016/j.corsci.2008.09.036

Martinez, S., & Stern, I. (2002). Thermodynamic characterization of metal dissolution and inhibitor adsorption processes in the low carbon steel/mimosa tannin/sulfuric acid system. *Applied Surface Science*, *199*(1-4), 83–89. doi:10.1016/S0169-4332(02)00546-9

Mu, G., Li, X., Qu, Q., & Zhou, J. (2006). Molybdate and tungstate as corrosion inhibitors for cold rolling steel in hydrochloric acid solution. *Corrosion Science*, *48*(2), 445–459. doi:10.1016/j.corsci.2005.01.013

Murulana, L. C., Kabanda, M. M., & Ebenso, E. E. (2016). Investigation of the adsorption characteristics of some selected sulphonamide derivatives as corrosion inhibitors at mild steel/hydrochloric acid interface: Experimental, quantum chemical and QSAR studies. *Journal of Molecular Liquids*, *215*, 763–779. doi:10.1016/j.molliq.2015.12.095

Nesic, S. (2007). Key issues related to modelling of internal corrosion of oil and gas pipelines–A review. *Corrosion Science*, *49*(12), 4308–4338. doi:10.1016/j.corsci.2007.06.006

Obot, I., & Obi-Egbedi, N. O. (2010). Inhibition of aluminium corrosion in hydrochloric acid using nizoral and the effect of iodide ion addition. *Chemistry (Weinheim an der Bergstrasse, Germany)*, *7*(3), 837–843.

Oguzie, E. E., Lia, Y., & Wang, F. H. (2007). Effect of 2-amino-3-mercaptopropanoic acid (cysteine) on the corrosion behaviour of low carbon steel in sulphuric acid. *Electrochimica Acta*, *53*(2), 909–914. doi:10.1016/j.electacta.2007.07.076

Oguzie, E. E., Unaegbu, C., Ogukwe, C. N., Okolue, B. N., & Onuchukwu, A. I. (2004). Inhibition of mild steel corrosion in sulphuric acid using indigo dye and synergistic halide additives. *Chemical Physics*, *84*, 363.

Okafor, P. C., Ikpi, M. E., Uwah, I. E., Ebenso, E. E., Ekpe, U. J., & Umoren, S. A. (2008). Inhibitory action of Phyllanthus amarus extracts on the corrosion of mild steel in acidic media. *Corrosion Science*, *50*(8), 2310–2317. doi:10.1016/j.corsci.2008.05.009

Ouakki, M., Galai, M., Cherkaoui, M., Rifi, E. H., & Hatim, Z. (2018b). Inorganic compound (apatite doped by Mg and Na) as a corrosion inhibitor for mild steel in phosphoric acidic medium. *Anal. Bioanal. Electrochem*, *10*(7), 943–960.

Ouakki, M., Galai, M., Rbaa, M., Abousalem, A. S., Lakhrissi, B., Rifi, E. H., & Cherkaoui, M. (2019). Quantum chemical and experimental evaluation of the inhibitory action of two imidazole derivatives on mild steel corrosion in sulphuric acid medium. *Heliyon*, *5*(11), e02759. doi:10.1016/j.heliyon.2019.e02759 PMID:31768434

Ouakki, M., Rbaa, M., Galai, M., Lakhrissi, B., Rifi, E. H., & Cherkaoui, M. (2018a). Experimental and quantum chemical investigation of imidazole derivatives as corrosion inhibitors on mild steel in 1.0 M hydrochloric acid. *J. Bio Tribo-Corrosion*, *4*(3), 35. doi:10.100740735-018-0151-2

Quartarone, G., Battilana, M., Bonaldo, L., & Tortato, T. (2008). Investigation of the inhibition effect of indole-3-carboxylic acid on the copper corrosion in 0.5 M H2SO4. *Corrosion Science*, *50*(12), 3467–3474. doi:10.1016/j.corsci.2008.09.032

Rbaa, M., Abousalem, A. S., Ebn Touhami, M., Warade, I., Bentiss, F., Lakhrissi, B., & Zarrouk, A. (2019a). Novel Cu (II) and Zn (II) complexes of 8-hydroxyquinoline derivatives as effective corrosion inhibitors for mild steel in 1.0 M HCl solution: Computer modeling supported experimental studies. *Journal of Molecular Liquids*, *290*(15), 111243. doi:10.1016/j.molliq.2019.111243

Rbaa, M., Benhiba, F., Obot, I. B., Oudda, H., Warad, I., Lakhrissi, B., & Zarrouk, A. (2019b). Two new 8-hydroxyquinoline derivatives as an efficient corrosion inhibitors for mild steel in hydrochloric acid: Synthesis, electrochemical, surface morphological, UV–visible and theoretical studies. *Journal of Molecular Liquids*, *276*, 120–133. doi:10.1016/j.molliq.2018.11.104

Rbaa, M., Galai, M., El Faydy, M., El Kacimi, Y., Ebn Touhami, M., Zarrouk, A., & Lakhrissi, B. (2017c). Synthesis and Characterization of New Benzimidazoles Derivatives of 8-hydroxyquinoline as a Corrosion Inhibitor for Mild Steel in 1.0 M Hydrochloric Acid Medium. *Anal. Bioanal. Electrochemistry*, *9*(7), 904–928.

Rbaa, M., Galai, M., El Kacimi, Y., Ouakki, M., Touir, R., & Lakhrissi, B. (2017a). Adsorption Properties and Inhibition of Carbon Steel Corrosion in a Hydrochloric Solution by 2-(4, 5-diphenyl-4, 5-dihydro-1h-imidazol-2-yl)-5-methoxyphenol. *Portugaliae Electrochimica Acta*, *35*(6), 323–338. doi:10.4152/pea.201706323

Rbaa, M., Galai, M., Faydy, E. L., El Kacimi, M., Ebn, Y., Touhami, M., Zarrouk, A., & Lakhrissi, B. (2017b). Synthesis, inhibition effects and thermodynamic studies of novel substituted quinolines on the corrosion of mild steel in 1 M HCl solution. *J. Mater. Environ. Sci*, *8*(10), 3529–3549.

Shaban, S. M., Abd-Elaal, A. A., & Tawfik, S. M. (2016). Gravimetric and electrochemical evaluation of three nonionic dithiol surfactants as corrosion inhibitors for mild steel in 1 M HCl solution. *Journal of Molecular Liquids*, *216*, 392–400. doi:10.1016/j.molliq.2016.01.048

Shahin, M., Bilgic, S., & Yilmaz, H. (2002). The inhibition effects of some cyclic nitrogen compounds on the corrosion of the steel in NaCl mediums. *Applied Surface Science*, *195*(1-4), 1–7. doi:10.1016/S0169-4332(01)00783-8

Shetty, S. K., & Shetty, A. N. (2017). Eco-friendly benzimidazolium based ionic liquid as a corrosion inhibitor for aluminum alloy composite in acidic media. *Journal of Molecular Liquids*, *225*, 426–438. doi:10.1016/j.molliq.2016.11.037

Singh, A. K., & Quraishi, M. A. (2010). The effect of some bis-thiadiazole derivatives on the corrosion of mild steel in hydrochloric acid. *Corrosion Science*, *52*(4), 1373–1385. doi:10.1016/j.corsci.2010.01.007

Singh, S. K., Tambe, S. P., Gunasekaran, G., Raja, V. S., & Kumar, D. (2009). Electrochemical impedance study of thermally sprayable polyethylene coatings. *Corrosion Science*, *51*(3), 595–601. doi:10.1016/j.corsci.2008.11.025

Song, X. Q., Wang, Y. W., Zheng, J. R., Liu, W. S., & Tan, M. Y. (2007). Synthesis and spectroscopic properties of lanthanide nitrate complexes with a new amide-based quinoxaline-2,3-dione ligand. *Mol Biomol Spectrosc*, *68*(3), 701–704. doi:10.1016/j.saa.2006.12.048 PMID:17419093

Suzuki, T., Nishihara, H., & Aramaki, K. (1996). The synergistic inhibition effect of octylmercapto-propionate and 8-quinolinol on the corrosion of iron in an aerated 0.5 M Na2SO4 solution. *Corrosion Science*, *38*(8), 1223–1234. doi:10.1016/0010-938X(95)00172-G

Tawfik, S. M. (2015). Corrosion inhibition efficiency and adsorption behavior of N,N-dimethyl-4-(((1-methyl-2-phenyl-2,3-dihydro-1H-pyrazol-4-yl) imino) methyl)-N alkylbenzenaminium bromide surfactant at carbon steel/hydrochloric acid interface. *Journal of Molecular Liquids*, *207*, 185–194. doi:10.1016/j.molliq.2015.03.036

Tazouti, A., Galai, M., Touir, R., Touhami, M. E., Zarrouk, A., Ramli, Y., Saraçoğlu, M., Kaya, S., Kandemirli, F., & Kaya, C. (2016). Experimental and theoretical studies for mild steel corrosion inhibition in 1.0 M HCl by three new quinoxalinone derivatives. *Journal of Molecular Liquids*, *221*, 815–832. doi:10.1016/j.molliq.2016.03.083

Verma, C., Ebenso, E. E., Bahadur, I., Obot, I. B., & Quraishi, M. A. (2015b). 5-(Phenylthio)-3H-pyrrole-4-carbonitriles as effective corrosion inhibitors for mild steel in 1 M HCl: Experimental and theoretical investigation. *Journal of Molecular Liquids*, *212*, 209–218. doi:10.1016/j.molliq.2015.09.009

Verma, C., Quraishi, M. A., & Singh, A. (2015a). 2-Amino-5-nitro-4,6-diarylcyclohex-1-ene-1,3,3-tricarbonitriles as new and effective corrosion inhibitors for mild steel in 1 M HCl: Experimental and theoretical studies. *Journal of Molecular Liquids*, *212*, 804–812. doi:10.1016/j.molliq.2015.10.026

Wang, C., Lai, C., Xie, B., Guo, X., Fu, D., Li, B., & Zhu, S. (2018). Corrosion inhibition of mild steel in HCl medium by S-benzyl-O, O¢-bis (2-naphthyl) dithiophosphate with ultra-long lifespan. *Results Phys*, *10*, 558–567. doi:10.1016/j.rinp.2018.07.002

Wang, H. L., Fan, H. B., & Zheng, J. S. (2003). Corrosion inhibition of mild steel in hydrochloric acid solution by a mercapto-triazole compound. *Chemical Physics*, *77*, 655.

Yadav, M., Kumar, S., Purkait, T., Olasunkanmi, L. O., Bahadur, I., & Ebenso, E. E. (2016). Electrochemical, thermodynamic and quantum chemical studies of synthesized benzimidazole derivatives as corrosion inhibitors for N80 steel in hydrochloric acid. *Journal of Molecular Liquids*, *213*, 122–138. doi:10.1016/j.molliq.2015.11.018

Zhang, L., He, Y., Zhou, Y., Yang, R., Yang, Q., Qing, D., & Niu, Q. (2015). A novel imidazoline derivative as corrosion inhibitor for P110 carbon steel in hydrochloric acid environment. *Petroleum*, *1*(3), 237–243. doi:10.1016/j.petlm.2015.10.007

Chapter 7
Triazepines Compounds as Novel Synthesized Corrosion Inhibitors:
State of Play and Perspectives

Khaoula Alaoui
Laboratory of Materials Engineering and Environment: Modeling and Application, Faculty of Science, Ibn Tofail University, Morocco

Ashraf Abousalem
Mansoura University, Egypt

Burak Tüzün
Cumhuriyet University, Turkey

Younes El Kacimi
https://orcid.org/0000-0002-8662-4644
Laboratory of Materials Engineering and Environment: Modeling and Application, Faculty of Science, Ibn Tofail University, Morocco

ABSTRACT

This chapter describes some recent good works in the field of metallic corrosion monitoring using Triazepine carboxylate inhibitors in hydrochloric acidic media. The different sections of this chapter cover electrochemical measurements and theoretical investigations. This review reveals Triazepine carboxylate compounds as very good inhibitors for mild steel in hydrochloric medium. The inhibition action of all Triazepine carboxylates compound studied was performed via adsorption on mild steel surface. Comparison between several adsorption isotherms reveal that the adsorption was spontaneous and followed Langmuir isotherm in HCl for all inhibitors and all studied temperatures. Kinetic and thermodynamic parameters for all inhibitors led to suggest the occurrence of chemical mechanism and also the spontaneity of the adsorption process on mild steel surface. The corrosion inhibition mechanism was also compared and discussed with the light of some Triazepine carboxyale compounds constituents.

DOI: 10.4018/978-1-7998-2775-7.ch007

INTRODUCTION

The triazepines consist of a group of four triazacycloheptatrienes. The four possible isomers (a, b, c and d) are presented in Fig. 1. Triazepines have attracted a great deal of attention as starting materials in the synthesis of fused heterocyclic systems with prospective pharmacological activities.

Triazepines are very interesting, both from the pharmacological and chemical points of view.

Condensed heterocyclic 1,2,4-triazepines were found to have salidiuretic, renal vasodilator, antioxidant, analgesic and immunomodulating activities. Additionally, the fusion of pyrimidine to the triazepine moiety results in improved pharmacological properties as antiviral, antifungal, and antidiabetic agents, and the products also act as inhibitors in cancer chemotherapy.

Similarly, fused triazepine derivatives having a bridgehead nitrogen atom within the molecule display interesting biological properties.

Furthermore, (Serrar, 2017) was focusing on developing synthetic routes to heterocyclic derivatives with potential biological activities, she herein report a safe, facile, fast and one-pot reaction synthesis of a novel substituted triazepines carboxylate compounds.

The new proposed synthetic routes seem to be of interest, since they have been compared with existing methods. Especially, the fact that the obtained structures present considerable antioxidant properties in different systems, gives them the potential of being therapeutic agents in many pathological diseases, since oxidative stress is implicated in many physiological systems.

In the light of this information, and considering that organic molecules containing heteroatoms, such as nitrogen (N), oxygen (O), sulfur (S) and phosphorus (P), have proven practically and theoretically to act as efficiently corrosion inhibitors in a wide range of acidic solutions (Bentiss, 2015; Popova, 2007).

In this paper, we have reported the works of some Triazepine carboxylate compunds which investigated as corrosion inhibitors. These articles were using deferent techniques: potentiodynamic polarization, electrochemical impedance spectroscopy (EIS), and gravimetric measurement. The chapter summarizes the corrosion inhibition of some triazepine derivatives as corrosion inhibitors on metal surfaces such as mild steel and carbon steel, in acidic medium. The information presented in this review is useful for diverse industrial fields and for other applications where corrosion inhibitors for metals are needed.

BACKGROUND

Mild steel is applied to approximately the entire industrial field for the reason of its excellent properties. However, its corrosion usually results in enormous financial losses and potential security problems where the protection technology has attracted the interest of several researchers (Jiang, 1995; Popova, 2007; Alaoui, 2018). Among these technologies, chemical inhibitors take a significant role for iron corrosion protection for the reason of their economy and convenience (El Kacimi, 2018; Tang, 2005; Oguzie, 2004). Today, new compounds have many bonds or heteroatoms in their structures were used. Many organic compounds were employed for iron corrosion in acidic media (Serrar, 2017; Sasikumar, 2015).

This review chapter describes some recent good works in the field of metallic corrosion monitoring using Triazepine carboxylate inhibitors in hydrochloric acidic media.

The different sections of the review chapter cover: Electrochemical measurements such as (The potentiodynamic polarization curves and electrochemical impedance spectroscopy investigations used to evaluate the corrosion inhibition performance of Triazepine carboxylate compounds as very good inhibitors for mild steel in Hydrochloric medium.

MAIN FOCUS OF THE CHAPTER

In order to review the use of some triazepine compounds as corrosion inhibitors used in different medium. Comparison of all Triazepine reviewed was done according to to the three sections listed below:

- Section 1:
    - Triazepine corrosion Inhibitors performances using:
        - weight-loss;
        - potentiodynamic polarization curves and
        - electrochemical impedance spectroscopy.
- Section 2:
    - Corrosion Kinetic Study and temperature effect
    - Thermodynamic parameters of the adsorption process
    - Adsorption isotherm analysis and thermodynamic parameters
- Section 3:
    - Molecular modeling simulation

Steel and Its Alloys

Iron is a strong metal that is fairly cheap, the combination of iron with other elements provides material properties that are acceptable for many applications such as mild steel and carbon steel which is used in manufacturing machine tools, automobiles, surgical equipment, electric power, and even building parts. One of the major defects of steels is their modest resistance to corrosion, especially in Industrial processes when we used acidic environments in order to clean and descale the undesirable corrosion products from the steel surface. The protection against this alteration can be realized by the use of inhibitors as many authors investigated Triazepine compounds which tested the most using carbon or mild steel. In order to clearly show the differences in steel composition, even though these steels are sometimes named the same, Table 1 was made for the materials found in the literature compared in this review chapter.

Table 1. Chemical compositions in wt.% of different iron-based materials examined for this review.

	C	Si	Mn	Cr	Mo	Ni	Al	Cu	Co	V	W	P	S	Ti	Fe
Mild steel 1	0.11	0.24	0.47	0.12	0.02	0.1	0.03	0.14	<0.012	<0.003	0.06	-	-	-	Balance
Mild steel 2	0.076	0.026	0.192	0.05	-	-	0.023	-	-	-	-	0.012	-	-	Balance
Carbon steel 3	0.37	0.23	0.68	0.077	-	0.059	-	0.16	<0.009	-	-	-	0.016	0.011	Balance
Carbon steel 4	0.36	0.27	0.66	0.21	0.02	-	0.06	0.22	-	-	-	0.015	0.02	-	Balance

EXPERIMENTAL TECHNIQUES

Weight Loss and Electrochemical Methods

Recently, several modern techniques were showing up, but the mainly used methods in the investigation of inhibition performance are weight loss method, the electrochemical method (potentiodynamic polarization and electrochemical impedance spectroscopy (EIS)) which uses a cell with three electrodes.

Molecular Dynamic Modelling

Recently, molecular dynamics simulation constituted an advancement part in corrosion inhibition studies as it is employed at the atomistic level to explore the interaction of inhibitor molecules on metal surfaces (Tang, 2010). The simulation of adsorption process of inhibitor molecules on Iron surfaces were investigated by molecular dynamics simulation using (Material Studio 2017) software from Accelrys Inc, USA. The most stable planes of crystalline metals is the one with densely packed, it is commonly used in molecular dynamics simulations to represent the corroding substrate metal (Salarvand, 2017; Obot, 2013). The simulations were performed in a simulation box of dimensions (3.44 nm x 4.05 nm x 5.34 nm). The molecular dynamics simulations were carried out with periodic boundary conditions to model a representative part of the interface devoid of any arbitrary boundary effects. Firstly, the surface of (pure Fe crystal) was cleaved, and relaxed by minimizing its energy by means of molecular mechanics. Next, the surface of cleaved plane Fe (110) was enlarged to a super cell of an appropriate size, and then a vacuum slab with zero Thickness was built above these surfaces. Since corrosion inhibition in aqueous solution is an electrochemistry process, the use of water molecules and different ions is essential The effect of chemical species involves in the adsorption system was taken into consideration, thus the system in the molecular dynamics simulations incorporated 250 H_2O, 5 H^+, 5 Cl^- and one inhibitor molecule in each case (Fouda, 2017). Finally, the most stable orientation of inhibitor molecules on metal surface was searched by Monte-carlo search simulations. The molecular dynamics simulations were performed at 298 K (controlled by the Andersen thermostat) using a canonical ensemble (NVT) with a time step of 1.0 fs and a simulation time of 500 ps. For the whole simulation procedure, the COMPASS (Condensed-phase Optimized Molecular Potentials for Atomistic Simulation Studies) force field (Fouda, 2017) was used, because it allows the accurate and simultaneous prediction of structural, conformational, vibrational, and thermo-physical properties for a broad range of chemical species, including organic molecules, metals, metal oxides, and metal halides (Sun, 1998).

TRIAZEPINE COMPOUNDS

Triazepines and their derivatives are found to be associated with various biological activities, such as cancer, viral infections (HIV), and cardiovascular disorders. The fusion of heterocyclic rings to different faces of the heptatomic nucleus have been shown to enhance the activity profiles. The synthesis and characterization of a family of triazepine derivatives are of particular interest to the chemical and biological communities and hence the chemistry and biochemistry of triazepine derivatives have been intensively investigated. However, the synthetic derivatives of triazepine also form a varied and important group of medicines. Some known derivatives of imidazotriazepines have shown activity as muscle relaxants, while others have antifungal, anti-diabetic, antimicrobial, antiviral, anticancer and analgesic properties. Triazepine is a seven-membered heterocyclic compound with three nitrogen atoms, where the position of the nitrogen atoms can vary (Elatar, 2015).

In the light of this information, and considering that organic molecules containing heteroatoms, such as nitrogen (N) and oxygen (O), have proven practically and theoretically to act as efficiently corrosion inhibitors in a wide range of acidic solutions. The efficiency of these inhibitors can be attributed to their high polarizability and lower electronegativity; so that these atoms and the functional groups can cover large metallic surface areas and easily electrons transfer to the empty orbitals of atoms. In addition, nitrogen-containing organic inhibitors is good anticorrosion materials for metals in hydrochloric acid, while compounds having sulfur atoms act as good inhibitors in sulfuric acid. Compounds holding nitrogen and sulfur behave as perfect corrosion inhibitors for both media. Several authors have decided to use Triazepine molecules and/or to synthesize novel triazepine molecules in order to use them as effective corrosion inhibitors.

Figure 1. Structures of (a) 1,2,3-triazepine (b) 1,2,4-triazepine (c) 1,2,5-triazepine (d) 1,3,5-triazepine

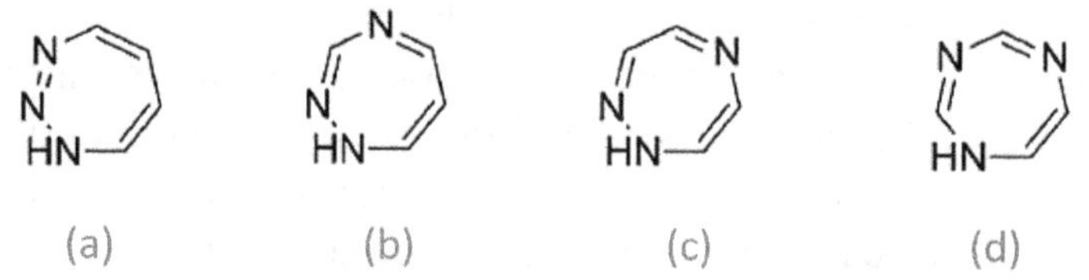

The following paragraph is dedicated to the review of triazepine used as an inhibitor.

EXPERIMENTAL STUDIES

The maximum of Triazepine corrosion inhibitors is used as corrosion inhibitors in hydrochloric acid (HCl) since it is the most applied in industries. A list of various Triazepines compounds that have been examined is given in Table 2. The additional effect of these molecules at different concentrations on the corrosion of metal was studied firstly by measurement of weight loss, potentiodynamic polarization, electrochemical impedance spectroscopy (EIS), the results obtained from these measurements were summarized in Table 3.

As we can see from Table 3, all the inhibitors have inhibition efficiency fluctuates between 86.9% as lowest efficiency and 97.4% as the highest corrosion inhibition efficiency although the different aggressive solutions used. According to the author's discussions, the intensity–potential curves indicated that all studied Triazepine compounds operate as mixed-type inhibitors because the displacement in potential is less than 85 mV/E_{corr} and their inhibition efficiencies increase with their concentrations to attain a maximum at 10^{-3} M. It is established also that the mechanism of the anodic and cathodic reactions changed by the addition of triazepine carboxylate derivative compounds. Also, they reported from the polarization curves that the current densities I_{corr} decrease by the addition of studied inhibitors which mean an increase in the blocked fraction of electrode surface by adsorption and consequently achieve high inhibition efficiencies. In other case, they postulated that The Nyquist plots gotten by EIS measurements are not perfect semicircles and explained this by the dispersion of frequency as a result of the roughness and inhomogeneous of electrode surface. Also, they confirmed that these structures adsorbed into the electrode interface by replacement of water molecules with organic inhibitor molecules. Furthermore, it is found that the triazepine compounds act by an adsorption on the metallic surface and their inhibition depends on the nature of the alkyl substituents. Table 3 resumes the effect of triazepines carboxylate derivatives on mild steels and carbon steel corrosion in acidic medium and using different electrochemical techniques such as potentiodynamic polarization curves and electrochemical impedance spectroscopy.

(Benhiba, 2020), interpreted that the P1 and P2 compounds have good corrosion inhibitor performance of mild steel in 1.0 M HCl medium. It appears that the corrosion rate decreases which leads to increasing the protection efficiency with the raise of inhibitor concentration, which reaches a highest value of 89% for the P2 compound, at an optimal concentration of 10^{-3} M. This decrease in corrosion rate is due to the adsorption of triazepines molecules (P1 and P2) on the metal, this promotes the adsorption of film more powerful which ensures the protection of our metal. Benhiba et al. found also, that the P1 and P2 organic inhibitors in 1.0 M HCl inhibit corrosion of mild steel by adsorption on the metal/electrolyte interface. In hydrochloric acid, P1 and P2 inhibitors can exist in the protonated form in equilibrium with its corresponding molecular form. The physical interactions represent the first step of the adsorption of P7 and P8 compounds tested.

Indeed, the adsorption of cationic forms of inhibitors will be fostered by the concentration of anions (Cl^-) on the metal surface. That is, the positively charged inhibitory molecules are subjected to the electrostatic effect by Cl^- ions adsorbed. Whereas after the release of H_2 in the case of the cathodic reaction, the protonated inhibitor molecules return to their neutral form which will be chemisorbed on the metal surface by donor interactions between the free electron pairs of (N, O) heteroatoms and the unoccupied d-orbital of the iron atoms.

About the 6-methyl-7H[1,2,4]triazolo[4,3-b][1,2,4]triazepine-8(9H)-thione (P3) inhibitor, (El Bakri, 2017) found that the (P3) have a good inhibition properties for the corrosion of carbon steel in phosphoric acid (2.0 M H_3PO_4) at 303 K, and the inhibition efficiency, increases with increase of the inhibitor concentration. The inhibition efficiency of P3 decrease proportionally with increasing temperature (303–333K) and its addition to 2.0 M H_3PO_4 leads to increase of apparent activation energy (E_a) of corrosion process. The corrosion process is inhibited by the adsorption of 6-methyl-7H[1,2,4]triazolo[4,3-b][1,2,4]triazepine-8(9H)-thione (P3) on carbon steel surface and the adsorption of the inhibitor fits a Langmuir isotherm model at 303 K. The calculated value of ΔG_{ads} reveals that the adsorption mechanism of P3 on carbon steel surface is mainly electrostatic-adsorption. Based on the Tafel polarization results, P3 can be classified as mixed inhibitor. In continue of his works, (El Bakri, 2019) reveal that the inhibitive action of three triazepine derivatives, (P4, P5 and P6) for carbon steel in 1.0 M HCl is investigated. The results revealed that they are relatively good inhibitors. Potentiodynamic polarization data indicated that synthesized derivatives are of mixed type inhibitor. EIS data reveals the inhibition of corrosion on getting adsorbed on the carbon steel surface. Langmuir model is best fitted into the obtained results. Quantum chemical parameters suggested the inhibition of carbonsteel corrosion in 1.0 M HCl solution by the studied compounds is essentially due to the effective transfer of electrons from the inhibitors to Fe atom, which facilitates donor-acceptor interactions between the inhibitors and the metal.

Molecular dynamics simulation results reveal that the interaction energy values of the three inhibitor compounds with the Fe(110) surface obey the order P6 > P5 > P4, in accordance with the experimental inhibition efficiency. We hope that the present study will provide a distinctive perspective on developing new type of triazolotriazepine derivatives as corrosion inhibitors.

Table 2. Triazepine compounds used as corrosion inhibitors

Sample Number	Metals/Alloy	Chemical Structure	IUPAC Name	Medium	Reference
P1	Mild steel (2)	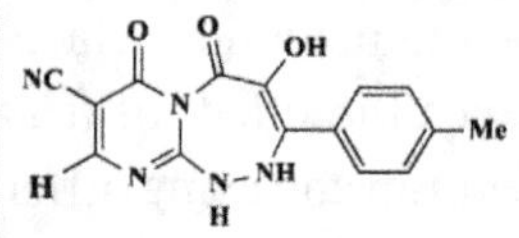	4-hydroxy-8-isocyano-9-phenyl-3-(p-tolyl)-1,2-dihydropyrimido[2,1-c]**[1,2,4]triazepine**-5,7-dione	1 M HCl	(Benhiba, 2020)
P2	Mild steel (2)	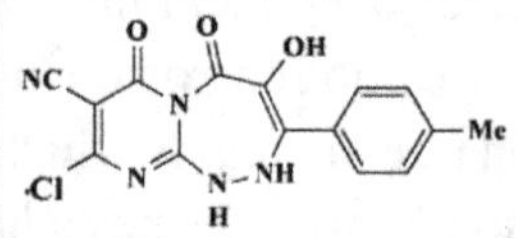	9-(4-chlorophenyl)-4-hydroxy-8-isocyano-3-(p-tolyl)-1,2-dihydropyrimido[2,1-c]**[1,2,4]triazepine**-5,7-dione	1 M HCl	(Benhiba, 2020)
P3	Carbon steel (3)	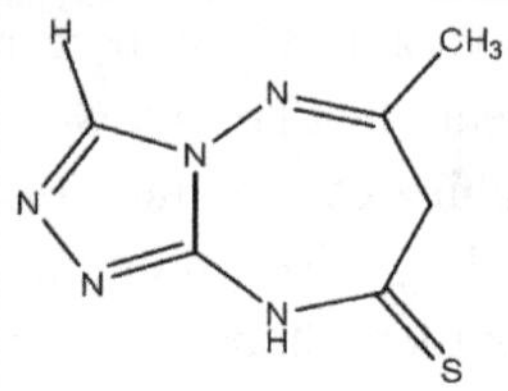	6-methyl-7H[1,2,4]triazolo[4,3-b]**[1,2,4]triazepine**-8(9H)-thione (MTTT)	2 M H_3PO_4	(El Bakri, 2017)
P4	Carbon steel (4)	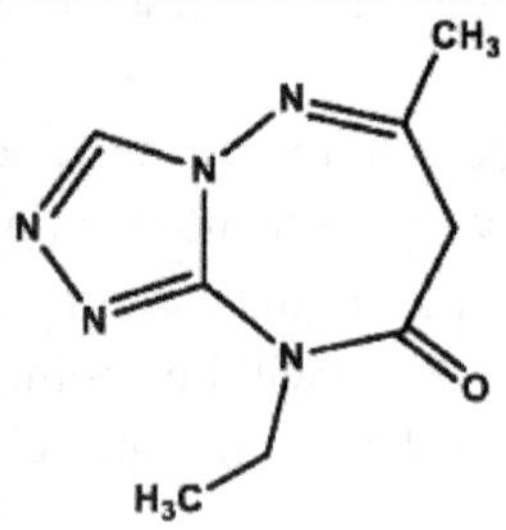	9-ethyl-6-methyl-7H-1,2,4-triazolo[4,3-b]**[1,2,4]triazepine**-8(9H)-one	1 M HCl	(El Bakri, 2019)
P5	Carbon steel (4)	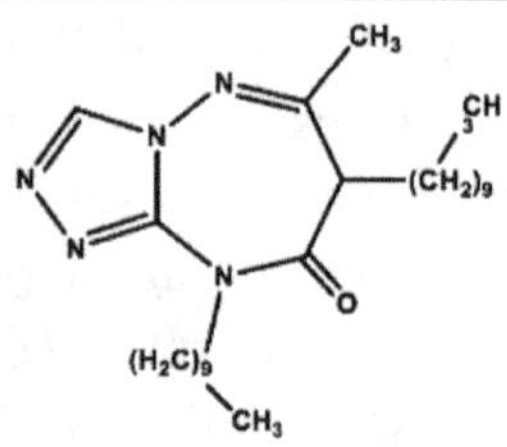	7,9-didecyl-6-methyl-7H-1,2,4-triazolo[4,3-b]**[1,2,4]triazepine**-8-one	1 M HCl	(El Bakri, 2019)
P6	Carbon steel (4)	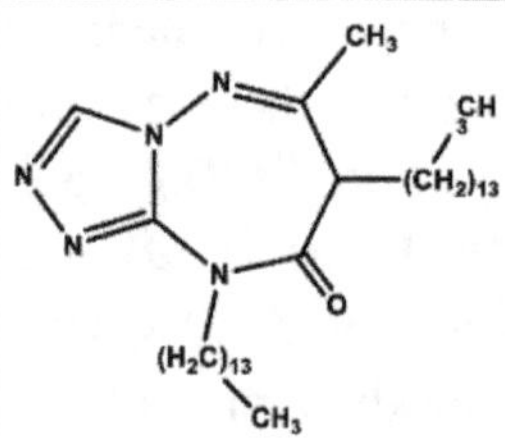	7,9-ditetradecyl-6-methyl-7H-1,2,4-triazolo[4,3-b]**[1,2,4] triazepine**-8-one	1 M HCl	(El Bakri, 2019)
P7	Mild steel (1)	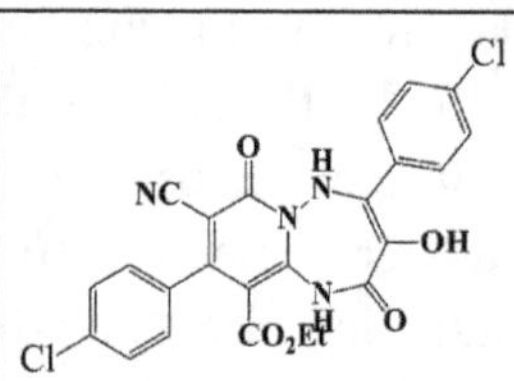	ethyl 4,9-bis(4-chlorophenyl)-8-cyano-3-hydroxy-2,7-dioxo-1,2,5,7-tetrahydropyrido[1,2-b]**[1,2,4]triazepine**-10-carboxylate	1 M HCl	(Alaoui, 2018; Alaoui, 2020)

continued on following page

Table 2. Continued

Sample Number	Metals/Alloy	Chemical Structure	IUPAC Name	Medium	Reference
P8	Mild steel (1)	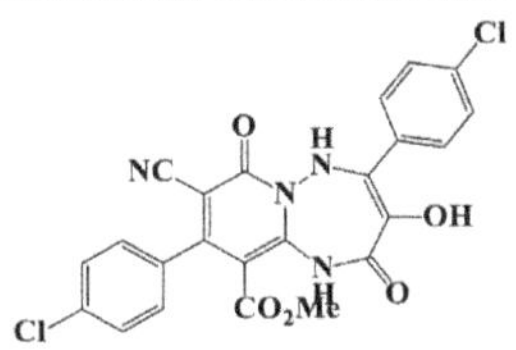	methyl 4,9-bis(4-chlorophenyl)-8-cyano-3-hydroxy-2,7-dioxo-1,2,5,7-tetrahydropyrido[1,2-b] **[1,2,4]triazepine**-10-carboxylate	1 M HCl	(Alaoui, 2018; Alaoui, 2020)
P9	Mild steel (1)	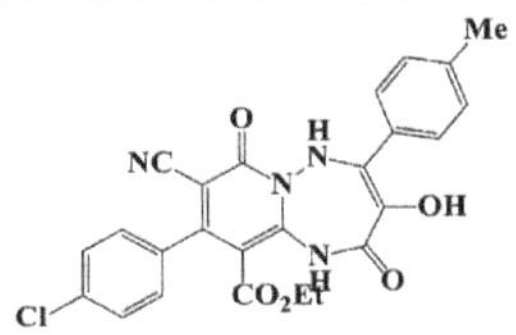	ethyl 9-(4-chlorophenyl)-8-cyano-3-hydroxy-2,7-dioxo-4-(p-tolyl)-1,2,5,7-tetrahydropyrido[1,2-b] **[1,2,4]triazepine**-10-carboxylate	1 M HCl	(Alaoui, 2018; Alaoui, 2020)
P10	Mild steel (1)	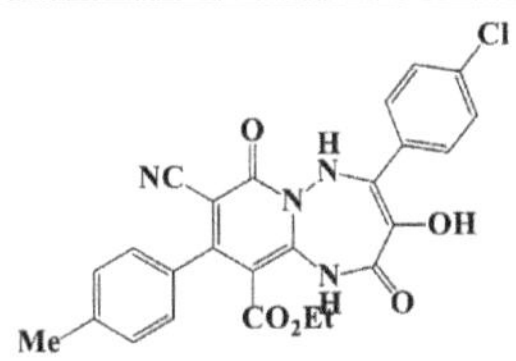	ethyl 4-(4-chlorophenyl)-8-cyano-3-hydroxy-2,7-dioxo-9-(p-tolyl)-1,2,5,7-tetrahydropyrido[1,2-b]**[1,2,4]triazepine**-10-carboxylate	1 M HCl	(Alaoui, 2019; Alaoui, 2020)
P11	Mild steel (1)	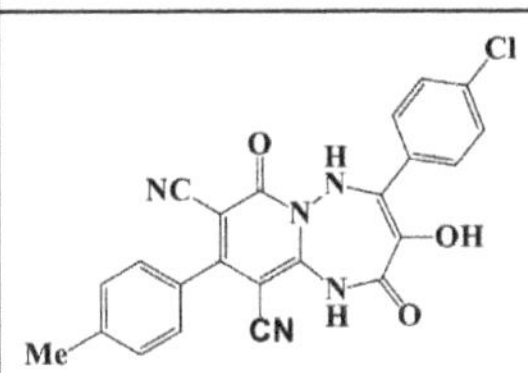	4-(4-chlorophenyl)-3-hydroxy-2,7-dioxo-9-(p-tolyl)-1,2,5,7-tetrahydropyrido[1,2-b]**[1,2,4]triazepine**-8,10-dicarbonitrile	1 M HCl	(Alaoui, 2019; Alaoui, 2020)
P12	Mild steel (1)	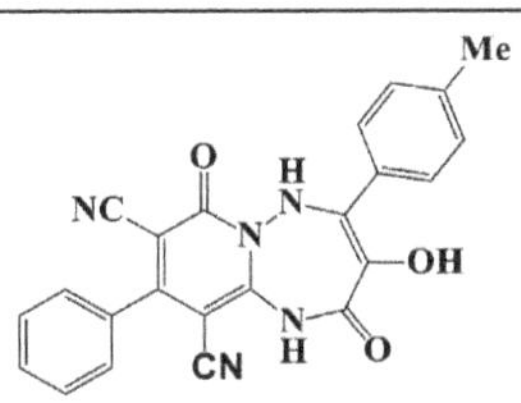	3-hydroxy-2,7-dioxo-9-phenyl-4-(p-tolyl)-1,2,5,7-tetrahydropyrido[1,2-b]**[1,2,4]triazepine**-8,10-dicarbonitrile	1 M HCl	(Alaoui, 2019; Alaoui, 2020)

a More data is available in the cited article (also regarding the structure of the derivatives).
b Composition given in Table 1, g – inhibition effectiveness, WL – weight loss, T – temperature, IT – immersion time.
c The units mpy (mils per year) in the cited article are converted to mm/y.

In order to facilitate the comparative analysis of six corrosion inhibitors, and taking into account their substitutions, we studied them in two distinct series. We have studied first in our previous work the adsorption and corrosion inhibition properties of three triazepines carboxylate substituted compounds, namely, ethyl 4,9-bis(4-chlorophenyl)-8-cyano-3-hydroxy-2,7-dioxo-1,2,5,7-tetrahydropyrido[1,2-b][1,2,4]triazepine-10-carboxylate (P7), methyl 4,9-bis(4-chlorophenyl)-8-cyano-3-hydroxy-2,7-dioxo-1,2,5,7-tetrahydropyrido[1,2-b][1,2,4]triazepine-10 -carboxylate (P8) and ethyl 9-(4-chlorophenyl)-8-cyano-3-hydroxy-2,7-dioxo-4-(p-tolyl)-1,2,5,7-tetrahydropyrido[1,2-b][1,2,4]triazepine-10-carboxylate (P9) on mild steel corrosion in 1.0 M of hydrochloride acid were investigated by using weight loss, electrochemical and theoretical investigations. It is found that these compounds act as excellent inhibitors and their inhibition efficiencies follow the order: P8 > P7 > P9 and depend on the type of the substituent. In addition, the possible relations between the experimental and theoretical parameters were discussed. It is found that the theoretical parameters confirmed those obtained by experimental studies.

Table 3. Parameters of Triazepines inhibitors performances at Optimum inhibitors concentration effect

Inhibitors	Metals/ Alloy [(a)]	Medium	Inhibition Efficiency [(b)]				Corrosion Kinetic Parameters [(c)]				Reference
			η_{wl} (%)	η_{pp} (%)	η_{EIS} (%)	η_{x} (%)	E_a ($KJ\ mol^{-1}$)	ΔH_a ($KJ\ mol^{-1}$)	ΔS_a ($J\ mol^{-1}\ K^{-1}$)	ΔG°_{ads} ($KJ\ mol^{-1}$)	
P1	Mild steel [(2)]	1 M HCl	87.0	88.0	90.0	88.3	96.6	94.0	76.82	-42.0	(Benhiba, 2020)
P2	Mild steel [(2)]	1 M HCl	89.0	91.0	91.0	90.3	197.8	105.2	85.13	-41.7	(Benhiba, 2020)
P3	Carbon steel [(3)]	2 M H_3PO_4	-	92.0	95.9	94.0	41.7	34.8	-45.394	30.7	(El Bakri, 2017)
P4	Carbon steel [(4)]	1 M HCl	-	94.4	96.9	95.7	86.5	83.9	60.52	-43.0	(El Bakri, 2019)
P5	Carbon steel [(4)]	1 M HCl	-	95.0	97.1	96.1	90.5	87.8	72.33	-44.0	(El Bakri, 2019)
P6	Carbon steel [(4)]	1 M HCl	-	97.0	97.8	97.4	91.5	88.9	71.91	-45.0	(El Bakri, 2019)
P7	Mild steel [(1)]	1 M HCl	87.7	87.0	87.4	87.4	33.2	30.6	-102.2	-40.1	(Alaoui, 2018; Alaoui, 2020)
P8	Mild steel [(1)]	1 M HCl	89.1	89.8	89.4	89.4	36.9	34.3	-91.7	-43.5	(Alaoui, 2018; Alaoui, 2020)
P9	Mild steel [(1)]	1 M HCl	87.4	86.0	87.3	86.9	34.3	31.7	-98.3	-41.2	(Alaoui, 2018; Alaoui, 2020)
P10	Mild steel [(1)]	1 M HCl	87.8	92.6	91.0	90.5	39.1	36.5	-86.8	-42.0	(Alaoui, 2019; Alaoui, 2020)
P11	Mild steel [(1)]	1 M HCl	94.7	94.6	93.7	94.3	51.4	48.8	-48.8	-44.8	(Alaoui, 2019; Alaoui, 2020)
P12	Mild steel [(1)]	1 M HCl	94.0	94.1	93.2	93.8	43.7	41.1	-73.6	-39.3	(Alaoui, 2019; Alaoui, 2020)

(a) Composition given in Table 1,

(b) More data is available in the cited article / Inhibition effectiveness, η_{wl} – weight loss, η_{pp} – potentiodynamic polarization, η_{EIS} – electrochemical impedance spectroscopy measurements.

(c) Corrosion kinetic parameters E_a - activation corrosion energy, ΔH_a - enthalpy of activation, ΔS_a – entropy of activation and ΔG°_{ads} - Free energy of adsorption.

Surface roughness morphology for mild steel before and after hydrochloric acid attack are investigated and discussed using 3D-profilometry images. The Scanning Electron Microscope (SEM) images of the corrosion product confirmed the protection offered by Triazepines carboxylate derivatives on the surface of the metal immersed in both media.

According to the literature, the protection efficiency of the organic compounds mainly depends on their size and their active centers (Saha, 2016). In addition, the obtained efficiency can be explained by the increase in the effective electron density at the inhibitor functional group by the presence of different substituents [19, 37]. The high inhibition efficiency of P8 and P7 compared to P9 caused by the presence of two –(Cl) group. The nucleophilic (electron releasing) character of –(Cl) is higher than the electrophonic (electron withdrawing). In addition, the high efficiency, protective film for P8 compared to P7 is due to the presence of methyl group -(CH_3) which are electron donating increases electronic density on triazepine carboxylate. So, the methyl groups tend to increase the electron density on the molecule, increasing the bond strength between the P7 and the metal surface.

Generally, the electrochemical reactions of iron in acidic solution are its anodic dissolution (1) and cathodic reduction of proton (2) reactions:

$$Fe \leftrightarrow Fe^{2+} + 2e^{-} \quad (1)$$

$$2H^{+} + 2e^{-} \leftrightarrow H_2 \quad (2)$$

It is found that the triazepine carboxylate derivatives addition hinders the acid attack on mild steel. In addition, it is observed that the triazepine carboxylate addition, displaces slightly the E_{corr} to the anodic directions. The entire tested compounds act like a mixed type inhibitors.

However, it is observed that these inhibitors have an effect on both, the cathodic and anodic slopes (β_c and β_a) signifying a modification on cathodic (equation (1)) and anodic (equation (2)) reactions mechanism. Furthermore, the change in β_a can be explained by the changes in the charge transfer coefficient for iron dissolution by the presence of an additional energy barrier created by inhibitors adsorption. Finally, it is seen also that the η_{PP} followed the same order: P8 > P7 > P9 such as found by the weight loss measurements.

For the second series of these inhibitors, attempts have been made to study the corrosion inhibition properties of three new triazepine carboxylate compounds namely Ethyl 9-(4-chlorophenyl)-8-cyano-3-hydroxy-2,7-diaxo-4-ptolylpyrido[1,2 b] [1, 2, 4] triazepine-10-carboxylate (P10), 9-(4-Chlorophenyl)-3-hydroxy-2,7-dioxo,4-ptolylpyrido[1,2-b] [1, 2, 4] triazepine-8,10-dicarbonitrile (P11) and 3-Hydroxy-2,7-dioxo-9-phenyl-4-p-tolylpyrido[1,2-b] [1, 2,4] triazepine-8,10-dicarbonitrile (P12) for mild steel in 1.0 M hydrochloric acid medium. The evaluation was carried out using mass loss, electrochemical impedance spectroscopy and polarization curves measurement. Impedance diagrams and Bode plots for uninhibited and inhibited systems were analyzed using Zview program. The fitted data observed trails in nearly the same pattern as the experimental results. It is showed that triazepine carboxylate compounds are very good inhibitors for mild steel corrosion in 1.0 M hydrochloric acid medium which act as mixed-type inhibitors. So, the inhibition efficiency was increased with inhibitor concentration in the order P11 > P12 > P10 which depended on their molecular structures. Electrochemical impedance spectroscopy showed that all compounds act by the formation of a protective film at the metal surface.

Besides, the recorded intensity-potential curves in the presence of inhibitors are characterized by the presence of anodic breakdown potential, $E_{b,}$ which sudden rise in current density takes place. Beyond this potential, the surface film is shifted from stable to unstable state. In addition, it is seen that E_b shifts to noble direction. The noble shift of E_b with concentrations reflected the increased inhibitors adsorption on metallic surfaces. This shift in E_b with the studied triazepine carboxylate concentrations reflected the formation of anodic protective film on the electrode surface.

Figure 2. Potentiodynamic polarization curves for mild steel in 1.0 M HCl in the absence and presence of various concentrations of P7, P8, P9, P10, P11 and P12 inhibitors.

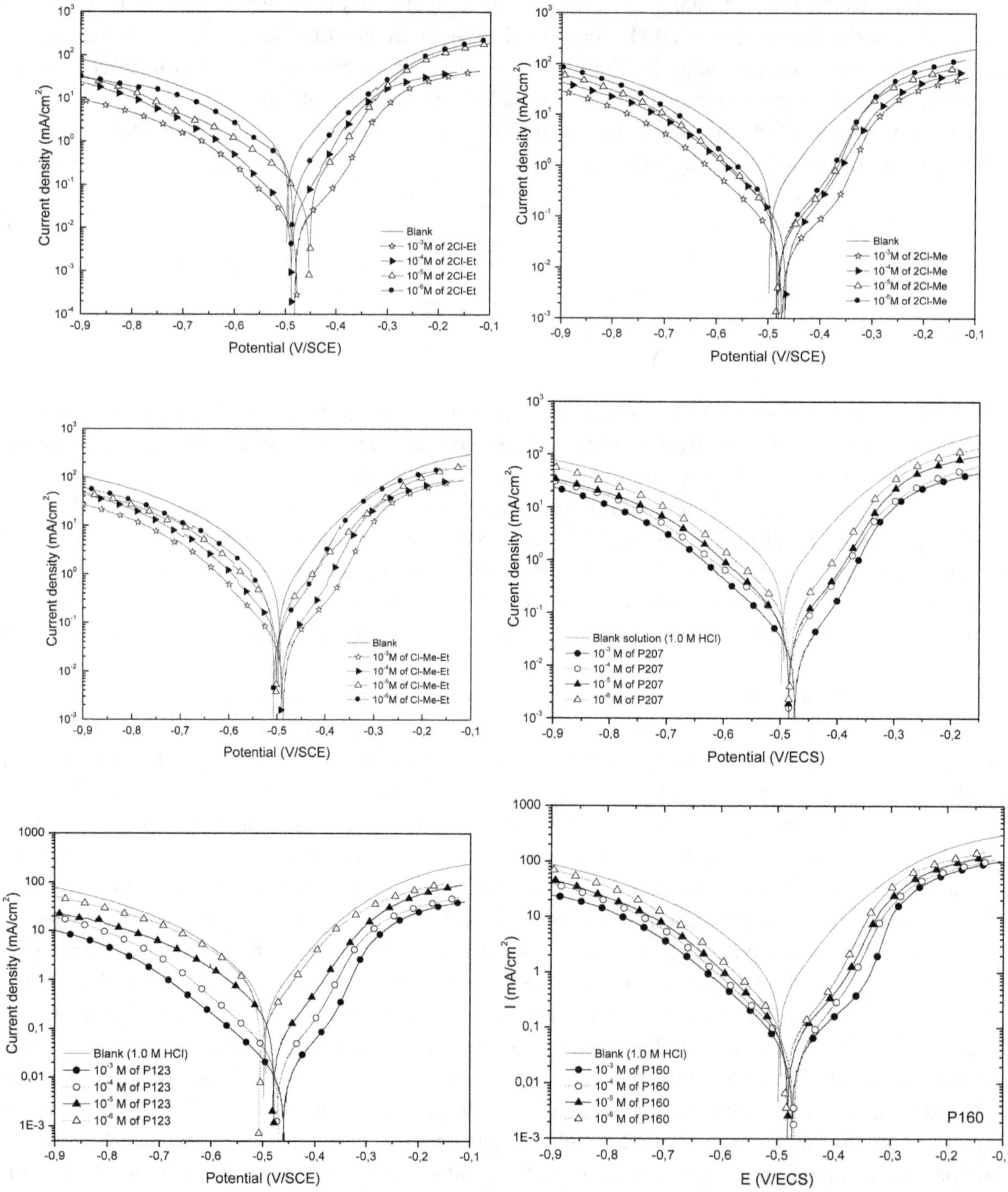

However, for potential higher than -300 mV_{SCE} versus E_{corr}, which defined as the desorption potential; the presence of triazepine carboxylate compounds did not change the current/potential characteristics. This comportment may be explained by the desorption of the inhibitor molecules from metal surface where its rate was superior to its adsorption rate and the corrosion rate increases with rising potential.

The presence of –(Cl) group and (-C°N) cyano group on triazepine carboxylate compounds can produce protonated species in HCl by the reaction of the amino group with HCl. This ammonium substituent may reduce the electronic density on the substituted triazepine carboxylate compounds.

On the other hand, according to the literature, at high concentration, the anions present in the corrosive solution can act as inhibitors or operate by blocking the pores of a passive film formed by the inhibitors molecules. Thus, the mechanism of the iron dissolution in HCl is like follows:

$$\mathrm{Fe} + \mathrm{Cl} \leftrightarrow [\mathrm{FeCl}^-]_{ads} \tag{3}$$

$$[\mathrm{FeCl}^-]_{ads} \leftrightarrow [\mathrm{FeCl}]_{ads} + e^- \tag{4}$$

$$[\mathrm{FeCl}]_{ads} \leftrightarrow [\mathrm{FeCl}^+]_{ads} + e^- \tag{5}$$

$$[\mathrm{FeCl}^+]_{ads} \leftrightarrow \mathrm{Fe}^{2+} + \mathrm{Cl}^- \tag{6}$$

The mechanism for hydrogen evolution is:

$$\mathrm{Fe} + \mathrm{H}^+ \leftrightarrow [\mathrm{FeH}^+]_{ads} \tag{7}$$

$$[\mathrm{FeH}^+]_{ads} + e^- \leftrightarrow [\mathrm{FeH}]_{ads} \tag{8}$$

$$[\mathrm{FeH}]_{ads} + \mathrm{H}^+ + e^- \leftrightarrow \mathrm{Fe} + \mathrm{H}_2 \tag{9}$$

In this case, it would be complicated for the protonated of the triazepine carboxylate derivative molecules to move toward the positively charged metal surface for the reason of the electrostatic repulsion. The anions are firstly adsorbed onto the metallic surface which allows the protonated of the triazepine derivative molecules to easily come near to the surface.

Figures 3 present the obtained plots from Electrochemical impedance spectroscopy for mild steel in 1.0 M HCl uninhibited solution and in the presence of different concentrations of triazepines compounds P7, P8, P9, P10, P11 and P12 at the E_{OCP}. It is noted that the impedance response was composed by one capacitive loop where its diameter significantly changed after all inhibitors additions and the maximum was obtained at 10^{-3} M of triazepine carboxylate compounds. This capacitive loop can be attributed to the charge transfer reaction. The obtained electrochemical parameters are listed in Table 3. This circuit contains the elements which can be named as; R_s is the solution resistance, R_{ct} is the charge transfer resistance and CPE_{dl} is the constant phase element of the double layer such as indicated above.

Excellent fit was obtained for all experimental data. For the description of the depressed loops with their centers below the real axis, a constant phase element (CPE) was used and expressed as follows:

$$Z_{CPE} = \frac{1}{A(i\omega)^n} \quad (10)$$

where A (Ω^{-1} s^n cm^{-2}) is the magnitude of the CPE, ω (rad s^{-1}) is the sine wave modulation angular frequency, $i^2 = -1$ is the imaginary number and n is an empirical exponent which presents the difference from the ideal capacitive behaviour. Depending on the values of n, the CPE can represent resistance (n = 0), capacitance (n = 1), inductance (n = -1) and Warburg impedance (n = 0.5). The double layer capacitance C_{dl} and the relaxation time constant (τ) values were obtained using the equations (11) and (12), respectively.

$$C_{dl} = \sqrt[n]{A(R_{ct})^{1-n}} \quad (11)$$

$$\tau = C_{dl}R_{ct} \quad (12)$$

Inspection of the results in Table 3 indicated that R_{ct} value increases while the C_{dl} value decreases with inhibitors concentrations. The change in the above parameters can be related to the gradual removal of water molecules (with high dielectric constant) by triazepine carboxylate molecules (with low dielectric constant) on metallic surface leading to the decrease in the number of the active sites of the corrosion reaction.

Moreover, the decrease in the C_{dl} with inhibitor concentrations was perhaps due to diminish in local dielectric constant and/or an augment in the thickness of 'e' double layer. On the other hand, the thickness of the formed layer (e) was evaluated from the Helmholtz model according to the following equation (13):

$$C_{dl} = \frac{\varepsilon_0 \varepsilon}{e} S \quad (13)$$

where ε_0, ε, e and S are the permittivity of space, the local dielectric constant, the film thickness and the surface area, respectively. Therefore, the decrease in the values of C_{dl} results an increase in the values of the thickness (e). In addition, the stability of 'n' (in the range of 0.81-0.90) indicated that the charge transfer kinetic controls the dissolution mechanism of mild steel in 1.0 M HCl without and with all triazepine carboxylate compounds studieds.

However, it is renowned also that the performance of inhibitors depends on their size and active centers. Thus, the best performance of compound P11 as corrosion inhibitor over compound P10 and

Figure 3. Nyquist plots for mild steel in 1.0 M HCl solution in the absence and presence of various concentrations of P7, P8, P9, P10, P11 and P12 at 298±2 K: (scater) experimental; (red line) fitted data.

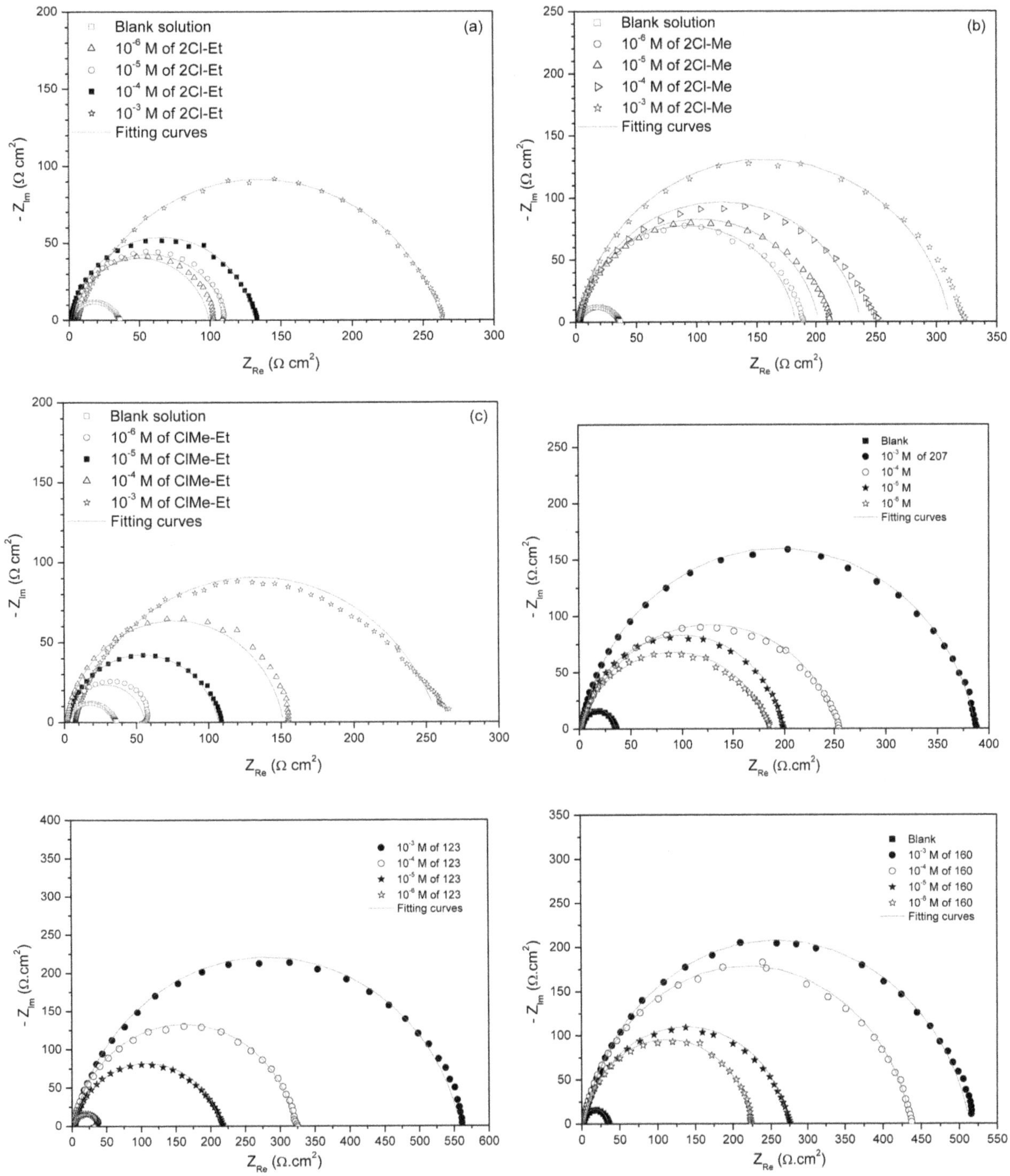

P11 may be attributed to the presence of nitrile (-C°N) cyano group and –(Cl) group. The nucleophilic (electron releasing) character of –Cl group is highest than the electrophonic (electron with drawing). So, P11 can produce protonated species in HCl by the reaction of the amino group with hydrochloride acid. This ammonium substituent may reduce the electronic density on the substituted triazepine carboxylate compounds. The same results were observed by other authors (Sun, 1998; El Faydy, 2018).

Corrosion Kinetic Study

In order to get information about the type of adsorption and the effectiveness of the Triazepine carboxylate compounds at higher temperature simulation of industrial process condition, polarization experiment was conducted in the range of 298 to 328±2 K without and with inhibitors. Electrochemical kinetic parameters such as corrosion potential (E_{corr}), corrosion current density (i_{corr}), anodic (β_a) and cathodic Tafel slope (β_c), determined from these experiments by extrapolation method, are presented in Figure 4.

Figure 4. Potentiodynamic polarization curves for mild steel in 1.0 M HCl with triazepine carboxylate compound P12 at various temperature.

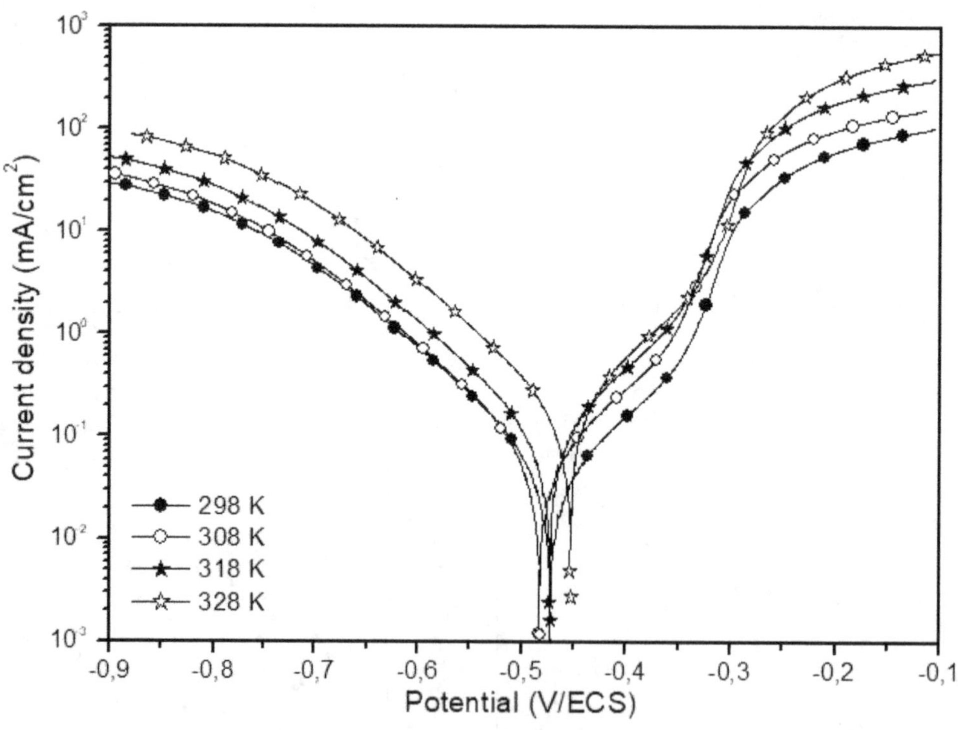

It is seen also that the triazepines carboxylate compounds is good inhibitors for the mild steel in 1.0 M HCl which the P11 carboxylate compounds present the best performance compared to others inhibitors studied. In addition, according to the substitution of each triazepine carboxylate compounds studied, it can be noted that the inhibition efficiency of inhibitors depends on the temperature and decreases with the rise of temperature from 298 to 328±2 K. This can be explained by the decrease of the strength of the adsorption process at elevated temperature and would suggest a physical adsorption mode.

It is seen also that the triazepine carboxylate compounds is good inhibitor for the mild steel in 1.0 M HCl, the presence of carboxylate compounds will provide a film barrier protecting steel surface from being exposed to acidic medium that can enhance the corrosion process. Increasing of $\eta\%$ will reduce the rate of corrosion process due to higher surface coverage *(θ)* from carboxylate compounds studied. The best inhibition efficiency reach is 94,6% at solution of 289 K with P11 triazepine carboxylate compounds. This is due to the concentration of inhibitor is the highest and can protect the mild steel surface from the aggressive HCl environment. At this condition, the surface coverage is the highest with 0.946 compare to the protection using others triazepine carboxylate compounds. Large surfaces areas of mild steel are covered by triazepine carboxylate compounds inhibitors film and this reducing the reaction sites available.

The adsorption isotherm type can provide additional information about the tested compounds properties. So, the coverage surface values (θ) can be easily determined from AC impedance, Tafel polarization or the linear polarization by the ratio η % /100. In the present study, the coverage surface values were evaluated from Tafel polarization. The Langmuir, Freundluich, Tempkin, Flory–Huggins and Frumkin, adsorption isotherms models considered were as described below:

$$\theta / _{1-\theta} = K_{ads} \times C_{inh} \quad (14)$$

$$\theta = K_{ads} \times C_{inh} \quad (15)$$

$$\exp\left(f \times \theta\right) = K_{ads} \times C_{inh} \quad (16)$$

$$\ln\left(\theta / _{C_{inh}}\right) = \ln K_{ads} + a \times \ln\left(1-\theta\right) \quad (17)$$

$$\left(\frac{\theta}{1-\theta}\right)\exp\left(-2 \times f \times \theta\right) = K_{ads} \times C_{inh} \quad (18)$$

where K_{ads} is the equilibrium constant of the adsorption process, C_{inh} is the inhibitor concentration, *f* is the factor of energetic inhomogeneity and the parameter '*a*' in Equation (17) is the water molecules number replaced by inhibitor molecules on metallic surface.

In this study, the Tempkin, Langmuir, Frumkin, Freundlich and Flory–Huggins isotherm were fitted. Their corresponding fitting curve results are shown in Table 4. It is noted that the inhibitor was adsorbed on metallic surface according to Langmuir isotherm model. This kind of isotherm involves

Table 4. Equilibrium constant and free energy of adsorption values for mild steel 1.0 M Hydrochloric acid in the presence of Triazepine carboxylate compounds at different concentration.

Isotherms	Linear Forms: Curves	Parameters	Triazepine Carboxylate Compounds					
			P7	P8	P9	P11	P12	P10
Langmuir	$\frac{C_{inh}}{\theta} = C_{inh} + 1/K_{ads}$	R^2	0.99989	0.99999	0.99963	0.99999	0.99995	0.99999
		K_{ads}(L mol^{-1})	147. 10^3	654. 10^3	66. 10^3	133×10^4	13×10^4	42×10^4
	$\frac{C_{inh}}{\theta} = f(C_{inh})$	Slope	1.14379	1.11229	1.13911	1.07228	1.05982	1,09689
Freundlich	$\ln\theta = \ln K_{ads} + \frac{1}{n}\ln C_{inh}$	R^2	0.96723	0.99521	0.93531	0.97755	0.79943	0.97681
		K_{ads}(L mol^{-1})	1.08	0.97	3.63	1.04	16.89	1.01
	$\ln\theta = f(\ln C_{inh})$	Slope (1/n)	0.03594	0.01232	0.18935	0.01525	0.34704	0.01723
		N	27.8	81.2	5.28	65.57	2.88	58.0
Temkin	$\theta = \frac{1}{f}\ln K_{ads} + \frac{1}{f}\ln C_{inh}$	R^2	0.95762	0.99397	09822	0.97806	0.84006	0.9743
		K_{ads}(L mol^{-1})	2.4 10^{16}	5.77 10^{39}	20.13 10^3	122.9×10^{30}	109.88×10^5	343×10^{26}
	$\theta = f(\ln C_{inh})$	Slope (1/*f*)	0.02749	0.0106	0.09007	0.01355	0.11622	0.01477
		f	36.4	94.3	11.1	73.80	8.60	67.7
Flory-Huggins	$\ln\left(\theta/C_{inh}\right) = \ln K_{ads} + a\times\ln(1-\theta)$	R^2	0.91635	0.98283	0.97889	0.98046	0.76608	0.95328
	$\ln\left(\theta/C_{inh}\right) = f(\ln(1-\theta))$	K_{ads}(L mol^{-1})	25.3 10^7	8.88 10^{14}	43.67 10^4	56.17×10^{10}	15.17×10^4	25.0×10^{10}
		Slope (a)	6.40134	12.2556	3.17271	7.36569	1.37621	8.24258
Frumkin	$\ln\left(C_{inh}\times\frac{1-\theta}{\theta}\right) = \ln K_{ads} - 2\times f(\theta)$	R^2	0.93312	0.99195	0.95097	0.97112	0.95993	0.96392
		K_{ads}(L mol^{-1})	8.16 10^{13}	7.22 10^{36}	13.5 10^5	50.4×10^{27}	92.13×10^3	72.06×10^{24}
	$\ln\left(C_{inh}\times\frac{1-\theta}{\theta}\right) = f(\theta)$	Slope (-2*f*)	27.28607	84.66614	5.91425	-60.0467	0.20589	-55.65795
		f	-13.6	-42.3	-2.9	30.02	-0.10	27.82

the assumption of no interaction between the adsorbed species on the electrode surface according to Elayyachy (2006). So, the adsorption constant, K_{ads}, is related to the free energy of adsorption, ΔG^*_{ads}, by the following equation (19):

$$K_{ads} = \frac{1}{55,55}\exp\left(-\frac{\Delta G^{\circ}_{ads}}{RT}\right) \tag{19}$$

where 55.55 value represents the water concentration in solution by mol L^{-1}, R is the universal gas constant and T is the absolute temperature.

The values of K_{ads}, R^2, ΔG^*_{ads} and other parameters are calculated. The high obtained value of K_{ads} reñects the high adsorption ability of this inhibitor on the mild steel surface. In addition, the negative value of the free energy of adsorption, ΔG^*_{ads}, indicates the strong interaction between inhibitor molecules and the mild steel surface according to (Avci, 2008).

Thermodynamic Parameters of the Activation Corrosion Process

In order to gain more information about the adsorption type of Triazepine carboxylates compounds and its effectiveness at higher temperatures, potentiodynamic polarization curves for mild steel in 1.0 M

HCl without and with various concentration of Triazepine carboxylates compounds were used in the temperatures range from 298 to 328±2 K. The activation parameters for the corrosion reaction can be regarded as an Arrhenius-type process, according to the following equation:

$$i_{corr} = A_r \exp\left(-E_a / RT\right) \tag{20}$$

where (E_a) is the apparent activation corrosion energy, R is the universal gas constant, and A_r is the Arrhenius pre-exponential factor. The apparent activation energies (E_a) in the absence and in the presence of Triazepine carboxylates compounds for mild steel are calculated by linear regression between ln(j_{corr}) and *1/T* (*Figure 5*), and the results are given in *Table 4*. All the linear regression coefficients are close to 1, indicating that the steel corrosion in hydrochloric acid can be elucidated using the kinetic model.

A corresponding increase in the corrosion rate occurs because of the greater area of metal that is consequently exposed to the acid environment according to (Szauer, 1981). The enthalpy of activation (ΔH_a) and the entropy of activation (ΔS_a) for the intermediate complex in the transition state for the corrosion of mild steel in 1.0 M HCl in the absence and in the presence of Triazepine carboxylates compounds were obtained by applying the alternative formulation of Arrhenius equation below:

$$i_{corr} = \left(\frac{RT}{Nh}\right)\exp\left(\frac{\Delta S_a}{R}\right)\exp\left(-\frac{\Delta H_a}{RT}\right) \tag{21}$$

where h is the Plank's constant and N is the Avogadro's number. (*Figure 6*) shows a plot of ln(j_{corr}/T) versus $1/T$. A straight line are obtained with a slope of ($-\Delta H_a/R$) and an intercept of (ln $R/Nh + \Delta S_a/R$) from which the values of ΔH_a and ΔS_a were calculated (*Table 3*). The positive values of ΔH_a in the absence and the presence of Triazepine carboxylates compounds reflect the endothermic nature of the steels dissolution process. One can also notice that E_a and ΔH_a values vary in the same way as shown in *Table 3*, indicating that the corrosion process is a unimolecular reaction. This result permits verifying the known thermodynamic equation between the E_a and ΔH_a.

$$E_a = \Delta H_a - T\Delta S_a \tag{22}$$

Therefore, when all of the details are worked out one ends up with an expression that again takes the form of an Arrhenius exponential multiplied by a slowly varying function of T. The precise form of the temperature dependence depends upon the reaction, and can be calculated using formulas from statistical mechanics involving the partition functions of the reactants and of the activated complex. Nevertheless, in order to carry simple calculations, Eq. (9) was rearranged to become:

$$\ln\left(\frac{i_{corr}}{T}\right) = \left(\frac{-\Delta H_a}{R}\right)\left(\frac{1}{T}\right) + \left[\ln\left(\frac{R}{Nh}\right) + \left(\frac{\Delta S_a}{R}\right)\right] \tag{23}$$

A plot of *ln* (i_{corr}/T) against *1000/T* should give a straight line with a slope of ($-\Delta H_a/R$) and intercept of [ln *(R/Nh)* + ($\Delta S_a/R$)], as shown in *Figure 6*. ΔH_a and ΔS_a were calculated and tabulated in *Table 3*.

Figure 5. Arrhenius plots of mild steel in 1.0 M HCl without and with performant triazepine carboxylate inhibitors studied.

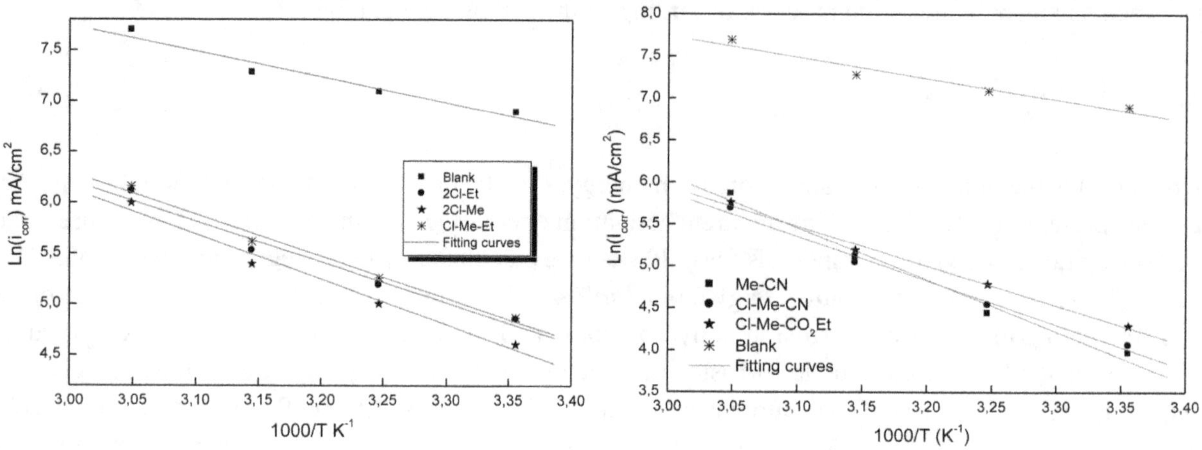

The activation energy E_a increases in the presence of the inhibitor (the same patterns was observed for all triazepine carboxylate compounds studied). Increases in E_a with the presence of triazepine carboxylate compounds indicate that a physical (electrostatic) adsorption occurred in the first stage. The increase in E_a may be resulted either due to physical adsorption that occurs in first stage (Noor, 2008) or due to decrease in the adsorption of inhibitor molecules on the steel surface with increase in temperature according to (Alaoui, 2016). The E_a value for all inhibitors studied was greater than 20 kJ mol^{-1} in both the presence and absence of the Triazepine carboxylates compounds, which reveals that the entire process is controlled by the surface reaction.

Figure 6. Transition Arrhenius plots for mild steel in 1.0 M HCl solution without and with performant triazepine carboxylate inhibitors studied.

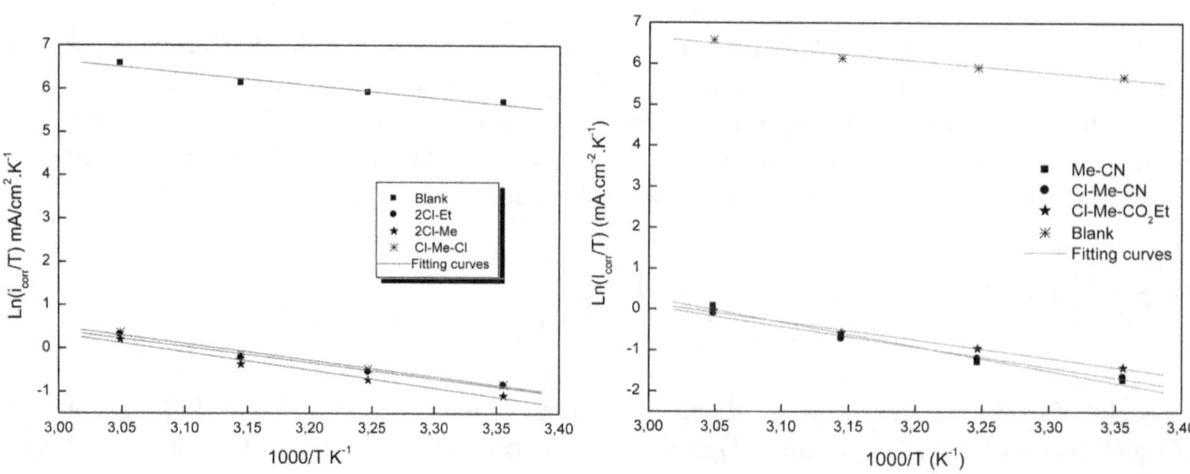

The values of ΔH_a^* are nearly the same and are higher in the presence of the inhibitor. This indicates that the energy barrier of the corrosion reaction increased in the presence of the inhibitor without changing the mechanism of dissolution.

The positive values of ΔH_a^* for both corrosion processes with and without the inhibitor reveal the endothermic nature of the steel dissolution process and indicate that the dissolution of steel is difficult according to Behpour, 2008).

According to the recorded values of enthalpy of activation (Table 3), exothermic adsorption behaviour were detected depending on applied temperatures. The value of enthalpy of activation $\left(\Delta H_{ads}^*\right)$ is listed in table 3.

The large negative value of $\left(\Delta S_a^*\right)$ for mild steel in 1.0 M HCl implies that the activated complex is the rate-determining step, rather than the dissociation step. In the presence of the triazepine carboxylate inhibitors, the value of $\left(\Delta S_a^*\right)$ increases and is generally interpreted as an increase in disorder as the reactants are converted to the activated complexes. The values of activation entropy $\left(\Delta S_a^*\right)$ are higher for inhibited solutions than that for the uninhibited solution (*Table 3*). The positive increment of $\left(\Delta S_a^*\right)$ suggests that an increase in randomness occurred on going from reactants to the activated complex. This observation is in agreement with the findings of other workers.

Monte Carlo Simulation

Monte Carlo simulation was employed in our study to explore the interaction between the investigated inhibitor species and metal was studied at atomic level according to (ElBelghiti, 2016). It is apparent from inspection of this figure that the studied compounds are located very close to the metal surface. The Molecular Dynamic parameters including the adsorption energy (dEad/dNi) are given in Table 5. The dEad/dNi are large negative values, this reflects the spontaneity of the adsorption process. It is also noted that P8 gains the largest negative magnitude of adsorption suggesting that P8 strongly adsorb on metal surface. P7 comes next in the order of interaction energy on metal surface, this due to the steric hindrance of the (–Et) groups and the lower donating ability of (–Et) groups in P7 compared to its counterpart Me in P8. Unlike the results found in literature which report on the higher adsorption ability of inhibitors in aqueous phases than in vacuum according to (Sasikumar, 2015; Zhang, 2004), this is not the case in our investigation, we find that in aqueous phase the values of adsorption energies are lower than those computed in vacuum, This probably suggests that the presence of water molecules shielded the active sites (heteroatoms, especially Oxygen) in inhibitors through dipole-dipole interaction or formation of hydrogen bond, and this to some extend hinder the inhibitors to take proper position on metal surface, i.e., lower binding ability unlike the vacuum, where the only interaction in inhibitors with metal surface according to (Zhang, 2004). The values of interaction energies (dEad/dNi) in both vacuum and aqueous phase decreases in the same order as follows: P8 > P7 > P9, which are reproducible with order of the %IE calculated from the practical experiments.

It was observed that the heteroatoms of the compounds adsorbed on the iron at first. Then, the remaining centers of the compounds have moved gradually close to the metal surface. Therefore, as seen in Fig. (7), the studied molecules *P11*, and *P12* adsorbed on the Fe surfaces with an almost flat orientation,

Figure 7. Most stable molecule dynamic orientation of inhibitors on Fe (110) in 1.0 HCl simulated: P7, P8, P9, P10, P11 and P12

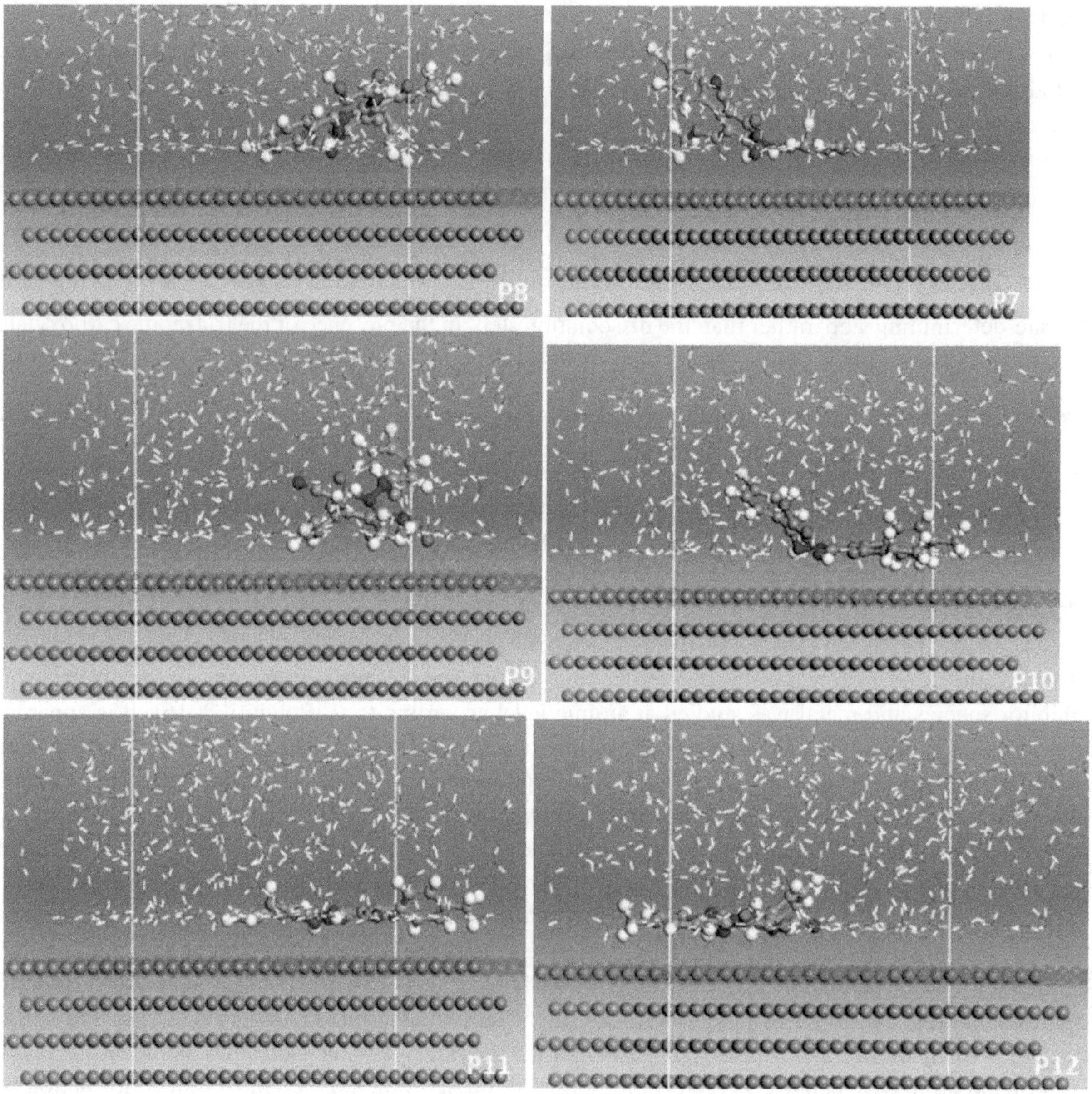

where compound *P10* is not fully oriented in flat position on metal surface. The parallel configuration as in compound *P11* and *P12* supports the maximum surface coverage and the higher binding interaction.

The values of (E_{ads}) are large and negative implies that the binding between the investigated inhibitors and the metal surface is spontaneous, and the more negative values of the adsorption energy can be attributed to the stable and strong interaction of the inhibitor on the metal surface (Saha, 2016). It is observed in Table 5 that the binding energy of *P11* on the iron surface is higher than the energy determined for H_3O^+ and Cl^-. This supports the inhibitors are stable on the iron surface than other competing

Table 5. List of molecular dynamic (MD) parameters from Monte Carlo simulations for the investigate compounds on Fe (110)

MD Parameter	P7	P8	P9	P10	P11	P12
Total energy	-362.81	-370.49	-342.03	-4269.70	-4341.42	-4278.84
Adsorption energy	-236.11	-237.49	-227.80	-4516.58	-4494.50	-4419.44
Rigid adsorption energy	-255.76	-256.26	-246.40	-4322.61	-4411.15	-4349.13
Deformation energy	19.66	18.77	18.60	-193.96	-83.36	-70.31
Compound: dE_{ad}/dN_i	-236.11	-237.49	-227.80	-226.23	-229.13	-200.18
H_2O: dE_{ad}/dN_i	--	--	--	-12.62	-13.80	-13.97
H_3O^+: dE_{ad}/dN_i	--	--	--	-123.05	-149.78	-134.55
Cl^-: dE_{ad}/dN_i	--	--	--	-153.05	-134.27	-108.51

species. It has been reported that the organic inhibitors with unoccupied orbitals promote the formation of a chelate on the metal surface by accepting electrons from a *d*-orbital of the metal during such strong adsorption process (Singh, 2010). Therefore, the compound *P10* and *P11* can form a stable chelate with the iron after strong adsorption, and this leads to a good corrosion inhibition efficiency.

Pair Correlation Function

The interaction of the N atoms of compounds in the model system with Fe (110) was investigated through the structural analysis of the MD simulation results in order to uncover the mechanistic of binding process of inhibitor on metal surface at the atomic level according to (Fouda, 2017). Pair Correlation Functions (PCFs), g(r), are derived from the output of Monte Carlo simulation trajectories according to (Fouda, 2017). In general, the peaks of PCF curves within 3.5 Å is caused by the chemical bonds, and that appear outside 3.5 Å is from the non-bonding interaction such as Coulomb and Van der Waals (VDW) interactions. The interaction models of investigated compounds on Fe (110) are performed in aqueous and vacuum model systems are shown in Figure 7. It is obvious from these Figure 8 to 13 that the interaction of hetero atoms incorporated in the molecular structure of investigated compounds occur almost at distance no more than 3.5 Å, this suggests that these atoms contribute in the chemical bonding with metal surface, while the interaction with average distance outside 3.5 Å for some regions on the molecular structure of compounds, this suggests that such portions of compounds interact with metal surface via electrostatic interaction.

CONCLUSION

Triazepine carboxylate compounds reviewed are acting as good corrosion inhibitors in acidic media and their inhibition efficiency increases with increasing concentration, and best performance is for (P11) (P11) compounds, namely "ethyl 9-(4-chlorophenyl)-8-cyano-3-hydroxy-2,7-dioxo-4-(p-tolyl)-1,2,5,7-tetrahydropyrido[1,2-b]triazepine-10-carboxylate". The inhibition efficiency trend of inhibitors obtained from electrochemical and molecular dynamic simulation are in good agreement. All the inhibitors follow the Langmuir adsorption isotherm with slopes of almost unity, monolayer interaction between the adsorbed

Figure 8. PCFs of interaction of heteroatoms in P7 with Fe (110) in Aqueous phase.

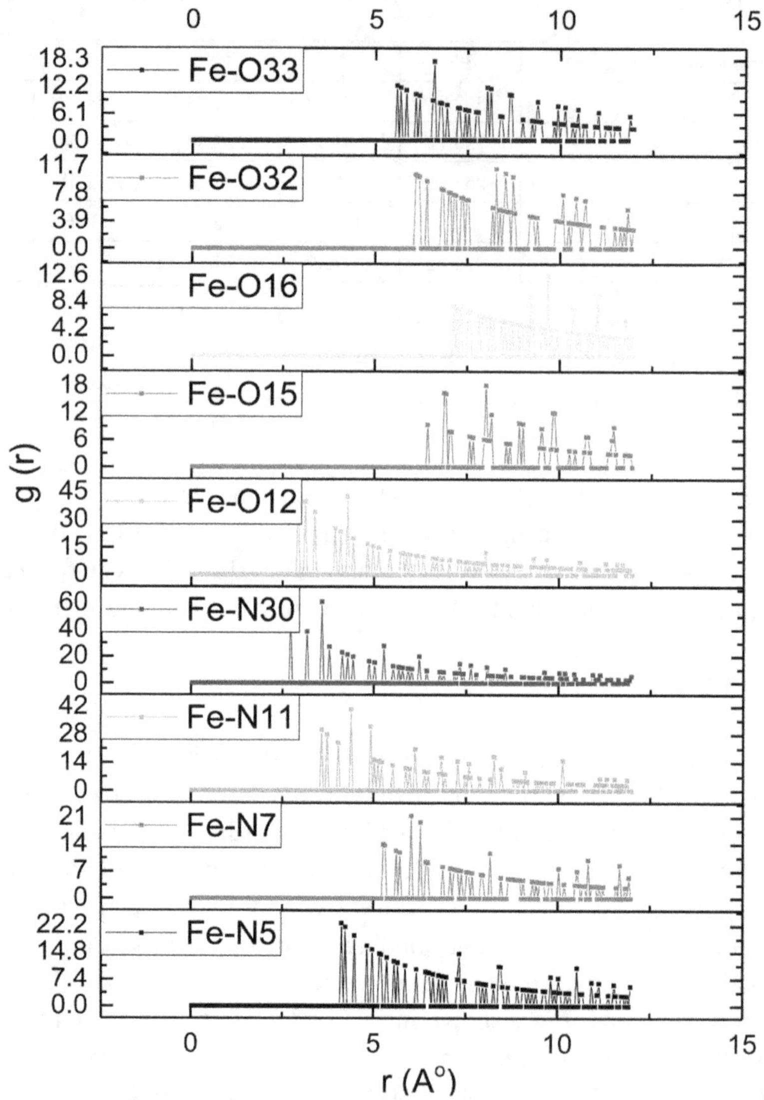

species. All the Triazepine inhibitors can be adsorbed on metals surface by chemical mechanism and also the spontaneity of the adsorption process. Molecular dynamic simulations supported the experimental results that showed P7, P8, P9, P10, P11 and P12 as efficient corrosion inhibitors. On the other hand, it could be interesting also to study the effect of these compounds on the corrosion of another metal in another acidic medium such as H_2SO_4 and H_3PO_4. This study confirmed the performance and the other application of the Triazepine carboxylate compounds.

Figure 9. PCFs of interaction of heteroatoms in P8 with Fe (110) in Aqueous phase.

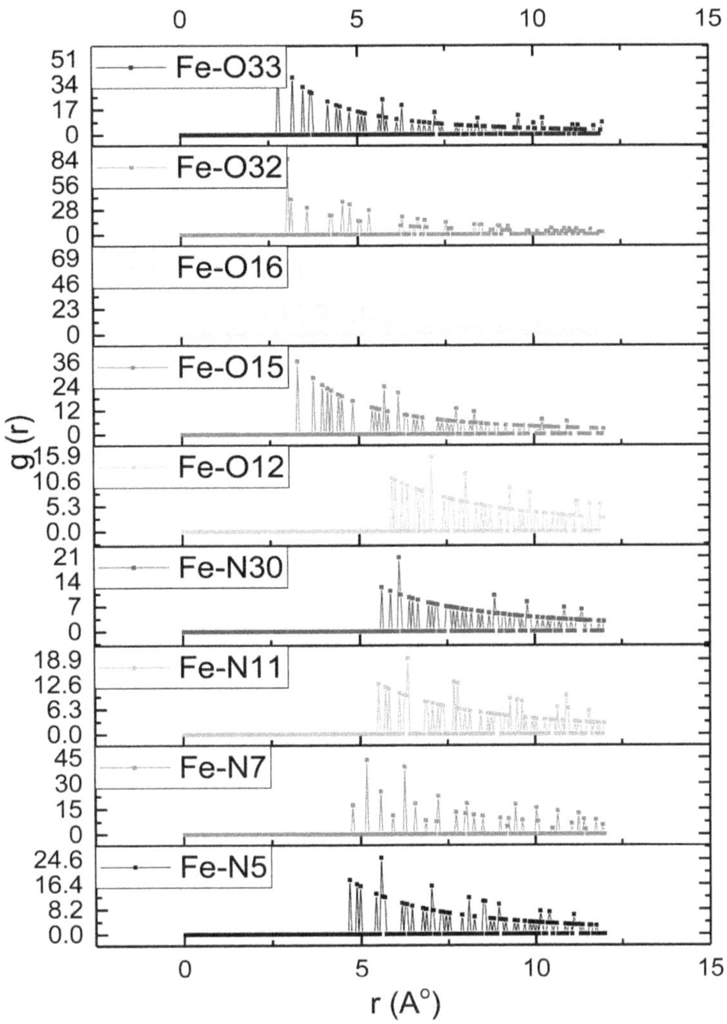

Figure 10. PCFs of interaction of heteroatoms in P9 with Fe (110) in Aqueous phase.

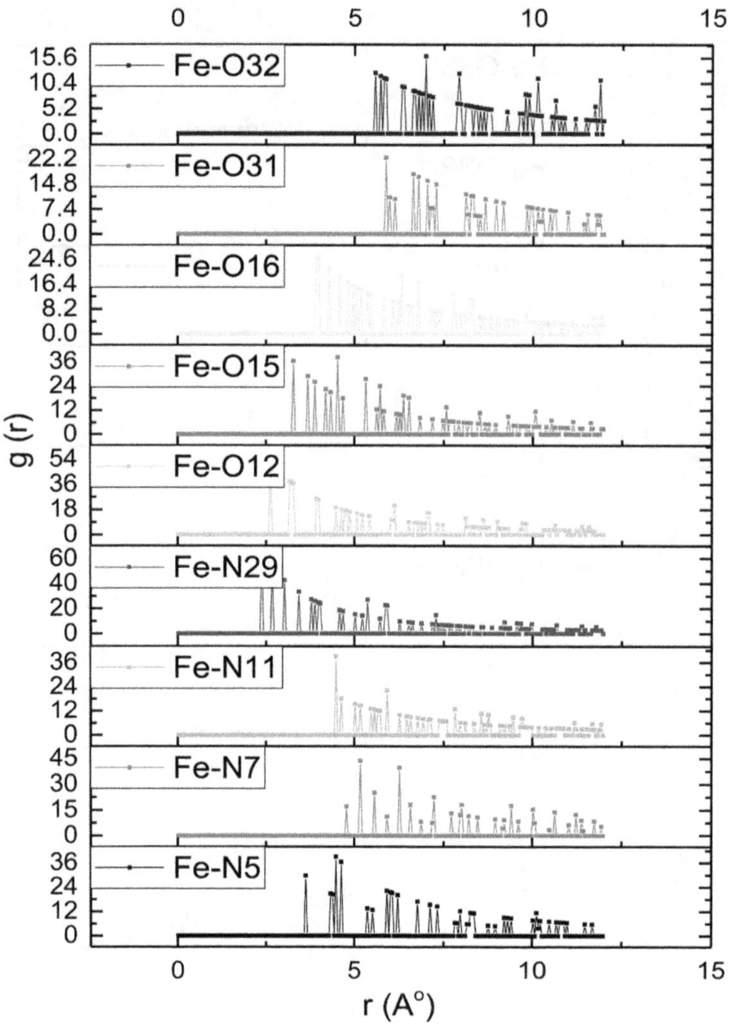

Figure 11. PCFs of interaction of heteroatoms in P10 with Fe (110) in Aqueous phase.

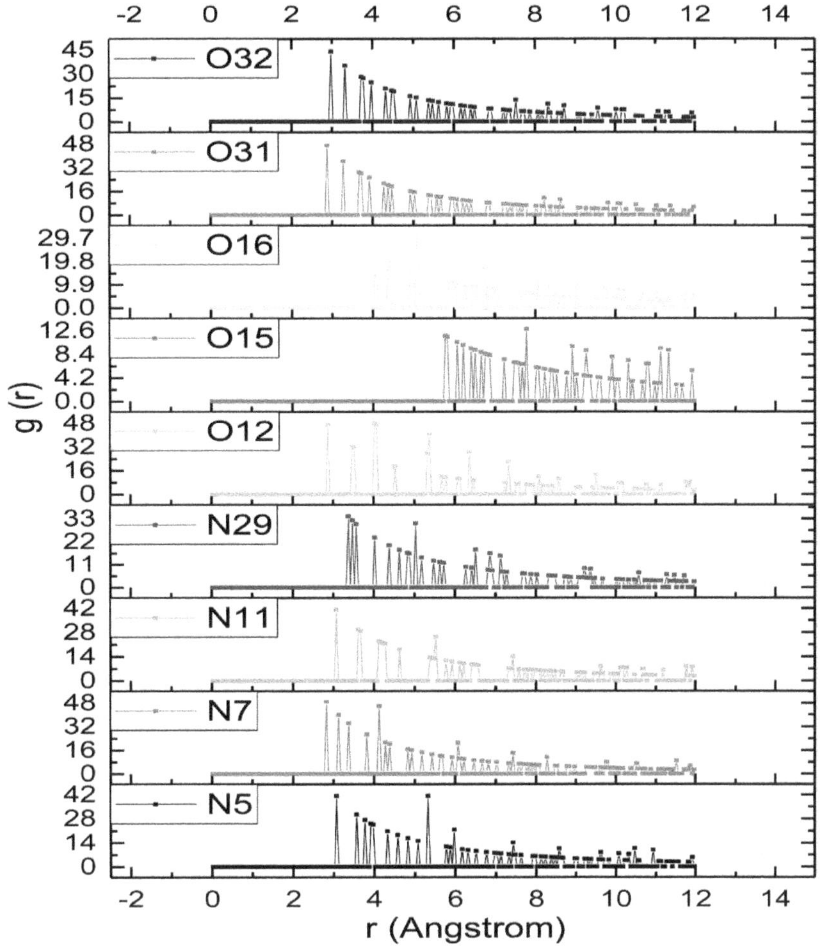

Figure 12. PCFs of interaction of heteroatoms in P11 with Fe (110) in Aqueous phase.

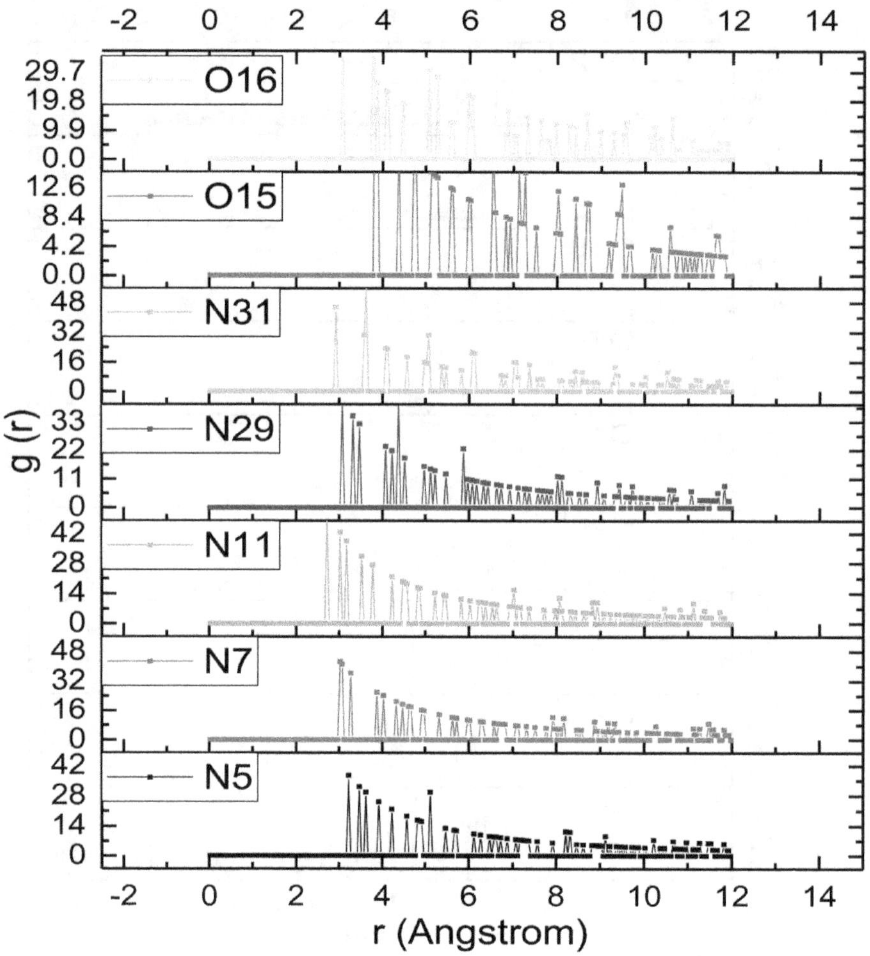

Figure 13. PCFs of interaction of heteroatoms in P12 with Fe (110) in Aqueous phase.

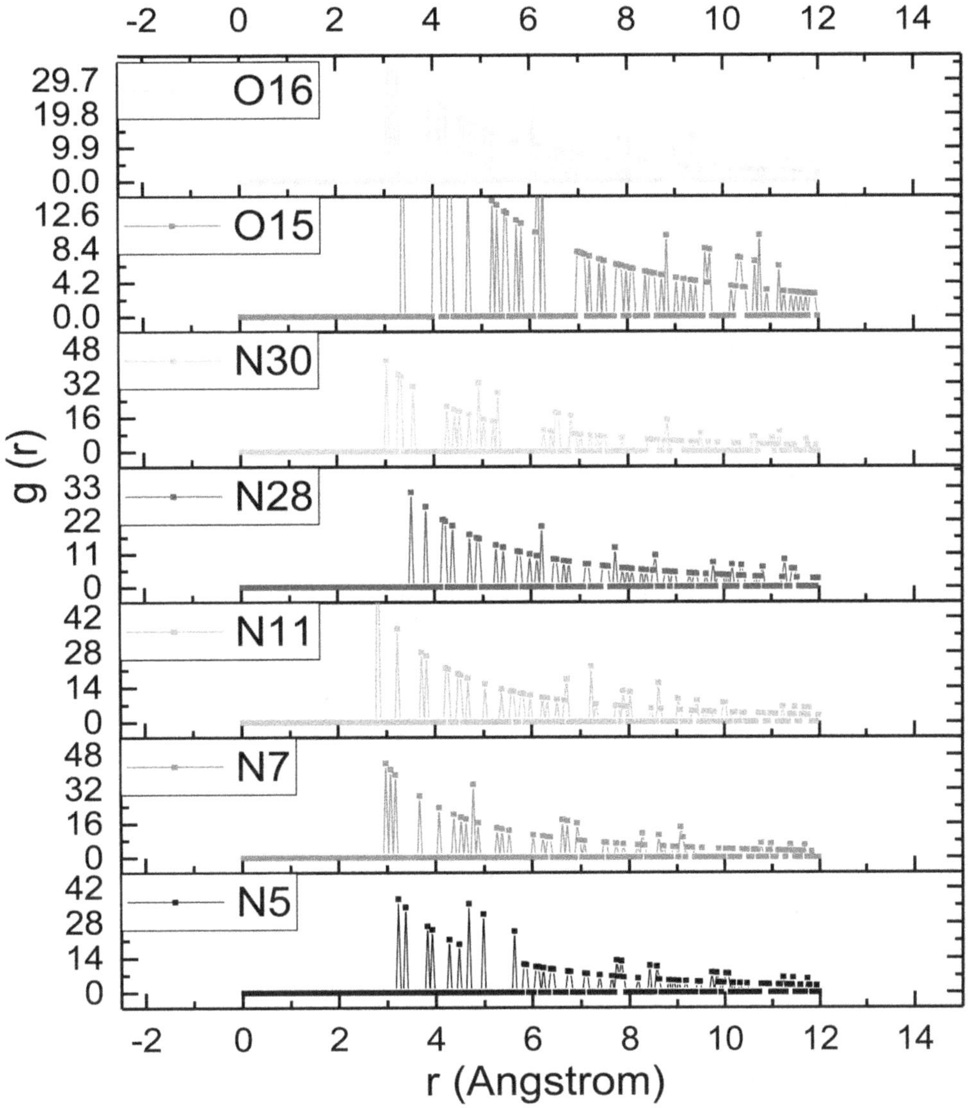

REFERENCES

Alaoui, K., El Kacimi, Y., Galai, M., Dahmani, K., Touir, R., El Harfi, A., & Ebn Touhami, M. (2016). Poly (1-phenylethene): As a Novel Corrosion Inhibitor for Carbon Steel / Hydrochloric Acid Interface. *Anal. Bioanal. Electrochem*, *7*, 830.

Alaoui, K., El Kacimi, Y., Galai, M., Serrar, H., Touir, R., Kaya, S., Kaya, C., & Ebn Touhami, M. (2020). New triazepine carboxylate derivatives: Correlation between corrosion inhibition property and chemical structure. *Int J Ind Chem*, *11*(1), 23–42. doi:10.100740090-019-00199-5

Alaoui, K., Ouakki, M., Abousalem, A. S., Serrar, H., Galai, M., Derbali, S., Nouneh, K., Boukhris, S., Touhami, M. E., & El Kacimi, Y. (2019). Molecular Dynamics, Monte-Carlo Simulations and Atomic Force Microscopy to Study the Interfacial Adsorption Behaviour of Some Triazepine Carboxylate Compounds as Corrosion Inhibitors in Acid Medium. *J Bio Tribo Corros*, *5*(1), 1. doi:10.100740735-018-0196-2

Alaoui, K., Ouakki, M., Abousalem, A. S., Serrar, H., Touhami, M. E., & Kacimi, Y. E. (2020). Chemical, Electrochemical, Theoretical and Surface Morphology Studies of Triazepine Carboxylate Compounds as Corrosion Inhibitors for Mild Steel in Hydrochloric Acid Medium. *J Bio Tribo Corros*, *6*(2), 54. doi:10.100740735-020-00352-z

Alaoui, K., Touir, R., Galai, M., Serrar, H., Ouakki, M., Kaya, S., Tüzün, B., Boukhris, S., Ebn Touhami, M., & El Kacimi, Y. (2018). Electrochemical and Computational Studies of Some Triazepine Carboxylate Compounds as Acid Corrosion Inhibitors for Mild Steel. *J Bio Tribo Corros*, *4*(3), 37. doi:10.100740735-018-0154-z

Alaoui, K., Touir, R., Galai, M., Serrar, H., Ouakki, M., Kaya, S., Tüzün, B., Boukhris, S., Ebn Touhami, M., & El Kacimi, Y. (2018), Electrochemical and Computational Studies of Some Triazepine Carboxylate Compounds as Acid Corrosion Inhibitors for Mild Steel. *Journal of Bio- and Tribo-Corrosion, 4*, 37.

Avci, G. (2008). Inhibitor effect of N,N¢-methylenediacrylamide on corrosion behavior of mild steel in 0.5 M HCl. *Materials Chemistry and Physics*, *112*(1), 234–238. doi:10.1016/j.matchemphys.2008.05.036

Behpour, M., Ghoreishi, S. M., Soltani, N., Salavati, N. M., Hamadanian, M., & Gandomi, A. (2008). Electrochemical and Theoretical Investigation on the Corrosion Inhibition of Mild Steel by Thiosalicylaldehyde Derivatives in Hydrochloric Acid Solution. *Corrosion Science*, *50*(8), 2172–2181. doi:10.1016/j.corsci.2008.06.020

Benhiba, F., Serrar, H., Hsissou, R., Guenbour, A., Bellaouchou, A., Tabyaoui, M., Boukhris, S., Oudda, H., Warad, I., & Zarrouk, A. (2020). Tetrahydropyrimido-Triazepine derivatives as anti-corrosion additives for acid corrosion: Chemical, electrochemical, surface and theoretical studies. *Chemical Physics Letters*, *743*, 137181. doi:10.1016/j.cplett.2020.137181

Bentiss, F., Lebrini, M., & Lagrenee, M. (2005). Thermodynamic characterization of metal dissolution and inhibitor adsorption processes in mild steel/2,5-bis(n-thienyl)-1,3,4-thiadiazoles/hydrochloric acid system. *Corrosion Science*, *47*(12), 2915–2931. doi:10.1016/j.corsci.2005.05.034

El Bakri, Y., El Aoufir, Y., Bourazmi, H., Harmaoui, A., Sebhaoui, J., Tabyaoui, M., Guenbour, A., Oudda, H., Lgaz, H., El Hajjaji, F., Ben Ali, A., Ramli, Y., & Essassi, E. M. (2017). Corrosion control of carbon steel in phosphoric acid by 6-methyl-7H-1,2,4-triazolo[4,3-b][1,2,4]-triazepine-8(9H)-thione: Electrochemical studies. *Journal of Materials and Environmental Science*, *8*(8), 2657–2666.

El Bakri, Y., Lei, G., El Hassane, A., & El Mokhtar, E. (2019). Electrochemical, DFT and MD simulation of newly synthesized triazolotriazepine derivatives as corrosion inhibitors for carbon steel in 1 M HCl. *Journal of Molecular Liquids*, *274*, 759–769. doi:10.1016/j.molliq.2018.11.048

El Faydy, M., Galai, M., Rbaa, M., Ouakki, M., Lakhrissi, B., Ebn Touhami, M., & El Kacimi, Y. (2018). Synthesis and Application of New Quinoline as Hydrochloric Acid Corrosion Inhibitor of Carbon Steel. *Anal. Bioanal. Electrochem*, *10*(7), 815–839.

El Kacimi, Y., Galai, M., Alaoui, K., Touir, R., & Ebn Touhami, M. (2018). Surface Morphology Studies and Kinetic-thermodynamic characterization of Steels treated in 5.0 M HCl medium: Hot-Dip Galvanizing application. *Anti-Corrosion Methods and Materials*, *65*(2), 176–189. doi:10.1108/ACMM-11-2016-1735

Elattar, K. M., Abozeid, M. A., Mousab, I. A., & El-Mekabatya, A. (2015). Advances in 1,2,4-triazepines chemistry. *RSC Advances*, *5*(129), 106710–106753. doi:10.1039/C5RA21108E

ElBelghiti, M., Karzazi, Y., Dafali, A., Hammouti, B., Bentiss, F., Obot, I. B., Bahadur, I., & Ebenso, E. E. (2016). Experimental, quantum chemical and Monte Carlo simulation studies of 3,5-disubstituted-4-amino-1,2,4-triazoles as corrosion inhibitors on mild steel in acidic mediumJ. *Mol. Liq.*, *218*, 281–293. doi:10.1016/j.molliq.2016.01.076

Fouda, A. S., Ismail, M. A., EL-ewady, G. Y., & Abousalem, A. S. (2017). Evaluation of 4-amidinophenyl-2, 2¢-bithiophene and its aza-analogue as novel corrosion inhibitors for CS in acidic media: Experimental and theoretical study. *Journal of Molecular Liquids*, *240*, 372–388. doi:10.1016/j.molliq.2017.05.089

Fouda, M. A., Ismail, M. A., EL-ewady, G. Y., & Abousalem, A. S. (2017). Ismail, G.Y. Elewady, A.S. Abousalem, Evaluation of 4-amidinophenyl-2,2?-bithiophene and its aza-analogue as novel corrosion inhibitors for CS in acidic media: Experimental and theoretical study. *Journal of Molecular Liquids*, *240*, 372–388. doi:10.1016/j.molliq.2017.05.089

Jiang, Z., Wang, J., Hu, Q., & Huang, S. (1995). The influence of 1-(2-pyridylazo)-2-naphthol (PAN) on the corrosion of titanium in 10 N sulfuric acid solution. *Corrosion Science*, *37*(8), 1245–1252. doi:10.1016/0010-938X(95)00029-J

Noor, E. A., & Al-Moubaraki, A. H. (2008). Thermodynamic study of metal corrosion and inhibitor adsorption processes in mild steel/1-methyl-4[4¢(-X)-styryl pyridinium iodides/hydrochloric acid systems. *Materials Chemistry and Physics*, *110*(1), 145–154. doi:10.1016/j.matchemphys.2008.01.028

Obot, I. B., Obi-Egbedi, N. O., Ebenso, E. E., Afolabi, A. S., & Oguzie, E. E. (2013). Experimental, quantum chemical calculations, and molecular dynamic simulations insight into the corrosion inhibition properties of 2-(6-methylpyridin-2-yl) oxazolo [5, 4-f][1, 10] phenanthroline on mild steel. *Research on Chemical Intermediates*, *39*(5), 1927–1948. doi:10.100711164-012-0726-3

Oguzie, E. E., Okolue, B. N., Ebenso, E. E., Onuoha, G. N., & Onuchukwu, A. I. (2004). Evaluation of the inhibitory effect of methylene blue dye on the corrosion of aluminium in hydrochloric acid. *Materials Chemistry and Physics*, *87*(2-3), 394–401. doi:10.1016/j.matchemphys.2004.06.003

Popova, A., Christov, M., & Vasilev, A. (2007). Inhibitive properties of quaternary ammonium bromides of N-containing heterocycles on acid mild steel corrosion. Part II: EIS results. *Corrosion Science*, *49*(8), 3290–3302. doi:10.1016/j.corsci.2007.03.012

Popova, A., Christov, M., & Zwetanova, A. (2007). Effect of the molecular structure on the inhibitor properties of azoles on mild steel corrosion in 1 M hydrochloric acid. *Corrosion Science*, *49*(5), 2131–2143. doi:10.1016/j.corsci.2006.10.021

Saha, S. K., Hens, A., Murmu, N. C., & Banerjee, P. (2016). A comparative density functional theory and molecular dynamics simulation studies of the corrosion inhibitory action of two novel N-heterocyclic organic compounds along with a few others over steel surface. *Journal of Molecular Liquids*, *215*, 486–495. doi:10.1016/j.molliq.2016.01.024

Salarvand, Z., Amirnasr, M., Talebian, M., Raeissi, K., & Meghdadi, S. (2017). Enhanced corrosion resistance of mild steel in 1 M HCl solution by trace amount of 2-phenyl-benzothiazole derivatives: Experimental, quantum chemical calculations and molecular dynamics simulation studies. *Corrosion Science*, *114*, 133–145. doi:10.1016/j.corsci.2016.11.002

Sasikumar, Y., Adekunle, A. S., Olasunkanmi, L. O., Bahadur, I., Baskar, R., Kabanda, M. M., Obot, I. B., & Ebenso, E. E. (2015). Experimental, quantum chemical and Monte Carlo simulation studies on the corrosion inhibition of some alkyl imidazolium ionic liquids containing tetrafluoroborate anion on mild steel in acidic medium. *Journal of Molecular Liquids*, *211*, 105–118. doi:10.1016/j.molliq.2015.06.052

Serrar, H., Marmouzi, I., Benzekri, Z., Boukhris, S., Hassikou, A., Faouzi, M. E. A., & Souizi, A. (2017). Synthesis and Evaluation of Novel Pyrido[1,2-b][1,2,4]triazine-2,6-dione and Pyrido[1,2-b][1,2,4] triazepine-2,7-dione Derivatives as Antioxidant Agents. *Letters in Organic Chemistry*, *14*(4), 267–277. doi:10.2174/1570178614666170227124523

Singh, A. K., & Quraishi, M. A. (2010). The effect of some bis-thiadiazole derivatives on the corrosion of mild steel in hydrochloric acid. *Corrosion Science*, *52*(4), 1373–1385. doi:10.1016/j.corsci.2010.01.007

Sun, H., Ren, P., & Fried, J. R. (1998). The COMPASS force field: Parameterization and validation for phosphazenes. *Computational and Theoretical Polymer Science*, *8*(1-2), 229–246. doi:10.1016/S1089-3156(98)00042-7

Szauer, T., & Brandt, A. (1981). Adsorption of oleates of various amines on iron in acidic solution. *Electrochimica Acta*, *26*, 1257. doi:10.1016/0013-4686(81)85108-0

Tang, L., Lia, X., Lia, L., Qua, Q., Mua, G., & Liua, G. (2005). The effect of 1-(2-pyridylazo)-2-naphthol on the corrosion of cold rolled steel in acid media: Part 1: Inhibitive action in 1.0M hydrochloric acid. *Materials Chemistry and Physics*, *94*(2-3), 353–359. doi:10.1016/j.matchemphys.2005.05.015

Tang, Y., Yang, X., Yang, W., Wan, R., Chen, Y., & Yin, X. (2010). A preliminary investigation of corrosion inhibition of mild steel in 0.5 M H2SO4 by 2-amino-5-(n-pyridyl)-1, 3, 4-thiadiazole: Polarization, EIS and molecular dynamics simulations. *Corrosion Science*, *52*(5), 1801–1808. doi:10.1016/j.corsci.2010.01.028

Zhang, D., Gao, L., & Zhou, G. (2004). Inhibition of copper corrosion in aerated hydrochloric acid solution by heterocyclic compounds containing a mercapto group. *Corrosion Science*, *46*(12), 3031–3040. doi:10.1016/j.corsci.2004.04.012

Chapter 8
Corrosion Inhibition Study of Mild Steel in the Acidic Mediums Containing a Mineral Compound-Based Phosphate

Moussa Ouakki
https://orcid.org/0000-0002-6265-4734
Laboratory of Materials, Electrochemistry, and Environment, Faculty of Science, Ibn Tofail University, Morocco

Mouhsine Galai
Laboratory of Materials Engineering and Environment: Modeling and Application, Faculty of Science, Ibn Tofail University, Morocco

Mohammed Cherkaoui
Laboratory of Materials, Electrochemistry, and Environment, Faculty of Science, Ibn Tofail University, Morocco

Mohamed Ebn Touhami
Laboratory of Materials Engineering and Environment: Modeling and Application, Faculty of Science, Ibn Tofail University, Morocco

E. H. Rifi
Laboratory of Organic Synthesis and Extraction Processes, Faculty of Science, Ibn Tofail University, Morocco

Z. Hatim
Laboratoire d'Electrochimie et Traitement de Surface, Université Chouaib Doukkali, Morocco

ABSTRACT

The corrosion inhibition of mild steel in hydrochloric, sulfuric, and phosphoric acids solutions containing a mineral compound-based phosphate (apatite) was evaluated by potentiodynamic polarization and

DOI: 10.4018/978-1-7998-2775-7.ch008

electrochemical impedance spectroscopy. Potentio-dynamic polarization measurements reveal that the inhibition efficiency increased with the concentration of the apatite, which appears to be a cathodic type inhibitor in the three mediums. Electrochemical impedance spectroscopy confirms this result; indeed, the transfer resistance increases with apatite concentration. The authors note that the double layer capacitance decreases simultaneously suggesting the formation of an adsorbed layer on the mild steel surface. The inhibition mechanism has been elucidated by a thermodynamic study, which showed that the film was formed by physi-sorption. The adsorption model obeys to the Langmuir adsorption isotherm. The parameters of activation energy were evaluated and discussed.

1. INTRODUCTION

The use of acids is widely applied in the industrial field. However, acidic aqueous solutions generally constitute aggressive media for metals and alloys. To overcome this undesirable phenomenon, the organic inhibitors whose mode of action generally results from their adsorption on the surface of the metal are the most frequently used. Their choice depends on the type of acid, its concentration, the temperature and the metallic material exposed to the action of the acid solution. Thus, sulfur-containing inhibitors are effective in sulfuric acid (Agrawal, 1990; El Azhar, 2001; Quraishi, 1997), while those containing nitrogen, such as nitrogen-containing heterocycles (Galai, 2017; Serrar, 2018; Kadiri, 2018) and amino acids (Kertit, 1997) are effective in hydrochloric acid. The use of a corrosion inhibitor based on dicalcium phosphate-dihydrate (DCPD), can form a protective surface which inhibits or slows the rate of corrosion film; the effectiveness of this medium apatite was observed in the distribution of drinking water, but little study to understand how it works and why is effective (Holm,1991; Schock,1985), the presence of calcium in solution (Larsen, 1986) claims that a minimum concentration of calcium ions is required to form a protective film and limit the dissolution of iron, and he also claims that the Ca/P ratio should be at least 1/5 of the work (Boffardi, 1993), showed that the precipitation of calcium and phosphate ions depends on the concentrations of the two species and the pH of the solution. However, according to (Atkinson, 1971; He, 1996) calcium ions are not actively involved in the protection of corroded surfaces. G. Saha et al (Saha, 1986) have studied the adsorption of the complex iron phosphates that only the surface of the iron hydroxides is combined with phosphate ions; no trace of calcium was detected in the complex iron phosphates.

The aim of the present work is to study the inhibition efficiencies of $((Ca)_9(PO_4)_5(OH)(Ca\text{-}HP))$, on mild steel corrosion in 1 M HCl, H_2SO_4 0.5 M and 2.0 M H_3PO_4 using electrochemical measurements such as polarization curves and electrochemical impedance spectroscopy.

2. EXPERIMENTAL STUDY

2.1. Materials

The experiments were performed with mild steel (MS), the chemical composition (in wt%) was presented in Table 1.

Table 1. Chemical composition of the mild steel (MS)

C	Si	Mn	Cr	Mo	Ni	Al	Cu	Co	V	W	Fe
0.11	0.24	0.47	0.12	0.02	0.1	0.03	0.14	<0.0012	<0.003	0.06	Balance

The aggressive solutions (1.0 M HCl, 0.5 M H_2SO_4 and 2.0 M H_3PO_4) were prepared by dilution of 37% HCl, 98% H_2SO_4 and 85% H_3PO_4, respectively. The working electrode was abraded with emery paper (up to 1200 grit), cleaned in acetone and washed by distillated water and finally dried at hot air before any electrochemical measurement. The concentrations of inhibitor were ranged from 100 to 300 ppm. The compound used was a mineral compound-based phosphate ($(Ca)_9(PO_4)_5(OH)(Ca\text{-}HP)$). The experimental data of triplicate measurements are highly reproducible.

2.2. Electrochemical Measurements

Electrochemical study was realized in a three-electrode electrochemical cell, composed of a mild steel plate as a working electrode (with an exposed area of 1 cm^2), a platinum counter-electrode and a saturated calomel electrode (SCE) was used as a reference electrode. An electrochemical measurement has been done using a potentiostat/galvanostat/PGZ100 controlled by analysis software VoltaMaster4. Electrochemical experiments were analyzed using electrochemical software ORIGIN PRO 6.1. Electrochemical parameters were, then, extracted using EC-Lab V10.02 and ZView software. EIS measurements were done in the range of frequency 100 kHz to 100 mHz at an Open Circuit Potential (OCP) with 10 points per decade. The amplitude of the applied alternative current signal was 10 $mV.ms^{-1}$. All experiments were performed after 30 min immersion of mild steel in acidic solution in absence and presence of the different concentrations of the studied inhibitors. Modeling of the impedance plots gives the charge transfer resistance (R_{ct}), the double layer capacitance (C_{dl}) and other parameters. Inhibition efficiency ($\eta_{imp\%}$) is calculated by the following equation (Stern, 1957):

$$\eta_{imp}\% = (R_{CT} - R^{\circ}_{CT} / R_{CT}) \times 100$$

$$\theta = R_{CT} - R^{\circ}_{CT} / R_{CT}$$

Where R^0_{ct} and R_{ct} are the charge transfer resistance in absence and presence of the inhibitor, respectively, θ is the recovery rate.

Potentiodynamic polarization study was performed on mild steel specimens by automatically changing the electrode potential from −900 to -100 mV/ ECS versus OCP at a scan rate of 1 mVs^{-1}. The corrosion current density (i_{corr}) was calculated by extrapolating the linear Tafel segments of anodic and cathodic curves back to the corrosion potential. The inhibition efficiency was determined using the following equation (Bouckamp, 1993):

Figure 1. Polarization curves for mild steel in (a) 1 M HCl, in (b) 0.5 M H_2SO_4 and in (c) 2.0 M H_3PO_4 with various concentrations of apatite at 298 K.

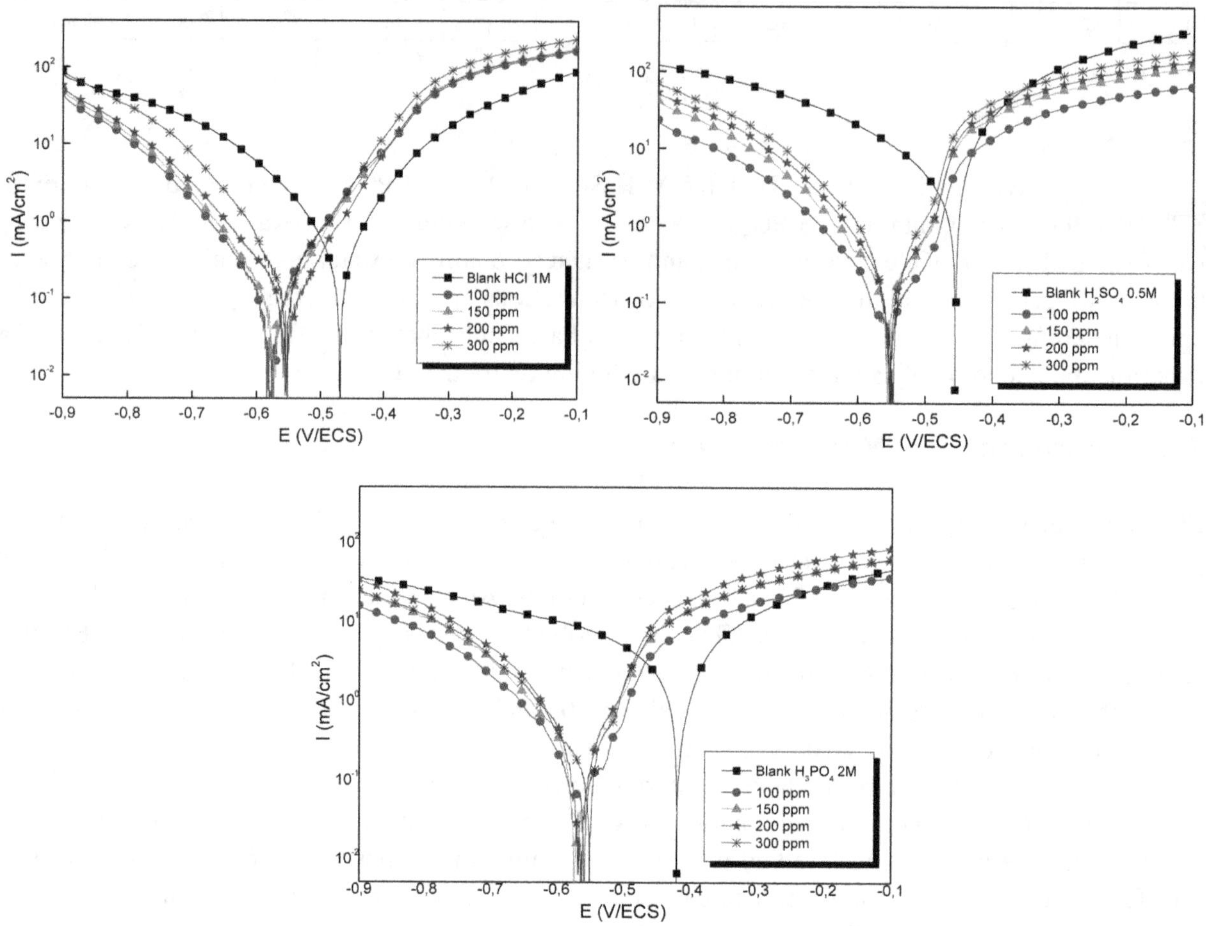

$$\eta_{pp}\% = \left[\left(i^{\circ}_{corr} - i_{corr}\right) \Big/ i^{\circ}_{corr} \right] \times 100$$

where i^0_{corr} and i_{corr} are the corrosion current densities values in absence and in presence of inhibitor, respectively.

2.3. Scanning Electron Microscopy (SEM)

The determination of the nature of the formed film on the surface of the metal, exposed to H_2SO_4 0.5 M, HCl 1M and 2.0 M H_3PO_4 solutions during 6h in absence and presence of the studied inhibitor, was realized by Scanning Electron Microscopy (Quanta FEG 450). These analyses were performed in the foundation of MAScIR-Rabat.

3. RESULTS AND DISCUSSION

3.1. Potentiodynamic Polarization Curves

The potentiodynamic polarization curves of mild steel in 1 M HCl, in 0.5 M H_2SO_4 and in 2.0 M H_3PO_4 with and without $((Ca)_9(PO_4)_5(OH)(Ca\text{-}HP))$ at 298K are shows in Figure 1. Electrochemical corrosion parameters, such as corrosion potential E_{corr} (mV/SCE), cathodic and anodic slopes β_c, β_a, corrosion current density i_{corr} and inhibition efficiency η_{pp}% are given in Table 2.

Table 2. Polarization data of mild steel in 1.0 M HCl, 0.5M H_2SO_4 and 2.0 M H_3PO_4 without and with Apatite at 298 K

Medium	Conc (ppm)	$-E_{corr}$ (mV/SCE)	i_{corr} (µA/cm²)	-βc mV/dec	βa mV/dec	IE%
1M HCl	Blank	469	762	150	126	-
	100	577	65	126	109	91,4
	150	572	125	131	112	83,6
	200	556	155	123	105	79,6
	300	554	174	125	107	77,1
0.5M H_2SO_4	Blank	451	1850	99	121	-
	100	553	176	123	70	90,4
	150	557	228	128	76	87,6
	200	554	261	126	77	85,8
	300	549	300	130	72	83,7
2M H_3PO_4	Blank	417	1519	173	133	-
	100	557	128	126	81	91,5
	150	573	144	131	103	90,5
	200	564	186	129	87	87,7
	300	549	207	125	62	86,3

Generally, the electrochemical reactions of iron in acidic solution are its anodic dissolution

Fe « $Fe^{2+} + 2e-$

and cathodic reduction of proton reactions:

$2H^+ + 2e-$ « H_2.

It's can be noted that the addition of studied compound prevents the acid attack of mild steel. In addition, it is observed that the addition of the inhibitor slightly shifts the Ecorr values towards the cathodic directions. According to Ouakki et al (Ouakki, 2018, 2020), the tested compound act as a cathodic type inhibitor in the three acids.

The given results in table 2 revealed that in i_{corr} values decrease with the decrease in inhibitor concentration, leading to an increase in the inhibition efficiency (Kadiri, 2018; Ouakki, 2018, 2020; Es-saadaoui,2019; Tarfaoui,2019).

However, it is observed that the inhibitor has an effect on both, cathodic and anodic slopes (βc and βa) signifying a modification on cathodic and anodic reactions mechanism. Furthermore, the change in βa can be explained by the changes in the charge transfer coefficient for iron dissolution by the presence of an additional energy barrier created in inhibitor adsorption (Abboud, 2007; Alaoui,2018).

3.2. *Electrochemical Impedance Spectroscopy (EIS)*

This technique furnishes details about surface characteristics and kinetics of the electrode processes. The mechanistic information of the system under subject is provided by the nature and shape of the

Figure 2. Nyquist impedance diagrams for mild steel in (a) 1 M HCl, in (b) 0.5 M H_2SO_4 and in (c) 2.0 M H_3PO_4 with various concentrations of apatite at 298 K.

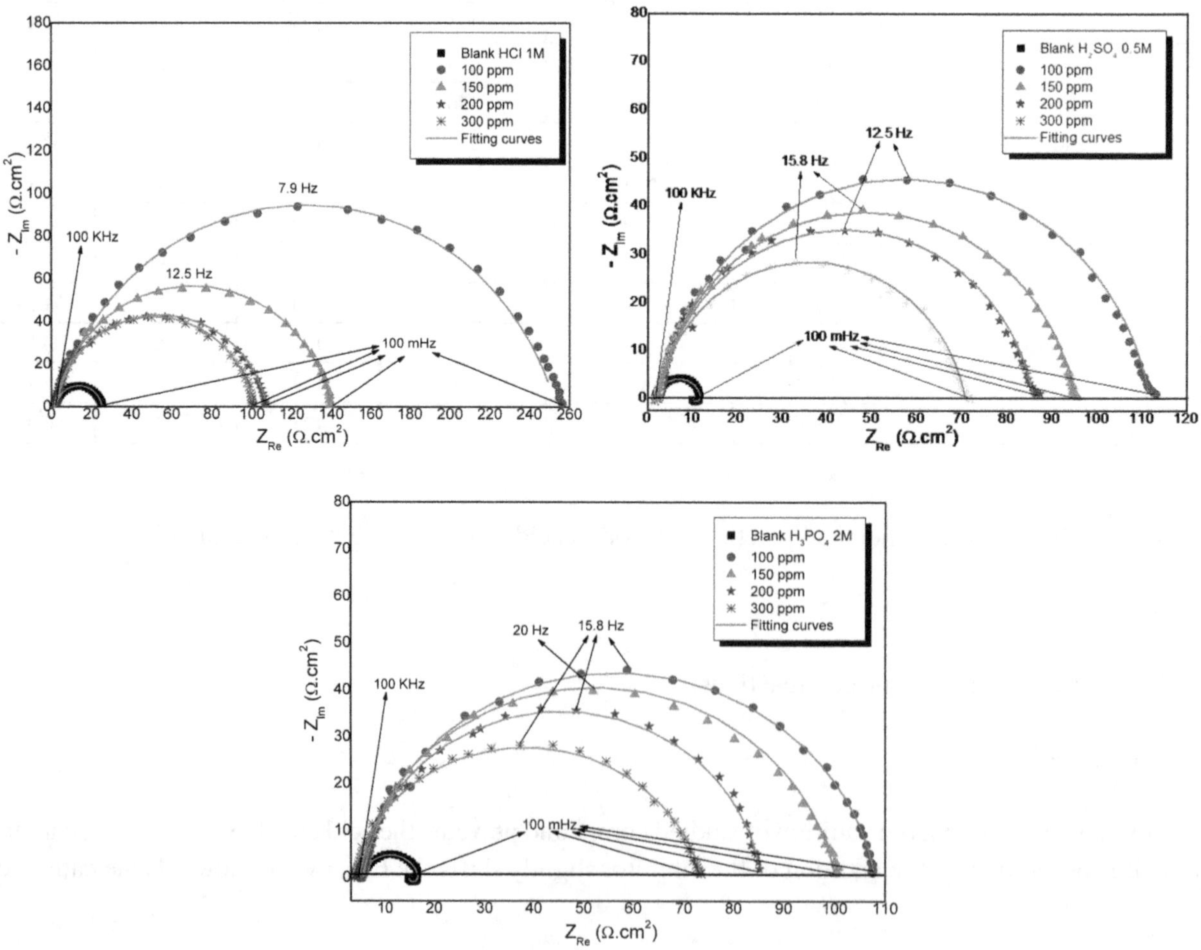

impedance curves (Asan, 2006). This technique is extensively employed to investigate the corrosion inhibition processes.

Nyquist plots of mild steel in 1 M HCl, in 0.5 M H_2SO_4 and in 2.0 M H_3PO_4 solutions at 298 K in the absence and presence of different concentrations of the studied compound are illustrated in Fig. 2. A depressed capacitive semi-circular shape was observed in the Nyquist plots with the summit of the semi-circular plot shifting to high frequency which is related to a single time constant (El Mehdi, 2003). But in H_2SO_4 0.5M and H_3PO_4 2.0M solutions without inhibitor, we note a first loop at high frequency (HF) attributed to the relaxation of the double layer capacitance in parallel with the charge transfer resistance R_{ct} inversely proportional to the corrosion rate and an inductive loop at low frequency (LF) (Kadiri, 2018; Ouakki, 2018, 2020; Lgaz, 2018). The LF inductive loop may be attributed to the relaxation process obtained by adsorption species as $SO_4^{2-}{}_{ads}$, $PO_4^{3-}{}_{ads}$ and H^+_{ads} on the electrode surface (Ouakki, 2019). It can be observed from Fig. 2 that the diameter of the semicircular Nyquist plot increases with increasing inhibitor concentration.

Figure 3. Electrochemical equivalent circuit used to fit the impedance spectra

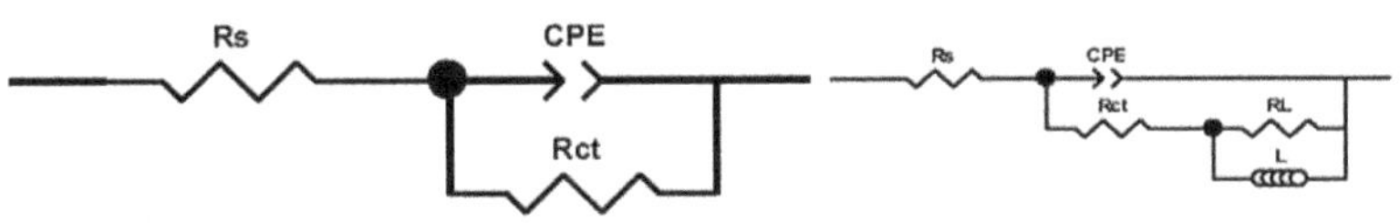

The characteristic values of these diagrams have been derived from the equivalent circuits presented in fig 3. The measured and simulated data fit well enough. According to the above-mentioned equivalent circuits, the solution resistance Rs, the charge transfer resistance R_{ct} and the constant phase element (CPE) were fitted and their parameters calculated according to different models were listed in Table 3.

The CPE was used as a substitute for the capacitor to fit more precisely the impedance behavior of the electric double layer. The impedance of the CPE is expressed as:

$$Z_{CPE} = Y^{-1}\left(j\omega\right)^{n}$$

where Y is the magnitude of the CPE; ω the angular frequency; n as the deviation parameter ($-1 < n < 1$) which is dependent on the surface morphology and often referred to the frequency dispersion as a result of the non-homogeneity or the roughness of the solid surface (Bommersbach, 2005; Finley, 1960).

Double layer capacitance (C_{dl}) values were obtained at maximum frequency (f_{max}) and calculated using the following equation:

$$C_{dl}=1/2\pi f_{max}R_{ct}$$

The results described below can be interpreted in terms of the equivalent circuit of the electrical double (Fig.3), which has been used previously to model the iron-acid interface (El Ouali, 2010). In

Table 3. Impedance parameters for corrosion of steel in 1.0 M HCl, 0.5M H_2SO_4 and 2.0 M H_3PO_4 without and with inhibitor at 298 K

Medium	Conc (ppm)	Rs ($\Omega.cm^2$)	Q (Ω^{-1} cm^2 s^{-n})	n	C_{dl} (μF cm^{-2})	R_{ct} ($\Omega.cm^2$)	R_L ($\Omega.cm^2$)	L (H)	Θ	η_{imp}%
1M HCl	Blank	1.9±0.2	270±0.5	0.897±0.02	170.9	23.25±0.4	-	-	-	-
	100	1.8±0.2	163±0.3	0.822±0.01	81.8	251.4±0.4	-	-	0.907	90,7
	150	1.4±0.2	168±0.6	0.873±0.02	97.5	138.6±0.4	-	-	0.832	83,2
	200	1.3±0.3	194±0.9	0.817±0.01	109.9	106.0±0.5	-	-	0.780	78,0
	300	1.8±0.2	247±1.2	0.902±0.02	165	98.7±0.5	-	-	0.764	76,4
0.5M H_2SO_4	Blank	1.6±0.2	430±0.7	0.830±0.03	180	11.7±0.3	2.17±0.1	0.48±0.02	-	-
	100	2.6±0.2	186±0.9	0.888±0.01	114	109±0.4	-	-	0.892	89,2
	150	2.1±0.2	187±0.3	0.885±0.01	119	93±0.4	-	-	0.874	87,4
	200	2.3±0.1	245±1.2	0.890±0.02	151	83.4±0.3	-	-	0.859	85,9
	300	1.8±0.1	277±1.3	0.878±0.03	160	68.8±0.5	-	-	0.829	82,9
2M H_3PO_4	Blank	5.2±0.3	243±0.8	0.880±0.02	209.0	9.7±0.5	2.0±0.1	0.26±0.01	-	-
	100	5.3±0.1	149±0.8	0.897±0.02	92.4	103±0.5	-	-	0.905	90,5
	150	4.6±0.2	133±0.6	0.896±0.01	108.4	95.5±0.4	-	-	0.898	89,8
	200	5.5±0.3	159±0.5	0.928±0.03	111.0	80.0±0.4	-	-	0.878	87,8
	300	3.2±0.2	262±1.2	0.850±0.02	129.6	70.4±0.5	-	-	0.862	86,2

this equivalent circuit, R_s is the solution resistance, R_{ct} is the charge transfer resistance and CPE is a constant phase element.

The most important data obtained from the equivalent circuit are presented in Table 3. It may be remarked that R_{ct} value decreases and C_{dl} increases with inhibitor concentration indicating that more inhibitor molecules are adsorbed on metallic surface and provided better surface coverage and/or enhanced the thickness of the protective layer at the metal/solution interface (Moradi, 2013; Tang, 2013). Values of inhibition efficiency η_{imp}% increase with the concentration of the inhibitor up to 90.7%, 89.2% and 90.5% for HCl, H_2SO_4 and H_3PO_4, respectively at the concentration of 100 ppm. The impedance study also gave the same efficiency trend as found in Tafel polarization method.

3.3. Adsorption Isotherm

The adsorption isotherm type can provide additional information about the tested compound properties. The fractional coverage surface (θ) can be easily determined from polarization curve. In the present study, the θ evaluated from EIS curve measurements (Table 3). in general, the most adsorption isotherms models considered were as described in reference (Khaled, 2008):

Temkin isotherm:

$$e^{f\theta} = K_{ads}C_{inh}$$

Langmuir isotherm:

Figure 4. Langmuir adsorption isotherm plot for mild steel in 1 M HCl, 0.5 M H_2SO_4 and 2.0 M H_3PO_4 at different concentrations of the tested compound at 298 K.

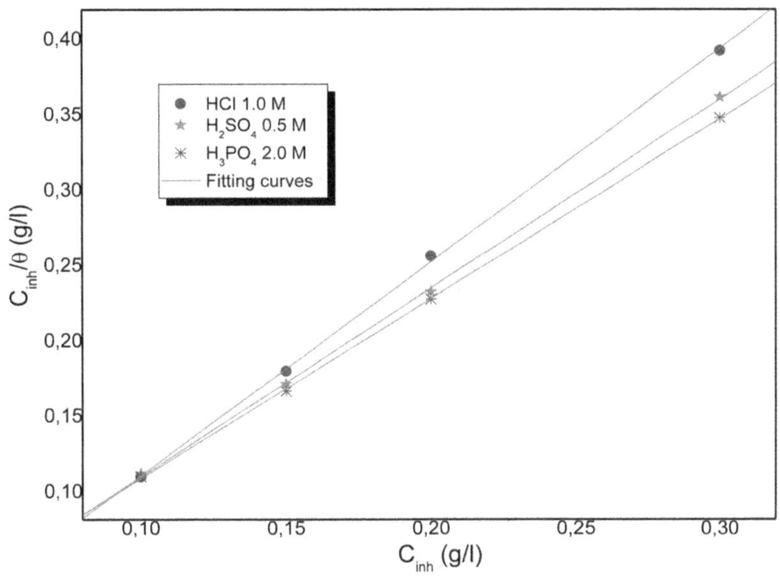

$$\frac{\theta}{1-\theta} = K_{ads} C_{inh}$$

Frumkin isotherm:

$$\frac{\theta}{1-\theta} e^{-2f\theta} = K_{ads} C_{inh}$$

Freundluich isotherm:

$$\theta = K_{ads} C_{inh}$$

where K_{ads} is the equilibrium constant of the adsorption process, C_{inh} is the inhibitor concentration and f is the factor of energetic inhomogeneity. The linear coefficient regression, r^2 and the slope, were used

Table 4. Equilibrium constant and free energy of adsorption values for mild steel in 1 M HCl, 0.5 M H_2SO_4 and 2.0 M H_3PO_4 in presence of the tested compound.

Medium	K_{ads} (L /g)	ΔG_{ads} (Kj/mol)	R^2	Slopes
1.0 M HCl	32.7	-42.8	0.99978	1.41589
0.5 M H_2SO_4	66.4	-44.6	0.99981	1.25151
2.0M H_3PO_4	99.0	-45.6	0.99992	1.19168

to choose the suitable isotherm (Table 4). K_{ads}, is related to the free energy of adsorption, ΔG_{ads}, by the following equation:

$$K_{ads} = \frac{1}{10^6} e^{-\frac{\Delta G_{ads}}{RT}}$$

where 10^6 represents the water concentration in solution in g.L^{-1}, R is the universal gas constant and T is the absolute temperature (298 K).

The best fit straight line (strong correlation with R^2 closer to 1) is obtained for the plot of C_{inh}/θ vs. C_{inh} in the three acids at 298 K with slopes around unity (Figure 4). This suggests that the adsorption of the tested compound on the metallic surface obeyed to the Langmuir's adsorption isotherm model, and exhibit single-layer adsorption characteristics. This kind of isotherm involves the assumption of no interaction between the adsorbed species on the electrode surface (Elayyachy, 2006). The values of ΔG_{ads} are calculated and are given in Table 4. The high K_{ads} values reflect the high adsorption ability of this inhibitor on metallic surface. In addition, the negative values of free energy of adsorption, ΔG_{ads}, indicate spontaneous adsorption of apatite molecules on the metallic surface and also the strong interaction between inhibitor molecules and the mild steel surface (Avci, 2008; Bayol, 2008). K_{ads} of this inhibitor in H_3PO_4 2.0M was higher than the K_{ads} in HCl 1M and H_2SO_4 0.5M, which means that apatite exhibits a stronger tendency to adsorb on steel surface in H_3PO_4 2.0M compared with that in 1.0 M HCl and H_2SO_4 0.5M. Generally, the standard free energy values of -20 KJ mol^{-1} or less negative are associated with an electrostatic interaction between charged molecules and charged metal surface (physical adsorption); those of -40 KJ mol^{-1}or more negative involve charge sharing or transfer from the inhibitor molecules to the metal surface to form a co-ordinate covalent bond (chemical adsorption) (Abiola, 2004; Ozcan, 2008). Based on the literature (Hongbo, 2002), the calculated ΔG_{ads} values in this work presented in Table 4 indicate that the adsorption mechanism of this apatite on mild steel surface in 1.0 M HCl, 0.5 M H_2SO_4 and 2.0 M H_3PO_4 solutions is chemical adsorption.

3.4. Effect of Temperature

Generally, temperature is a factor that has a direct influence on the kinetics of chemical reactions in solution. According to literature, temperature may cause a change in the behavior of steel in a corrosive environment and in the nature of metal/inhibitor interaction (Abdallah, 2002). In order to determine the effect of this variable on the inhibitory power of the compound $((Ca)_9(PO_4)_5(OH)(Ca\text{-}HP))$ on mild steel, we performed stationary electrochemical characterization in potentiostatic mode at different temperatures (298K to 328K).

The polarization plots obtained in HCl 1.0M, H_2SO_4 0.5M and H_3PO_4 2.0M without and with the addition of 100 ppm of the tested compound in the temperature range (298-328 K) are presented in the figures 5, 6 and 7. The values of the corrosion current densities (i_{corr}), corrosion potentials of the steel (E_{corr}) and the inhibition efficiencies as a function of temperature are given in table 5.

Based on the analysis of the results obtained from the figures (5, 6 and 7) and Table 5, we find that:

- The corrosion current density values increase with temperature in the whole studied range. This is probably due to the acceleration of steel corrosion kinetics with increasing temperature.

- The curves in the cathodic part are parallel, explaining that H^+ ions reduction on the steel surface is done according to the same pure activation mechanism in these three acidic mediums.

Figure 5. Potentiostatic polarization plots of mild steel in HCl 1.0M without and with the addition of 100 ppm of inhibitor at different temperatures

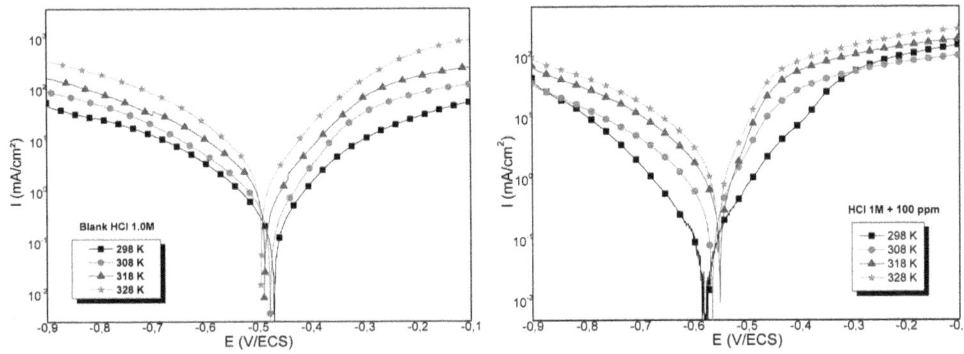

Figure 6. Potentiostatic polarization plots of mild steel in H_2SO_4 0.5M without and with the addition of 100 ppm of inhibitor at different temperatures

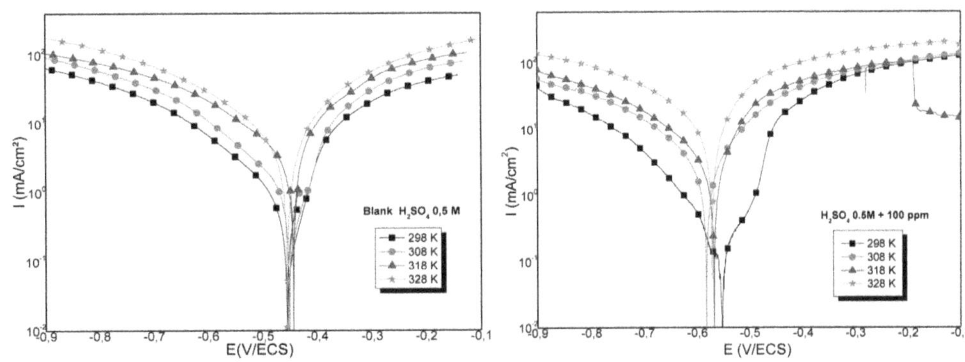

Figure 7. Potentiostatic polarization plots of mild steel in H_3PO_4 2.0M without and with the addition of 100 ppm of inhibitor at different temperatures

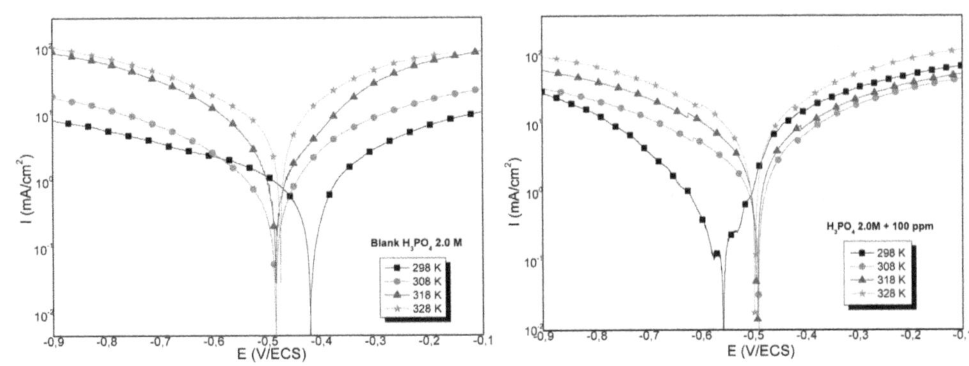

Table 5. Electrochemical characteristics of mild steel in 1 M HCl, 0.5M H_2SO_4 and 2.0 M H_3PO_4 with and without 100 ppm of inhibitor at different temperatures.

Medium		Temperature (K)	$-E_{corr}$ (mV/ECS)	i_{corr} (µA/cm²)	-βc (mv/dec)	βa (mv/dec)	E%
1M HCl	Blank	298	469	762	150	126	-
		308	475	1150	184	112	-
		318	488	1550	171	124	-
		328	492	2350	161	118	-
	100 ppm	298	577	65	126	109	91,4
		308	564	127	135	104	88,9
		318	550	212	147	112	86,3
		328	554	381	156	103	83,8
0.5M H_2SO_4	Blank	298	451	1850	99	124	-
		308	453	2250	92	114	-
		318	449	2480	96	102	-
		328	442	3340	102	97	-
	100 ppm	298	553	176	123	70	90,4
		308	582	267	136	80	88,1
		318	568	352	132	86	85,8
		328	573	559	124	90	83,2
2M H_3PO_4	Blank	298	417	1519	173	133	-
		308	484	3240	155	128	-
		318	485	4240	145	143	-
		328	475	4850	130	118	-
	100 ppm	298	557	128	126	81	91,5
		308	488	365	135	103	88,7
		318	491	594	141	109	85,9
		328	496	787	138	113	83,7

- Inhibition efficiency η_{PP} (%) of the tested compound decreases with the temperature increase. This can be explained by the positive contribution to the corrosion phenomenon. This can be also explained by a slight change of equilibrium adsorption- desorption in favor of the desorption process thus decreasing the inhibitory power of the molecule (Obot, 2010; Mu, 2004).

3.5. Thermodynamic Part

In order to gain more information about the adsorption type of apatite compound and its effectiveness at higher temperatures, potentiodynamic polarization curves for mild steel in the three acidic mediums without and with the compound were used in the temperatures range from 298 to 328 ± 2 K. The activation parameters for the corrosion reaction can be regarded as an Arrhenius-type process, according to the following equation (Ekanem, 2010):

Figure 8. Arrhenius plots for mild steel in acidic mediums with and without of 100 ppm of inhibitor.

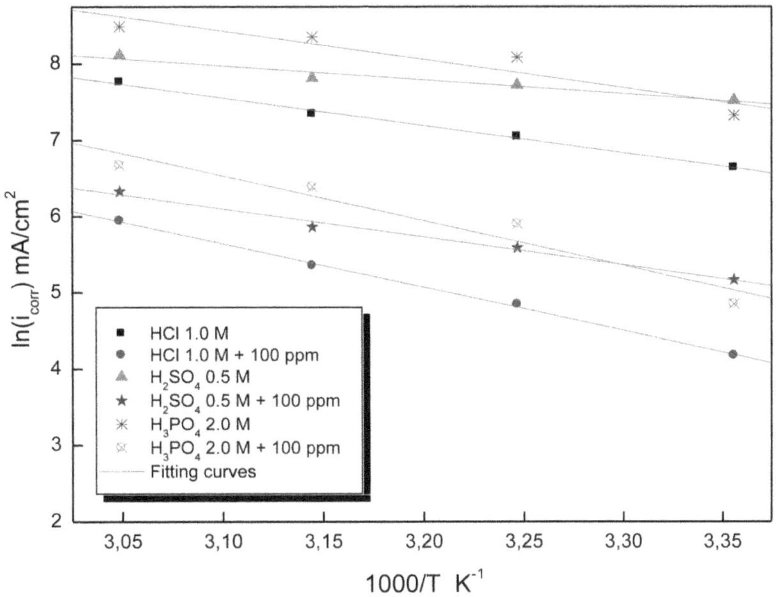

$$\ln i_{corr} = \text{Ln } A - \frac{Ea}{RT}$$

where (Ea) is the apparent activation corrosion energy; (R) is the universal gas constant, and (A) is the Arrhenius preexponential factor. The apparent activation energies (Ea) in the absence and the presence

Figure 9. Arrhenius plots for mild steel in acidic mediums with and without of 100 ppm of inhibitor.

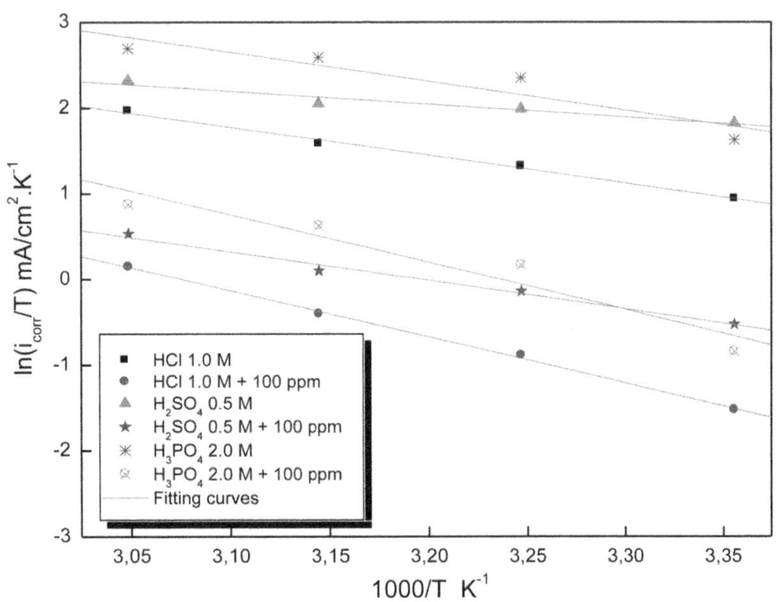

of tested compound for mild steel are calculated by linear regression between ln(i_{corr}) and 1000/T (Fig. 8), and the results are given in Table 6. All the linear regression coefficients are close to 1, indicating that the steel corrosion in hydrochloric acid can be elucidated using the kinetic model.

A corresponding increase in the corrosion rate occurs because of the greater area of metal that is consequently exposed to the acid environment (Szauer, 1981). The enthalpy and entropy of activation respectively (ΔHa) and (ΔSa) for the intermediate complex in the transition state for the corrosion of mild steel in the three acidic mediums in the absence and in the presence of tested compound were obtained by applying the alternative formulation of Arrhenius equation (Yadav, 2013):

$$\ln\frac{icorr}{T} = (\ln(\frac{R}{Nh}) + \frac{"Sa}{R}) - \frac{"Ha}{RT}$$

where (h) is the Plank's constant and (N) is the Avogadro's number. A plot of ln (i_{corr}/T) against 1000/T should give a straight line with a slope of (– ΔHa/R) and intercept of [ln (R/Nh) + (ΔSa/R)], as shown in Fig. 9. (ΔHa) and (ΔSa) were calculated and tabulated in Table 6.

Table 6. The values of activation parameters Ea, ΔHa and ΔSa for mild steel in 1M HCl, 0.5M H_2SO_4 and 2.0M H_3PO_4 in the with and without of 100 ppm of inhibitor

Medium	conc	Ea (KJ/mol)	ΔHa (KJ/mol)	ΔSa (J/mol.K)
1M HCl	Blank	29.8	27.2	-98.2
	100 ppm	47.3	44.7	-60.1
0.5M H_2SO_4	Blank	15.1	12.5	-140.5
	100 ppm	30.4	27.8	-108.7
2.0 M H_3PO_4	Blank	30.8	28.2	-88.0
	100 ppm	48.6	46.0	-48.6

For three mediums, as observed in Table 5, the Ea increased with 100 ppm of apatite in the three acids. This behavior may be interpreted as physical adsorption. Indeed, a higher energy barrier for the corrosion process in the inhibited solution is associated with physical adsorption or weak chemical bonding between the inhibitor's species and the steel surface (Gharda, 2017; Dahmani, 2017).

The values of (ΔHa) are higher in the presence of the inhibitor. This indicates that the energy barrier of the corrosion reaction increased in the presence of the inhibitor without changing the mechanism of dissolution. The positive values of (ΔHa) for both corrosion processes with and without the inhibitor reveal the endothermic nature of the steel dissolution process and indicate that the dissolution of steel is difficult (Geethanjali, 2015). According to the recorded values of enthalpy of activation (Table 6), exothermic adsorption behavior was detected depending on applied temperatures. The value of enthalpy of activation (ΔHa) is respectively (44.7 $KJ.mol^{-1}$), (27.8 $KJ.mol^{-1}$) and (46.0 $KJ.mol^{-1}$) in present of 100 ppm of the tested inhibitor for the HCl, H_2SO_4 and H_3PO_4. All values of Ea are higher than the values of ΔHa, indicating that the corrosion process must involve a vapor reaction, simply the hydrogen evolution reaction, associated with a decrease in the total reaction volume (Shaban, 2016; Murulana,2016).

Figure 10. SEM image of mild steel after 6 h of immersion in 1.0 M HCl solution Blank and solution with 100 ppm of inhibitor

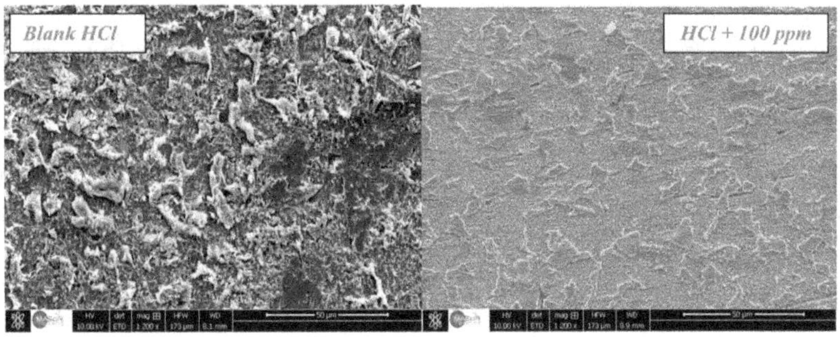

Figure 11. SEM image of mild steel after 6 h of immersion in 0.5 M H_2SO_4 solution Blank and solution with 100 ppm of inhibitor

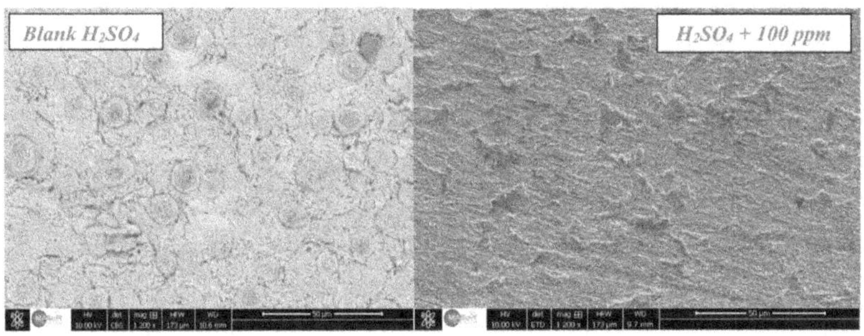

The large negative value of ΔSa for mild steel in the three acidic mediums implies that the activated complex is the rate-determining step, rather than the dissociation step. In the presence of the tested inhibitor, the value of (ΔSa) increases and is generally interpreted as an increase in disorder as the reactants are converted to the activated complexes (Krishnegowda, 2013; Olasunkanmi, 2015). The values of activation entropy (ΔSa) are higher for inhibited solutions than that for the uninhibited solution (Table 6).

Figure 12. SEM image of mild steel after 6 h of immersion in 2.0 M H_3PO_4 solution Blank and solution with 100 ppm of inhibitor

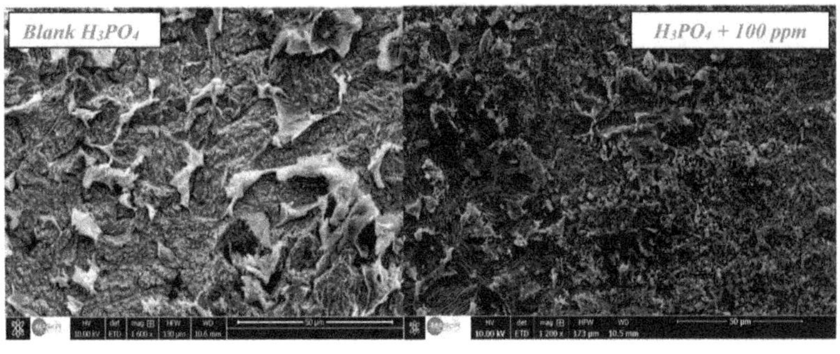

3.6. Surface Analysis by SEM

To know the chemical compositions and to characterize the surface of the steel before and after addition of inhibitors in the mediums acid 1.0 M HCl, 0.5 M H_2SO_4 and 2.0M H_3PO_4, we realized an analysis by the scanning electron spectroscopy (SEM). The observation of the images before the addition of the inhibitor shows that the surface of mild steel destroyed by the corrosion in the acid mediums (Fig. 10, 11 and 12). On the other hand, the observation of the images after the addition of the inhibitor shows that the mild steel surface shows a change of state, but is not thin, because the inhibitor carries the calcium and phosphorus atoms.

4. CONCLUSION

Concluding the experimental part, it was clearly demonstrated that all techniques used, are able to characterize and to follow the corrosion inhibition process promoted by mineral compound was investigated on mild steel corrosion in three acidic mediums 1.0 M HCl, 0.5 M H_2SO_4 and 2.0M H_3PO_4. Polarization measurements showed that the inhibitor is cathodic type in the three acidic mediums. The compound tested exhibited good inhibition properties for mild steel corrosion in the three acidic solutions and their inhibition efficiency is inversely proportional to their concentration. The results of EIS indicate that the values of C_{dl} tend to decrease and both R_{ct} and η_{imp}% tends to increase with increasing the inhibitor concentration. These results can be attributed to an increase of the thickness of the protective film formed on the mild steel surface. The adsorption characteristics of the inhibitor followed the isothermal adsorption model of Langmuir. The study of temperature effect on the inhibition efficiency shows that this one decreases with the temperature increase. This confirms that the adsorption of the inorganic compound occurs by an intermediate adsorption between physisorption and chemisorption. Surface morphology studies confirmed the mitigation of mild steel corrosion by the formation of a protective film on the surface of the steel.

BACKGROUND

Mild steel (MS) is a class of strong and tough steel that contains a low quantity of carbon. It is the most important engineering material particularly for structural, instrumental, industrial and automobile applications. But it is highly susceptible to corrosion especially in acidic media. Corrosion of metals, particularly M-steel, is in most cases an electrochemical reaction involving the metal and its environment. Hence, various corrosion protection technologies have been employed, among them; the use of inhibitors is one of the practical methods for preventing corrosion of MS especially in acid media. A wide variety of corrosion inhibitors have been used. Corrosion inhibitors block the active sites and enhance the adsorption process, thus decreasing the rate of dissolution of the equipment.

MAIN FOCUS OF THE CHAPTER

The main focus of this chapter is to investigate the inhibition potential and kinetic information of a mineral compound-based phosphate (apatite) against the corrosion of mild steel in three acidic mediums 1.0 M HCl, 0.5 M H_2SO_4 and 2.0M H_3PO_4. The inhibitory action was investigated using potentiodynamic polarization and electrochemical impedance spectroscopy. The purpose of the chapter is also to provide information about the adsorption mechanism of this compound onto mild steel surface. The effect of temperature is used to study of corrosion of steel in order to know the behavior of high-temperature inhibitors. Also, it allows to know how this factor can influence the inhibitory efficacy of the studied compound. In order to support our electrochemical results, SEM analyses were carried out to demonstrate the morphology of metallic surfaces corroded in the three acidic mediums in the absence and presence of the apatite.

REFERENCES

Abboud, Y., Abourriche, A., Saffaj, T., Berrada, M., Charrouf, M., Bennamara, A., Al Himidi, N., & Hannache, H. (2007). 2,3-Quinoxalinedione as a novel corrosion inhibitor for mild steel in 1 M HCl. *Materials Chemistry and Physics*, *105*, 1–5.

Abdallah, M. (2002). Rhodanine azosulpha drugs as corrosion inhibitors for corrosion of 304 stainless steel in hydrochloric acid solution. *Corrosion Science*, *44*, 717–728.

Abiola, O. K., & Oforka, N. C. (2004). Adsorption of (4-amino-2-methyl-5-pyrimidinyl methylthio) acetic acid on mild steel from hydrochloric acid solution (HCl)—Part 1. *Materials Chemistry and Physics*, *83*, 315.

Agrawal, R., & Namboodhiri, T. K. G. (1990). The inhibition of sulphuric acid corrosion of 410 stainless steel by thioureas. *Corrosion Science*, *30*, 37.

Alaoui, K., Touir, R., Galai, M., Serrar, H., Ouakki, M., Kaya, S., Tüzün, B., Boukhris, S., Ebn Touhami, M., & El Kacimi, Y. (2018). Electrochemical and Computational Studies of Some Triazepine Carboxylate Compounds as Acid Corrosion Inhibitors for Mild Steel. *Journal of Bio- and Tribo-Corrosion*, *4*, 37.

Asan, A., Soylu, S., Kıyak, T., Yıldırım, F., Oztas, S. G., Ancın, N., & Kabasakaloğlu, M. (2006). Investigation on some Schiff bases as corrosion inhibitors for mild steel. *Corrosion Science*, *48*, 3933.

Atkinson, R.J., Posner, A.M., & Quirk, J.P. (1972). Kinetics of isotopic exchange of phosphate at the L-FeOOH-aqueous solution interface. *Journal of Organic and Nuclear Chemistry, 34*, 2201-2211.

Avci, G. (2008). Inhibitor effect of N, N¢-methylenediacrylamide on corrosion behavior of mild steel in 0.5 M HCl. *Materials Chemistry and Physics*, *112*, 234.

Bayol, E., Gurten, A. A., Dursun, M., & Kayakırılmaz, K. (2008). Adsorption Behavior and Inhibition Corrosion Effect of Sodium Carboxymethyl Cellulose on Mild Steel in Acidic Medium. *Wuli Huaxue Xuebao*, *24*, 2236.

Boffardi, B. (1993). The chemistry of polyphosphate. *Materials Performance*, *32*, 50–53.

Bommersbach, P., Alemany-Dumont, C., Millet, J. P., & Normand, B. (2005). Formation and behaviour study of an environment-friendly corrosion inhibitor by electrochemical methods. *Electrochimica Acta*, *51*, 1076–1084.

Bouckamp, A. (1993). Faculty of Chem. Tech., Universidal of Twente, The Netherlands. User's Manual Equivalent Circuit, Ver, 4, 51.

Dahmani, K., Galai, M., & Cherkaoui, M., El hasnaoui, A., & El Hessni, A. (2017). Cinnamon essential oil as a novel eco-friendly corrosion inhibitor of copper in 0.5 M Sulfuric Acid medium. *J. Mater. Environ. Sci*, *8*, 1676.

Ekanem, U. F., Umoren, S. A., Udousoro, I. I., & Udoh, A. P. (2010). Inhibition of mild steel corrosion in HCl using pineapple leaves (Ananas comosus L.) extract. *Journal of Materials Science*, *45*, 5558–5566.

El Azhar, M., Mernari, B., Traisnel, M., Bentiss, F., & Lagrenée, M. (2001). Corrosion inhibition of mild steel by the new class of inhibitors [2,5-bis(*n*-pyridyl)-1,3,4-thiadiazoles] in acidic media. *Corrosion Science*, *43*, 2229.

El Mehdi, B., Mernari, B., Traisnel, M., Bentiss, F., & Lagrenee, M. (2003). Synthesis and comparative study of the inhibitive effect of some new triazole derivatives towards corrosion of mild steel in hydrochloric acid solution. *Materials Chemistry and Physics*, *77*, 489.

El Ouali, I., Hammouti, B., Aouniti, A., Ramli, Y., Azougagh, M., Essassi, E. M., & Bouachrine, M. (2010). Thermodynamic characterisation of steel corrosion in HCl in the presence of 2- phenylthieno (3, 2-b) quinoxaline. *J. Mater. Environ. Sci*, *1*(1), 1–8.

Elayyachy, M., El Idrissi, A., & Hammouti, B. (2006). New thio-compounds as corrosion inhibitor for steel in 1 M HCl. *Corrosion Science*, *48*, 2470.

Essaadaoui, Y., Galai, M., Ouakki, M., Kadiri, L., Ouass, A., Cherkaoui, M., Rifi, E. H., & Lebkiri, A. (2019). Study of the anticorrosive action of eucalyptus Camaldulensis extract in case of mild steel in 1.0 m HCL. *J. Chem. Technol. Metall*, *54*(2), 431–442.

Finley, H. F., & Hackerman, N. (1960). Effect of Adsorption of Polar Organic Compounds on the Reactivity of Steel. *Journal of the Electrochemical Society*, *107*, 259–263.

Galai, M., Rbaa, M., El Kacimi, Y., Ouakki, M., Dkhirech, N., Touir, R., Lakhrissi, B., & Ebn Touhami, M. (2017). Anti-corrosion Properties of some Triphenylimidazole Substituted Compounds in Corrosion Inhibition of Carbon Steel in 1. 0 M Hydrochloric Acid Solution. *Anal. Bioanal. Electrochem*, *9*, 80.

Geethanjali, R., & Subhashini, S. (2015). Thermodynamic Characterization of Metal Dissolution and Adsorption of Polyvinyl Alcohol-Grafted Poly(Acrylamide-Vinyl Sulfonate) on Mild Steel in Hydrochloric Acid. *Portugaliae Electrochimica Acta*, *33*(1), 35–48.

Gharda, N., Galai, M., Saqalli, L., Ouakki, M., Habbadi, N., Ghailane, R., Souizi, A., Ebn Touhami, M., & Peres-lucchese, Y. (2017). Synthesis, Structural Properties and Complex Corrosion Inhibition Cu (II) With Amino Acid (DL-α-Alanine). *Oriental Journal of Chemistry*, *33*, 1665–1676.

He, Q.H., Leppard, G.G., Paige, C.R., & Snodgrass, W.J. (1996). Transmission electron microscopy of a phosphate effect on the colloid structure of iron hydroxide. *Water Research, 30*, 1345-1352.

Holm, T. R., & Schock, M. R. (1991). Potential Effects of Polyphosphate Products on Lead Solubility in Plumbing Systems. *Journal - American Water Works Association*, *83*, 76–82.

Hongbo, F. (2002). *Synthesis and Application of New Type Inhibitors*. Chemical Industry Press.

Kadiri, L., Galai, M., Ouakki, M., Essaadaoui, Y., Ouass, A., Cherkaoui, M., Rifi, E. H., & Lebkiri, A. (2018). Coriandrum Sativum. L Seeds Extract as a Novel Green Corrosion Inhibitor for Mild Steel in 1.0 M Hydrochloric and 0.5 M Sulfuric Solutions. *Anal. Bioanal. Electrochem*, *10*(2), 249–268.

Kertit, S., Hammouti, B., Taleb, M., & Brighli, M. (1997). Peptidic compounds as corrosion inhibitor for iron in acid chloride solution. *Bulletin of Electrochemistry*, *13*, 241.

Khaled, K. F. (2008). Molecular simulation, quantum chemical calculations and electrochemical studies for inhibition of mild steel by triazoles. *Electrochimica Acta*, *53*, 3484.

Krishnegowda, P. M., Venkatesha, V. T., Krishnegowda, P. K. M., & Shivayogiraju, S. B. (2013). Acalypha torta Leaf Extract as Green Corrosion Inhibitor for Mild Steel in Hydrochloric Acid Solution. *Industrial & Engineering Chemistry Research*, *52*, 722–728.

Larsen, M. J., & Jensen, S. J. (1986). Solubility study of the initial formation of calcium orthophosphates from aqueous solutions at pH 5–10. *Archives of Oral Biology*, *31*, 565–572. PMID:3467681

Lgaz, H., Chung, I. M., Salghi, R., Ali, I. H., Chaouiki, A., El Aoufir, Y., & Khan, M. I. (2018). On the understanding of the adsorption of Fenugreek gum on mild steel in an acidic medium: Insights from experimental and computational studies. *Applied Surface Science*, *463*, 647–658.

Moradi, M., Duan, J., & Du, X. (2013). Investigation of the effect of 4,5-dichloro-2-n-octyl-4-isothiazolin-3-one inhibition on the corrosion of carbon steel in *Bacillus* sp. inoculated artificial seawater. *Corrosion Science*, *69*, 338–345.

Mu, G. N., Li, X., & Li, F. (2004). Synergistic inhibition between *o*-phenanthroline and chloride ion on cold rolled steel corrosion in phosphoric acid. *Materials Chemistry and Physics*, *86*, 59–68.

Murulana, L. C., Kabanda, M. M., & Ebenso, E. E. (2016). Investigation of the adsorption characteristics of some selected sulphonamide derivatives as corrosion inhibitors at mild steel/hydrochloric acid interface: Experimental, quantum chemical and QSAR studies. *Journal of Molecular Liquids*, *215*, 763–779.

Obot, I. B., & Obi-Egbedi, N. O. (2010). Theoretical study of benzimidazole and its derivatives and their potential activity as corrosion inhibitors. *Corrosion Science*, *52*, 657–660.

Olasunkanmi, L. O., Obot, I. B., Kabanda, M. M., & Ebenso, E. E. (2015). Some Quinoxalin-6-yl Derivatives as Corrosion Inhibitors for Mild Steel in Hydrochloric Acid: Experimental and Theoretical Studies. *Journal of Physical Chemistry*, *119*, 16004–16019.

Ouakki, M., Galai, M., Cherkaoui, M., Rifi, E. H., & Hatim, Z. (2018). Inorganic compound (apatite doped by Mg and Na) as a corrosion inhibitor for mild steel in phosphoric acidic medium. *Anal. Bioanal. Electrochem.*, *10*(7), 943–960.

Ouakki, M., Galai, M., Cherkaoui, M., Rifi, E.H., & Hatim, Z. (2020). Study on the corrosion inhibition of mild steel in two acidic mediums (HCL 1m and H2SO4 0.5m) containing apatite. *J. Chem. Technol. Metall, 2*.

Ouakki, M., Galai, M., Rbaa, M., Abousalem, A. S., Lakhrissi, B., Rifi, E. H., & Cherkaoui, M. (2019). Quantum chemical and experimental evaluation of the inhibitory action of two imidazole derivatives on mild steel corrosion in sulphuric acid medium. *Heliyon*, *5*, e02759. PMID:31768434

Ozcan, M., Solmaz, R., Kardas, G., & Dehri, I. (2008). Adsorption properties of barbiturates as green corrosion inhibitors on mild steel in phosphoric acid. *Colloid Surf. A*, *57*, 325.

Quraishi, M. A., Wajid Khan, M. A., Ajmal, M., Muralidharan, S., & Venkatakrishnaiyer, S. (1997). Influence of molecular structure of substituted benzothiazoles on corrosion inhibition and hydrogen permeation through mild steel in sulphuric acid. *British Corrosion Journal*, *32*, 72.

Saha, G., & Kurmaih, N. (1986). Technical Note: The Mechanism of Corrosion Inhibition by Phosphate-Based Cooling System Corrosion Inhibitors. *Corrosion*, *42*, 233–235.

Schock, M.R., & Wagner, I. (1985). Internal corrosion of water distribution systems. *DVGW-Forschungstelle*, 700.

Serrar, H., Galai, M., Benhiba, F., Ouakki, M., Benzekri, Z., Boukhris, S., Hassikou, A., Souizi, A., Oudda, H., & Ebn Touhami, M. (2018). Two derivatives of 7-amino-thiazolo[3,2-a] pyrimidine as inhibitors of mild steel corrosion in 1.0 m hcl solution: part i: synthesis of inhibitors and electrochemical study. *J. Chem. Technol. Metall*, *53*(2), 324–335.

Shaban, S. M., Abd-Elaal, A. A., & Tawfik, S. M. (2016). Gravimetric and electrochemical evaluation of three nonionic dithiol surfactants as corrosion inhibitors for mild steel in 1 M HCl solution. *Journal of Molecular Liquids*, *216*, 392–400.

Stern, M., & Geary, A. L. (1957). Electrochemical Polarization: I. A Theoretical Analysis of the Shape of Polarization Curves. *Journal of the Electrochemical Society*, *104*, 56–63.

Szauer, T., & Brandt, A. (1981). On the role of fatty acid in adsorption and corrosion inhibition of iron by amine—Fatty acid salts in acidic solution. *Electrochimica Acta*, *26*, 1257–1260.

Tang, Y., Zhang, F., Huc, S., Cao, Z., Wu, Z., & Jing, W. (2013). Novel benzimidazole derivatives as corrosion inhibitors of mild steel in the acidic media. Part I: Gravimetric, electrochemical, SEM and XPS studies. *Corrosion Science*, *74*, 271–282.

Tarfaoui, K., Ouakki, M., Galai, M., Nehiri, M., Ebn Touhami, M., Barhada, N., & Ouhssine, M. (2019). Valorization of the essential oil of Zingiber officinale by its Use as inhibitor against the corrosion of the carbon steel in acid medium hydrochloric acid 1M. *Mediterr. J. Chem*, *8*(1), 17–24.

Yadav, M., Kumar, S., Sinha, R. R., & Behera, D. (2013). Experimental and Quantum Chemical Studies on the Corrosion Inhibition Performance of Benzimidazole Derivatives for Mild Steel in HCl. *Eng. Chem. Res*, *52*, 6318.

Section 5

Corrosion Methods and Control in Civil Engineering and Other Specific Industrial Applications of Stainless Steel

Chapter 9
Corrosion Resistance Methods for Stainless Steel: A Review

Sudesna Roy
https://orcid.org/0000-0003-3129-0891
School of Mechanical Engineering, KIIT University, India

Pramod kumar Mandal
School of Mechanical Engineering, KIIT University, India

ABSTRACT

304 grade stainless steel is known to be important in most structural applications due to its high mechanical strength, hardness, and machinability. It is considered a versatile steel that has good formability, can be welded as it is, and is non-magnetic austenitic structure. The corrosion resistance is also excellent at room temperature for most corrosive acids and alkalis. However, its corrosion resistance decreases at higher temperatures when exposed to water for prolonged periods of time. It is sensitive to pitting, crevice and stress corrosion cracking at elevated temperatures. In some cases, the resistance is improved by addition of corrosion inhibitor that negatively affects its formability and welding advantages. Therefore, other methods of corrosion protection are desired. This chapter provides in-depth review of corrosion protections materials and methods that have been used for protecting 304SS in different specific applications. It also provides systemic analysis of the potentiodynamic polarization method to compare the corrosion potential of different materials.

1. INTRODUCTION

Stainless steels play an important role in this competitive modern world, having about 2% of the whole steel production only. It has till date has crossed 10^9 MT production in the world (Ronald,2007). It is widely used for structural applications in buildings, transportation, ship building, mining, and energy and packaging industries. Among the different grades of steel, 304stainless steel is known to be important

DOI: 10.4018/978-1-7998-2775-7.ch009

in both industry and academic research. This is due to its adequate mechanical properties viz. tensile strength of 515 MPa and yield strength of 205 MPa, 40% ductility and 92 HRB, as well as good physical properties, like density of 8.00 g/cm^3, melting temperature of 1450°C, modulus of elasticity 193 GPa, electrical resistivity of 720 nΩm, and thermal conductivity of 16.3 W/m-K (Metals Handbook,1990) and (Harvey,1982). It also exhibits excellent heat resistance properties and can withstand up to 870°C working environment with no morphological changes. Its formability is suitable for both cold working, i.e. below <400°C, as well as hot working condition, i.e. above 700°C. Here cold working means fabrication below re-crystallization temperature, while hot working means fabrication above the re-crystallization temperature. Although 304SS is known to exhibit remarkable work hardening under several conditions, which makes it suitable for structural applications, it has also has lower stabilizing capacity of microstructure in austenitic phase. Hence, under high stress conditions i.e. plastic deformation, phase transformation by strain-induced martensitic transformation is noted to occur (Nakajima,2010).

The practical applications of 304SS are aplenty. It is a widely used for industrial applications which includes automotive parts of vehicles, fasteners, sleeve, bolts, and fixture etc, and in field of construction include railing guard, wire rope mesh etc. It is also widely used in the food industry for processing of different types of food containers like that needed for transferring food from one place to other and also tableware like cutlery, knifes, plates etc. It is also particularly used in the dairy industry as a safe and hygienic carrier for carrying milk and dairy products. Several types of chemical storage equipment's, trays, plates, stools etc are used for carrying chemical substance and other drugs from one place to other place.

It is widely used in chemical industries for container, trays, cups, ladles, vessels etc for storing hazardous chemical solutions. It is popularly used in heat exchangers (support columns, guiding bar, bolts etc) and steam boiler plant (pipes, tubes, water storage etc). It is also widely used in marine applications for fabrication of structural and functional ship parts, seats, railing guard etc. It is however, most widely used in household utensils like plate, tray, cup, spoon, refrigeration appliances, door lock, exterior design, other storing containers, etc. (Gedge,2008), (Baddoo,2008). Stainless steel is an alloy steel that has other alloying elements besides carbon and iron. Table 1 shows a typical example of composition of 304SS.

In 304SS, the composition of chromium and nickel plays a vital role for stabilizing microstructure properties and passivity. The amount of chromium is optimized to improve its corrosion resistance by passivation. Passivation is the formation of an impervious layer of chromium oxide that protects it from corrosion. Whereas nickel is an austenite phase stabilizer, that makes the steel non-magnetic for use in common day-to-day situations. Manganese is also an austenite stabilizer. The addition of manganese also improves its hot working properties and boosts its toughness, strength and hardenability. Phosporous and sulphur are used to improve its machinability, however, in low amounts (<0.05 wt %) as they also degrade its corrosion properties. Nitrogen and oxygen is naturally occurring and is usually present in most steel in small quantities and serves as austenite stabilizer.

The amount of carbon is at an average 0.06 wt% which classifies it under low carbon steel. However, based on the amount of carbon, there are a few variants of the 304SS i.e. 304L and 304H, where the carbon varies from 0.03 to 0.1 wt%, respectively. 304L stainless steel has low carbon contents, and makes this widely used for large and heavy gauge application. Benefits of using this graded steel is that there are no requirements of post-weld annealing and better suits for welding purposes. On the other hand, 304H steel with higher carbon contents in their composition, and corrosion resistance against chemically activated and sea shore (salty) is less, hence there are more chances of getting corroded. However, having higher contents of carbon in their composition, it better suits for the condition where good mechanical properties needed e.g.; clamp, hinges, pressure vessels application etc. This ensures that the low and high carbon variant 304 stainless steel can be used for heavy gauge applications, machine parts, structural components and higher temperature gradients, respectively but not suitable for corrosion resistance application (Prasad,2012) and (Sen,2011). Some of these key differentiating points between 304L and 304H have been discussed in the Table 2.

Although the addition of the other alloying elements (Cr, Mo, Ti, etc) is designed to improve the corrosion resistance, it however, negatively affects its formability and welding properties. The disadvantages being as follows:

1. Easily corrodes when it comes in contact with atmospheric gases i.e. sulphur, nitrogen and chlorine.
2. Low cutting performance.
3. High maintenance cost.
4. Low resistance against fire than concrete.
5. It doesn't resist fluctuation of load, leads to tension in structure.
6. Difficulty in molding processing.

Although the composition of 304SS is tuned for corrosion resistance properties, the corrosion behavior of such austenitic steels has been found to be less than optimum for various working conditions (Guocheng,2008). It is susceptible to pitting and crevice corrosion in the presence of chlorine and even stress corrosion cracking above 60°C (Al-Fadhalah,2018).

Moreover, under the action of bacterium (marine environment) it has been reported that the stability of the passivating layer does not remain intact and leads to localized breakdown at the site of attack (Fang,2015), (Tsutsumi,2007). The studies done on aerobic and anaerobic bacteria show that localized breakdown occurred forming micro pits that increase in size linearly with time (Tsutsumi,2006).

Pitting and crevice corrosion are the two most important types of corrosion processes for almost all grades of steel. Both pitting and crevice are localized corrosion processes. In pitting, the surface of steel when exposed to some destabilizing environment is locally able to breakdown a microscopic area (in form of tiny holes or pits) where aggressive corrosion occurs. Whereas, in crevice sharp corners and design flaws introduce stagnant corrosive media that again locally breaks down the passive film. These types of corrosion appear at localized gaps where flow of air and water are not available. They are also common in metal to non-metal contact component; example may be coupling, gasket, washer and joints of parts. Corrosion on 304SS steel surfaces may be related to temperature, pH, chloride and ions concentration present in the solution (VeraCruz,1998), (Widodo,2013) and (Maeir,2010).

This types of corrosion is dangerous than normal uniform corrosion, it creates small holes at the localized portion. These tiny holes of corroded surface are impossible to detect using naked eyes, so for measuring and monitoring corrosion various types of characterization methods are available. Usually high magnification range microscopy is sufficient to visually detect corrosion. This includes characterization methods like FE-SEM (field emission scanning electron microscopy) and optical/laser confocal microscopy for initial inspection of the corrosion site.

Overall corrosion has negative impacts on steel manufacturing industry in forms of scraps (waste). As per reports of United States gross domestic product (GDP), the total expenditure cost of corrosion related to steel was estimated about to $276 billion per year, which was 3.1% of the U.S. (GDP) (Meligi,2010). Compared to 316SS and 321SS grades of steel, 304SS has higher corrosion resistance capacity due to the higher amount of Cr in its composition. However, it is not possible to resist corrosion in acidic media, seashore and hydrogen conversion environments. Therefore to control the corrosion in 304SS, selective surface protection of steel is usually adopted. In this context lot of research had been conducted to alter the surface properties that reduce corrosion.

The surface modification techniques may be (i) additive in the form of coatings or (ii) subtractive in the form of laser, ion implantation etc. This chapter deals with the corrosion preventive methods, specially adapted for inhibiting corrosion in 304SS steel that are published by research groups. The US Department of Energy (DOE) has published a report by Argonne National Laboratory on the status on bipolar plate materials (Kopasz,2017). Now bipolar plates are an integral part of the proton exchange membrane (PEM) fuel cell that connects and separates the electrodes in the fuel stack. They have identified stainless steel, aluminum and titanium as the preferred materials for bipolar plate formation (Hassan,2019). Of these materials, steel and particularly 304SS graded is of particular importance due to its low cost and more availability.

Other graded stainless steels like 316 have more manufacturing cost than the acceptable limit set by DOE. In this context, 304SS is first structurally formed and then coated with the corrosion resistant material. The price can further be reduced by opting for non-precious coating materials like rare earth elements and their compounds. In this respect there are many studies that have been done on 304SS using both different coating materials as well as deposition method. The next section assimilates these different corrosion protection materials for various applications as originally published. These can be as simple as screen painting and use of corrosion inhibitors or more exclusive like vapor deposition and functionalized surface modification techniques. Before delving deep into subject of reviewing the corrosion protection methods, lets us also look into the basics of corrosion and the characterization and measurement methods that are available for characterizing corrosion.

2. THEORETICAL ASPECTS OF CORROSION

2.1. Basic Electrochemistry

Corrosion is the mechanism of oxidation/reduction by which the metal looses an electron and becomes electropositive, according to the equation 1.

$$M ® M^{+x} + Xe^{-} \quad (1)$$

Electrochemistry of corrosion process includes a fluidic medium, like water or acid to act as an electron acceptor or polarizer. There are a number of different types of corrosion, depending on its environment and cause. The corrosion which can be identified by simply visual inspection are uniform, pitting, crevice and galvanic corrosion. The types that can be identified by special tool (measurement/monitoring methods discussed below) are erosion, cavitation, fretting and intergranular corrosion. And finally the types that require microscopic investigations are exfoliation, dealloying, stress corrosion cracking and corrosion fatigue (Hameed,2016).

However, independent of the environment the mechanism of corrosion is similar. The corrosion system is similar to an electrochemical cell which has an anode that looses electrons, and a cathode that accepts electrons. The anode is usually the sacrificial metal that gets corroded. Now due to this flow of electrons the cell has an electrical potential difference between the electrodes. As soon as the electrodes are externally connected by a circuit, the flow of current can be measured in the cell. This is very useful in monitoring the electrochemical reactions in a cell.

The electrochemical reactions in a cell can be accurately measured and monitored to evaluate the corrosion reactions that occur. There are many corrosion prevention techniques that appropriately use corrosion inhibitors, galvanic couples, anodic and cathodic protection and smart designs to keep corrosion at bay. However, corrosion cannot be completely ignored and forgotten. In this context corrosion monitoring is the most useful technique to prevent massive failures. There are number of methods to compare corrosion potential of different materials that are generally used. Linear polarization by potentiodynamic scan is a particularly useful tool in this area.

2.2. Characterization Methods

Visual inspection of corroded parts usually does not show any visible marked changes on the steel surface. The part is usually inspected under a high magnification optical/laser microscope. Pitting or crevice corrosion is evident by the presence of tiny microscopic holes on the surface of the steel. For a closer look, the specimens may be metallographically prepared and examined by electron microscope methods like scanning electron microscopy and transmission electron microscopy. SEM provides idea about the physical as well as the chemical changes that may have occurred in conjunction with advanced microscopy techniques like back scattered diffraction or wavelength dispersive spectroscopy. However, these advanced microscopy or spectroscopy methods do not explore the corrosion potential. In this context, advanced potentiodynamic scanning method, electron impedance spectroscopy are the methods that are usually employed.

2.3. Potentiodynamic Scan

A potentiodynamic scan is a useful method in characterizing or defining the corrosion resistance of a material. Here, the corrosion resistant material is used as an electrode in an electrochemical cell. A potentiostat applies a weak current which acts as a driving force for the electrochemical reactions in the cell. This driving force dictates which electrochemical reaction takes place at the cathode and anode as well as their rate. The half cell potential for given electrochemical reaction is given by the Nernst Equation, shown in equation 2.

$$E = E_0 - \frac{RT}{nF}\ln\frac{product\ of\ activity\ of\ products}{product\ of\ activity\ of\ reactants} \tag{2}$$

The potentiodynamic scan is used to interpret data about the corrosion reactions that occur in the cell. It is most importantly used to assess the current density, i usually expressed in terms of A/cm^2. The current density is given by the Tafel equation, equation 3, where the over potential, η is given as

$$\eta = \beta \log\frac{i}{i_0} \tag{3}$$

This is used to construct a Tafel plot (or simply polarization curve) of applied current density, i (usually in logarithmic) against the applied potential (η), where β is the Tafel slope and i_0 is the exchange current density. This has a cathodic reaction ($E < E_{corr}$) and an anodic reaction ($E > E_{corr}$).

A typical Tafel plot of a potentiodynamic scan can be used to derive the corrosion current density (i_{corr}) and corrosion potential ($E_{corr)}$ as seen in figure 2. Tafel slopes β_c and β_a for the cathodic and anodic reactions, respectively, and the corrosion rate. Tafel extrapolation is probably one of the simplest methods that is used in electrochemistry to predict corrosion performance of materials. The Figure 2 shows an actual Tafel plot generated by a potentiodynamic polarization scan of a polished sample (304SS) in 3.5 wt% NaCl solution. The OCP or corrosion potential of the sample is 0.38 V. The anodic Tafel coefficient is the slope of the extrapolated anodic current, β_a is 0.067 V/dec and the cathodic Tafel coefficient is the slope of the extrapolated cathodic current, β_c is 0.28 V/dec. The corrosion current is the intersection of the extrapolated cathodic and anodic currents, shown in the figure. The corrosion current in this case of 304SS is 1.76 x 10^{-4} A/cm^2 (Mandal,2020).

2.4. Electrochemical Impedance Spectroscopy (EIS)

In general electrochemical impedance spectroscopy is used for investigating the electrical properties of different types of materials i.e., ionic, semiconductor or insulator. It provides information about their conductivity, dielectric constant and capacitance. This technique is based on measuring the impedance (resistance and capacitance) of the material by application of a sinusoidal AC current (Zhu,2019). Here a sinusoidal AC potential is applied and the resulting current through the cell is measured. The resulting current is analyzed in terms of Fourier transform to obtain the impedance. The impedance, Z of the system is a complex number which is calculated in terms of the phase shift, φ between the input and output waves and the frequency (ω). If we consider, Z' as the real part and Z" is the imaginary part of the expression, the plot of Z on the Z' and Z" plane is called as the Nyquist plot. The Nyquist plot is essential in deciphering even the minute changes in the system that can be quantified.

In corrosion, EIS can detect real time change in the Z of the system due to the presence of the corrosion by-products as they evolve. However, parallel microstructural confirmation through optical/electron/laser microscopy is needed to understand the corrosion mechanism. The Nyquist plot is continually monitored over the duration of the experiment, and the relation of the impedance, Z, i.e. linearity is monitored. The EIS of a ferritic-martensitic steel shows the Nyquist plot for real time monitoring for over 1000 hrs (Encinas-Sanchez,2019). The article shows that for P91 steel when exposed to a corrosive environment (molten salt bath), the initial reaction shows not only the corrosion process and mechanism, but indicates instability of the salt bath at the measured temperatures (Encinas-Sanchez,2019).

2.5. Measurement of Corrosion

Measurement of corrosion is most important in assessing the severity of failures due to corrosion. Moreover, in recent times its serves as a method to evaluate the performance of various corrosion resistance materials and develop the correct material for the appropriate use. The corrosion rate (mm/year), given in equation 4, for a electrochemical reaction is calculated from the combination of the corrosion density equation and Faraday's equation,

$$Corrosion\,rate = \frac{weight\,of\,material\,removed \times exposed\,surface\,area}{density} \quad (4)$$

Where the weight of material removed is calculated from Equation 5,

$$Weight\,of\,material\,removed = \frac{M \times Q}{n \times F} \quad (5)$$

Where, M is the molecular weight, Q is the total charge passed (= i_{corr} x time), n is the number of electrons transferred in the reaction and F is the Faraday's constant.

For 304SS, the calculated corrosion rate extrapolated from the Tafel plot (Figure 1) is 8.6 mm/year in a 3.5 wt% NaCl solution. It is important to stress that the corrosion rate is calculated or predictive in nature and will change with the corrosive media, experimental procedure, etc.

Usually i_{corr} and E_{corr} are compared for different materials to obtain a comparative idea of corrosion for them. Typically the electrode potential $E_{corr,}$ is related to the oxidizing power of the electrolyte and i_{corr} is related to the polarization of the electrode (material). A material with a lower i_{corr} has better corrosion resistance. Also corrosion rate is an overall indicator of the rate of material removal in a year. A lower corrosion rate translates to better corrosion resistance. However, this is not particularly useful for specific types of corrosion like pitting.

2.4. Online Corrosion Monitoring

The above discussion for corrosion analysis, as the reader will note evaluates a single Tafel or Nyquist plot for evaluation that generates data for the technical analysis. It is therefore imperative to establish steady state or equilibrium condition to maintain that the corrosion reactions are indeed over in the measurement time line. Although it is a good method for establishing and comparing corrosive nature of different materials and system, it does not provide dynamic testing flexibility. Online monitoring provides that platform for real time corrosion measurement, that is both dynamic (changes with time) and flexible.

Although corrosion analysis provides specific parameters like open circuit potential, Tafel coefficients, corrosion current density for investigating an equipment's condition in highly corrosive environments such as in metallurgical, oil-gas, petrochemical, water treatment, chemical, pharmaceutical, environmental and food industries. They usually have difficult to analyze these data, since it is either under water or under soil or in molten stages. This makes corrosion measurements difficult to investigate through offline mode. This has therefore given rise to the need of an online corrosion monitoring system.

Online corrosion monitoring involves continually measuring the corrosion process by attaching a probe to the system. The probe gives a weak signal (converted graphically) through which corrosion analysis is conducted over a long period of time. The probes involved are usually working electrode and reference electrode. Usually two or sometimes even three probes are strategically placed over large areas, over which the signal is collected. The probes are usually made of the same material as the corroding metal to get better correlation with actual data. The corrosion of the working probes in the corrosive media, leads to generation of a weak signal on application of a small potential. The signal is proportional to the i_{corr} at the corroding site. There are a few techniques that can be used in conjunction with online monitoring to study the corrosion behavior. The techniques may be used online to give the actual corrosion loss or off-line to predict the corrosion loss. Laser-induced breakdown spectroscopy (LIBS) is an example of such widely used online corrosion monitoring system (Zeng,2018) and (Xiong,2019). LIBS is an online method to study the corrosion behavior of metal, which is based on the application of high-power density laser pulses focused on a materials upper surface to stimulate the formation of transient plasma. After cooling the process emission spectrum produced during process gives the information of the sample properties (Zeng,2018). LIBS has been widely used method for online corrosion monitoring, since industries involving liquid stages of metals, gaseous form hence very difficult to rectify corrosion in offline mode (Zeng,2018). Electrochemical frequency modulation (EFM) on the other hand utilizes a technique much similar to that of EIS. It uses two sinusoidal waves of two different frequencies to the cell simultaneously. This is direct method that obtains the corrosion rates directly without the use of Tafel plots (Bosch,2001). So these kinds of method have to involve for cathodic protection of equipment's which is highly corrosive in such environments. Now a day's Online corrosion monitoring attracted researcher as well as scientist for further investigation, since this technique involved accurate corrosion protection in easy way through high quality digital images or graph with attached probes which is connected to system.

3. REVIEW OF MATERIALS

A comparison of the different coating materials and their deposition technique on 304stainless steel has been listed in Table 3 along with their corrosion parameters as listed by the authors (references shown). The materials may broadly be divided as polymer, metallic and non-metallic materials. It is not only the selection of the anti-corrosive material that is important but also the coating method that dictates its corrosion performance. The deposition methods have been discussed below.

3.1 Electroless Plating

Electroless plating is a chemical deposition method in which the substrate is dipped in a saturated solution containing the electrolyte. The coating material by virtue of thermodynamics deposits on the substrate uniformly. Zuo et al. (2008) deposited palladium through electroless plating method on stainless steel surface. Through analysis they revealed that films grown were blackish in color and surface was smooth, dense good adhesion as well as no defects found on it. Chemical analysis suggested that corrosion resistance of palladium films showed good corrosion resistance than bare one, but difficulties with palladium thin films was that it could not be resistible at higher concentration of chloride ions(chemically activated area) for long range.

3.2 Electrodeposition Method

Electrodeposition is used to coat the substrate with a thin layer of a material by application of a small current through the electrolyte. This has more uniformity and control over the substrate thickness through controlling the deposition parameters (like current, zeta potential, pH, etc). The system has somewhat flexibility to deposit metals as well as polymeric and ceramic coatings by this method. This method has been conveniently used by many to produce reliable corrosion resistant coatings.

Subathira et al. (2010) deposited polyaniline/polypyrrole composite coatings by electrochemical or electrodeposition method at constant potential and revealed that coating layer obtained was smooth and had good adhesion. Corrosion test showed that stability of the coating decreases at higher concentrations and higher deposition times lead to rapid increase in current density (i_{corr}) .corrosion potential (E_{corr}) value showed the anodic protection against corrosion. Ren et al. (2010) doped polyaniline (PANI) with dodecylbenzesulfonate anions by electrodeposition process on stainless steel substrate. Electrochemical measurement at atmospheric condition of H_2SO_4 and HCl shows that corrosion potential increases and decreases the corrosion current density in acidic condition.

In metals deposited by electrochemical process, nickel is probably the most widely used material. Li et al. (2019) had deposited nickel on stainless steel substrate by electrodeposition method and revealed through chemical analysis at NaCl acidic environment that nickel coated sample showed corrosion potential (E_{corr}) in increasing order and corrosion current density in decreasing order than bare stainless steel. Here the researchers reported that the nickel coated sample showed higher corrosion resistance (2 orders in magnitude) as compared to uncoated 304SS. In another article, Jiang et al. (2018) deposited $Ni(OH)_2$ reinforced PANI coating through electrodeposition method and conducted electrochemical analysis creating NaCl acidic environment. They reported that $Ni(OH)_2$ reinforced polyaniline coated surface was found less porous and more smooth to enhance the hydrophobic nature of stainless steel. Potentiodynamic scans show an excellent corrosion potential and polarization resistance of $Ni(OH)_2$ reinforced polyaniline than simple polyaniline coating, since it forms an active-passive layer and facilitates barrier against corrosion when immersed in acidic medium for long range.

In ceramic coatings, hydroxyapetite is probably most widely used material to prevent biocorrosion of metallic implants (variants of 304 and 316 grade SS are also commonly used as implant materials). Huda et al. (2018) deposited hybrid composite of HAP/rGO/Pd nanocomposite on 304SS by electrodeposition method. The samples when immersed in synthetic serum ($CaCl_2$, Na_2HPO_4) for 2 days through potentiodynamic polarization test revealed that corrosion potential and polarization resistance was high as well as corrosion current density was lower for the coated sample as compared to uncoated 304SS. This suggests positive attribute of 304 SS towards corrosion barrier in synthetic serum environments. In another article by Eljaouhari et al. (2019) zinc phosphate-polypyrrole (ZP/PPy) coatings were deposited by electrodeposition method and electrochemical analysis was conducted on 3% NaCl acidic environments. The potentiodynamic polarization analysis revealed that ZP/PPy coated on 304SS shows positive corrosion potential as well as low corrosion current density than zinc phosphate or polypyrrole coated samples alone. The double layer of ZP/PPy coatings provides excellent corrosion resistance in acidic medium.

In nanocomposites produced by this method, Wei et al. (2018) had grown WSe_2/TiO_2 nano-composite on stainless steel by anodic oxidation and electrochemical deposition. Analysis of the results show that WSe_2/TiO_2 nanocomposite coated on stainless steel had excellent open circuit potential than bare stainless steel and alone TiO_2 coated sample. Cathodic protection was improved for 304SS coated with WSe_2/TiO_2 nano-composite. These have been found to be hugely applicable in marine applications.

3.3 Alloyed and Addition Method

Alloying by chemical or mechanical methods, involves addition of two or more elements to the base material. This is found to improve not only the mechanical properties of the base material, but also its corrosion resistance. The alloying of stainless steel by addition of chromium to improve its passivation is probably the most widely known. Some other significant articles are discussed here.

Zhang et al. (2001) alloyed yttrium over substrate and revealed that yttrium alloyed substrate had significantly enhanced the corrosion resistance (passive layer) than bare stainless steel in H_2SO_4 acidic environmental condition as well as no changes in the mechanical properties of yttrium alloyed sample. This coated sample had also showed good erosion rate and provide scratch free surface (due to impingement of solid particle on stainless steel upper surface). On the other hand, Shi et al. (2018) added different contents of yttrium on 304SS to improve its corrosion resistance in 3.5% NaCl solution at 298K. They have reported that 0.007 wt% and 0.049 wt% yttrium added on surface of sample showed less barrier capability from corrosion while 0.019% contents yttrium showed best corrosion resistance property against pitting corrosion than others. They also reported that at first the corrosion resistance decreased, then gradually increased and finally corrosion resistance decreased at the higher concentrations of yttrium. The reduction of corrosion resistance at higher amounts of yttrium may be related to poisoning. In another article by Xuegang et al. (2015) CeO_2-Cr modified Ti coatings on stainless steel by pack cementation method and revealed after chemical analysis on NaCl acidic environment that open circuit potential was minimum for bare stainless steel than Ti(Titanium) coated and CeO_2-Cr modified with titanium(Ti). This modified layer on stainless steel showed excellent passivation and corrosion current density (i_{corr}) than bare stainless steel and Ti coated on surface of 304SS. In a electroplymerization method, Gao et al. (2016) deposited PANI doped phosphomolybdic acid and showed through EIS at H_2SO_4 and atmospheric condition, that the impedance value of coated samples was higher than uncoated sample, which indicate corrosion resistance was improved more than bare stainless steel. The phosphomolybdic acid was re-doped and deposited on stainless steel sample which do not reflects any diffusion in the coated sample morphology. Tafel plot also showed lower corrosion current density and higher corrosion potential value of coated sample high of the coated samples as compared to the uncoated ones which indicate good corrosion resistance than bare one.

Pan et al (2016) deposited niobium by high energy micro arc alloying technique. They have reported that niobium coated stainless steel exhibits low corrosion current density and higher corrosion potential than bare one and also coated sample showed 180 times more polarization resistance thus suggesting for better corrosion resistance than uncoated one. The niobium coating probably formed a passive anti-corrosive layer. They also evaluated the corrosion through both EIS as well as potentiodyanic scan to show the results to be similar. This shows the effectiveness of using simple techniques like Tafel plots that has the capability to produce results similar to dynamic complicated techniques like EIS.

Sol-gel is another chemical method that is commonly used to produce coatings of almost all materials. Many publications have been reported with sol-gel method to produce a myriad of coating materials. Miranda et al. (2016) deposited silicon oxide on 304 stainless steel sheets using the sol-gel and dip-coating processes. Corrosion test performed in NaCl acidic condition show that corrosion current density of coated sample was low (by only 1 order) as compared to bare one and uncoated sample showed poor corrosion resistance due to the presence of chloride ion which acts as barrier for corrosion resistance at acidic medium. Akhtar et al. (2018) deposited silane coatings on 304SS by sol-gel method and through electrochemical analysis keeping 0.6 NaCl acidic conditions. They have reported that the corrosion potential of the silane coated stainless steel improved in acidic media than the uncoated one. The corrosion current density, extrapolated through a Tafel plot, was 3 orders lower than uncoated one. This attributed towards forming of a physical barrier against corrosion.

Dinu et al. (2018) deposited chromium nitride and oxy-nitride chromium coatings by reactive cathodic arc method. The electrochemical analysis was performed at 0.10 M NaCl + 1.96M H_2O_2 acidic media for pro-longed hours. It shows that chromium oxy-nitride coatings had the lowest corrosion current density, high corrosion potential with fewer pores on surface than the chromium nitride as well as the bare stainless steel substrate.

This review of different materials has been tabularized (Table 3) to show the effect that the material as well as the coating/deposition process has on its corrosion behavior. It is observed that the coating properties like microstructure, adhesion and composition control the corrosion behaviour of the material. For deposition processes like electroless coating, sol gel and dip coating, the corrosion performance is rather low and provides current density values only slightly better than that of bare steel (Zuo,2008), (Akhtar,2018) and (Miranda,2016). Whereas for other deposition processes like electrochemical deposition, cathodic arc and sol-gel provides atomic bonding between the substrate and coating material, and thus improves its performance considerably. Moreover, the microstructure features like grain boundaries, porosities and defects will affect the interaction and penetration of the corrosive environment to the substrate. Another important aspect, often neglected by researchers, is the coverage of the coating on the substrate material. This includes free edges and corners that are not easy to cover. Although most publications provide in-depth analysis on the corrosion they do not emphasize on this aspect.

In general, the materials may be divided as polymers, metals and non-metallic ceramic materials. For polymer based coating (Jing,2019), (Zuo,2008), (Eljhouri,2019) and (Jiang,2018) the deposition methods are usually electrodeposition, sol-gel and dip coating techniques. The corrosion density, as an indicator of the corrosion resistance of the material, is reported for PANI based coatings to be in the range of 10^{-6} to 10^{-7} A/cm^2 i.e. almost 2-5 orders higher in magnitude than that for 304SS (Jing,2019). On the other hand, metallic coatings that have been deposited on 304SS include palladium, nickel, yttrium, titanium, chromium and niobium, require deposition techniques like arc deposition, electrodeposition and vapor deposition methods Jing,2019), (Yangzhi,2016) and (Xuegang,2015). The corrosion current density for nickel deposited by electrodeposition in this case has been reported to be in the range of 10^{-7} A/cm^2 (Jing,2019). Whereas corrosion inhibitors like yttrium (Xuegang,2015) and niobium (Yangzhi,2016), have much lower current densities 1 to 2 orders lower. This indicates protective coatings provide better corrosion protection than addition of corrosion inhibitors. For non-metallic coatings mostly include a metallic layer and may be included as a composite coating. These are chromium nitride and oxy-nitride, silane, silica, titanium oxide and hydroxyapetite (HAP) coatings. In this series, silane shows the lowest corrosion current density at 10^{-9} A/cm^2 (Zhang,2001) and HAP with reduced graphene oxide and palladium nano-composite shows a current density at 10^{-8} A/cm^2 (Wei,2018). These corrosion current densities are again much higher, 5-6 orders higher than that for 304 SS grade steel.

5. CONCLUSION

304SS has been established widely used as the preferred material in various industries due to its high strength, ductility and hardness. It can be used as-it-is for welding purposes without having to go through any heat treatment procedures. It, however, has poor corrosion resistance that makes it unlikely to be used in aggressive environments like that of fuel cells and heat exchangers. The basics of electrochemistry covered in the chapter, helps to understand the corrosion process and its different characterization methods. In relevance to the present day evaluation of corrosion performance there are only a few characterization methods that are actually of importance. The potentiodynamic polarization method is the most popular and reliable method that has been commonly employed by most researchers to evaluate the corrosion reactivity of different materials. It generates a Tafel plot that is used to calculate the corrosion parameter, namely corrosion current density, i_{corr} (measured in $\mu A/cm^2$). Electrochemical impedance spectroscopy is also used in the same manner to give more real time corrosion performance data. This chapter provides an overview of these corrosion parameters, obtained from their published reports and tabulated together for easy comparison of both the anti-corrosive materials as well as some generally used deposition processes.

Many researchers have employed different corrosion protection materials on 304SS and have performed electrochemical analysis like linear potentiodynamic scans and EIS on them. This chapter has systematically reviewed all polymer, metallic and non-metallic coating materials through different deposition techniques that have been used by researchers. A tabulation of the different materials and their coating processes actually provides a quick comparison as well as an insight into the effectiveness of different materials and the coating methods.

A wide range of corrosion protection materials have been deposited by researchers ranging from polymers like polyanalline that provide excellent corrosion resistance at low cost deposition techniques. Polymer based materials are the most common as they deploy low temperature and straight forward procedures for deposition. Whereas, metallic (pure, alloyed and doped) require more sophisticated and advanced deposition methods to obtain high quality coatings. However, metallic coatings generally have better adhesion strength and provide better corrosion resistance even at the most corrosive/aggressive environment. The lowest corrosion current density is in the range of 10^{-8} to 10^{-9} A/cm^2 for non-metallic nano-composite coatings.

It is worthwhile to investigate the relation of the corrosion current density to the deposition process also. For processes where the interaction between substrate and coating material is not in the atomistic scale, i.e. there is no bonding or adhesion, the corrosion current density is very low. For processes like dip coating and electroless deposition, the corrosion current density values are almost equal to that of bare steel. However, for deposition processes like vapor deposition and electrochemical deposition the bonding strength is appreciable such that the corrosion current density values are almost 4-6 orders higher in magnitude.

REFERENCES

ASM International Handbook Committee. (1990). *Metals Handbook. ASM International.*

Bastian, M., & Frankel, G. S. (2010). Pitting Corrosion of Bare Stainless Steel 304 under Chloride Solution Droplets. *Journal of the Electrochemical Society*, *157*(10), C302–C312. doi:10.1149/1.3467850

Bosch, R., Hubrecht, J., Syrett, B., & Bogaerts, W. (2001). Electrochemical Frequency Modulation: A New Electrochemical Technique for Online Corrosion Monitoring. *Corrosion*, *57*(1), 57. doi:10.5006/1.3290331

Cruz, P. V., Nishikata, A., & Tsuru, T. (1998). Pitting corrosion Mechanism of Stainless steel under wet-dry exposure in Chloride-Containing Environments. *Corrosion Science*, *40*(1), 125–139. doi:10.1016/S0010-938X(97)00124-8

Eljaouhari, A., Chennah, A., Jaddi, S. B., & Ahsaine, H. (2019). Electrosynthesis of Zinc Phosphate-Polypyrrole coatings for improved corrosion resistance of steel. *Surfaces and Interfaces*, *15*, 224–231. doi:10.1016/j.surfin.2019.02.011

Fang, D., Jun-mao, Q., & Hao-hua, D. (2015). *Study on Thermal Physical Properties of 304 Stainless Steel*. The Minerals, Metals & Materials Society.

Gedge, G. (2008). Structural uses of stainless steel Buildings and civil engineering. *Journal of Constructional Steel Research*, *64*(11), 1194–1198. doi:10.1016/j.jcsr.2008.05.006

Guocheng, L., Haidong, C., Chunchun, X., & Zonghu, H. (2008). Effect of Strain and Chloride Concentration on Pitting Susceptibility for Type 304 Austenitic Stainless Steel. *Chinese Journal of Chemical Engineering*, *16*(2), 314–319. doi:10.1016/S1004-9541(08)60080-4

Howard, E. B., & Timothy, L. G. (Eds.). (1985). Metals Handbook. American Society for Metals.

Huda, F. S. G. A. (2018). HAP/rGO/Pd nanocomposite thin film coating on SST 304 - Synthesize, characterization, and properties investigations. *Journal of Alloys and Compounds*, *741*, 562–574. doi:10.1016/j.jallcom.2018.01.047

Indrani, S., & Emmanuel, A., N. S. K., & Eric, F. (2011). Microstructure and mechanical properties of annealed SUS 304H austenitic stainless steel with copper. *Materials Science and Engineering A*, *528*(13), 4491–4499.

Jiang, L., Junaid, A. S., Yangzhi, G., Hongbin, L., & Xiangkang, M. (2018). Electrodeposition of $Ni(OH)_2$ reinforced polyaniline coating for corrosion protection of 304 stainless steel. *Applied Surface Science*, *440*, 1011–1021. doi:10.1016/j.apsusc.2018.01.145

Jing, L., Lida, P., Qiang, F., Yingluo, Z., & Nan, G. (2019). Wettability and corrosion behavior of a Ni coating on 304 stainless steel surface. *Surface and Coatings Technology*, *357*, 740–747. doi:10.1016/j.surfcoat.2018.10.050

Khaled, Al-F., & Muhammad, A. (2018). Microstructure Refinement and Mechanical Properties of 304 Stainless Steel by Repetitive Thermomechanical Processing. *Metallurgical and Materials Transactions A, 49*(4).

Kopasz, J. P., Benjamin, T. G., & Schenck, D. *(2017). Bipolar Plate Workshop Summary Report.* doi:10.2172/1406358

Majid, H. A., & Rana, A. M. A. (n.d.). Tribocorrosion. *Advances in Tribology, 5*, 89-110.

Meligi-El, A. A. (2010). Corrosion Preventive Strategies as a Crucial Need for Decreasing Environmental Pollution and Saving Economics. *Recent Patents on Corrosion Science*, *2*(1), 22–33. doi:10.2174/1877610801002010022

Mihaela, D., Emile, S. M. M., Anca, C. P., Alina, V., Leslie, F. P., & Mariana, B. (2018). Enhancement of the Corrosion Resistance of 304 Stainless Steel by Cr–N and Cr(N,O) Coatings. *Coatings*, *8*(4), 132. doi:10.3390/coatings8040132

Miranda, J. R., Sánchez, M. A., García, E. G., Velazquez, D.Y. M., Á. de J., & Ramírez, M. (2016). Mechanical Properties of SiO2 Coatings for Corrosion Protection of 304 Stainless Steel. *Characterization of Metals and Alloys,* 109–116.

Nakajima, A., Akita, M., Uematsu, Y., & Tokaji, K. (2010). Effect of strain-induced martensitic transformation on fatigue behavior of type 304 stainless steel. *Procedia Engineering*, *2*(1), 323–330. doi:10.1016/j.proeng.2010.03.036

Nancy, B. (2008). Stainless steel in construction: A review of research, applications, Challenges and opportunities. *Journal of Constructional Steel Research*, *64*(11), 1199–1206. doi:10.1016/j.jcsr.2008.07.011

Noor, U. H., Bahadir, T., & Ali, M. S. (2019). A Competitive Design and Material Consideration for Fabrication of Polymer Electrolyte Membrane Fuel Cell Bipolar Plates. *Designs*, *3*(1), 13. doi:10.3390/designs3010013

Pan, T. J., Chen, Y., Zhang, B., Hu, J., & Li, C. (2016). Corrosion behavior of niobium coated 304 stainless steel in acid solution. *Applied Surface Science*, *369*, 320–325. doi:10.1016/j.apsusc.2016.02.088

Philip, D. H. (1982). Engineering Properties of Steels. American Society for Metals.

Pramod, M., Usha, N. K., Soobhankar, P., & Sudesna, R. (2019). Investigation of the effects of electrophoretic deposition parameters on 304SS steel coated with graphene oxide for PEMFC application. *Materials Today: Proceedings*. doi:10.1016/j.matpr.2019.12.360

Qinyi, W., Xiutong, W., Xiaobo, N., Xinran, L., Jing, S., Hong, L., Wencheng, W., Yanliang, H., ... Baorong, H. (2018). Characteristics and anticorrosion performance of WSe_2/TiO_2 nanocomposite materials for 304 stainless steel. *Surface and Coatings Technology*, *352*, 26–32. doi:10.1016/j.surfcoat.2018.08.004

Ren, Y. J., Chen, J., & Zeng, C. L. (2010). Corrosion protection of type 304 stainless steel bipolar plates of proton-exchange membrane fuel cells by doped polyaniline coating. *Journal of Power Sources*, *195*(7), 1914–1919. doi:10.1016/j.jpowsour.2009.10.003

Ronald, L. P., Clara, H., Doris, M. E., Paulo, R. R., & Angelo, F. P. (2007). A Short Review on Wrought Austenitic Stainless Steels at High Temperatures: Processing, Microstructure, Properties and Performance. *Materials Research*, *10*(4), 453–460. doi:10.1590/S1516-14392007000400021

Shan, Z., Xinyang, S., Xiaoyang, G., Jianrong, W., Naiqin, Z., & Junwei, S. (2019). Equivalent circuit model recognition of electrochemical impedance spectroscopy via machine learning. *Journal of Electroanalytical Chemistry (Lausanne, Switzerland)*, *855*, 113627. doi:10.1016/j.jelechem.2019.113627

Siva, P. K., Srinivasa, C. R., & Damera, N. R. (2012). A Review on Welding of AISI 304L Austenitic Stainless Steel. *Journal for Manufacturing Science and Production*, *14*(1), 1–11.

Subathira, A., & Meyyappan, R. M. (2012). Corrosion Protection of Stainless Steel by Polyaniline/Polypyrrole Composite Coating. *International Journal of Engineering Science and Technology, 2,* 7105-7111.

Sultan, A., & Asif, M. (2018). Enhancement of Anticorrosion Property of 304 Stainless Steel Using Silane Coatings. *Applied Surface Science*, *440*, 1286–1297. doi:10.1016/j.apsusc.2018.01.203

Teguh, D. W., & Kazuhiko, N. (2013). *Corrosion Resistance Enhancement of 304 Stainless steel Under droplet of Chloride Solution by Mechanical Surface Treatment*. The Minerals, Metals & Materials Society.

Tiancheng, Z. (2001). Improvement in the corrosion-erosion resistance of 304 stainless steel with alloyed yttrium. *Journal of Materials Science*, *36*(14), 3479–3486. doi:10.1023/A:1017920314397

Víctor, E. S., Maria, T. D. M., Isabel, L., & Gustavo, G. M. (2019). Electrochemical impedance spectroscopy (EIS): An efficient technique for monitoring corrosion processes in molten salt environments in CSP applications. *Solar Energy Materials and Solar Cells*, *191*, 157–163. doi:10.1016/j.solmat.2018.11.007

Weining, S., Shufeng, Y., & Jingshe, L. (2018). Correlation between evolution of inclusions and pitting corrosion in 304 stainless steel with yttrium addition. *Scientific Reports*, *8*(1), 4830. doi:10.103841598-018-23273-x PMID:29556014

Xianqiang, X., Liya, F., Yiming, X., Qiqiang, Z., Jiangshan, L., Yong, W., Guihua, C., Chenglin, W., Tianding, L., Shuai, F., & Guolin, Z. (2019). Online corrosion monitoring of photoelectrochemical cathodic protection of carbon steel using particle video microscope. *Optik (Stuttgart)*, 164091.

Xing, X., Han, Z., Wang, H., & Lu, P. (2015). Electrochemical corrosion resistance of CeO2-Cr/Ti coatings on 304 stainless steel via pack cementation. *Journal of Rare Earths*, *33*(10), 1122–1128. doi:10.1016/S1002-0721(14)60535-4

Yangzhi, G., Junaid, A. S., Hongbin, L., & Xiangkang, M. (2016). Anti-corrosive performance of electropolymerized phosphomolybdic acid doped PANI coating on 304SS. *Applied Surface Science*, *360*, 389–397. doi:10.1016/j.apsusc.2015.11.029

Yusuke, T., Atsushi, N., & Tooru, T. (2006). Monitoring of Rusting of Stainless Steels in Marine Atmospheres Using Electrochemical Impedance Technique. *Journal of the Electrochemical Society*, *153*(7), B278–B282. doi:10.1149/1.2202110

Yusuke, T., Atsushi, N., & Tooru, T. (2007). Pitting corrosion mechanism of Type 304 stainless steel under a droplet of chloride solutions. *Corrosion Science*, *49*(3), 1394–1407. doi:10.1016/j.corsci.2006.08.016

Zeng, Q., Pan, C., Li, C., Fei, T., Ding, X., Du, X., & Wang, Q. (2018). Online monitoring of corrosion behavior in molten metal using laser-induced breakdown spectroscopy. *Spectrochimica Acta. Part B, Atomic Spectroscopy*, *142*, 68–73. doi:10.1016/j.sab.2018.01.011

Zuo, Y., Junlei, T., Chongzhi, F., Yuming, T., & Jinping, X. (2008). An electroless plating film of palladium on 304 stainless steel and its excellent corrosion resistance. *Thin Solid Films*, *516*(21), 75. doi:10.1016/j.tsf.2008.04.049

APPENDIX: FIGURE AND TABLES

Figure 1. Tafel plot of 304 stainless steel tested in 3.5 wt% NaCl at ambient conditions

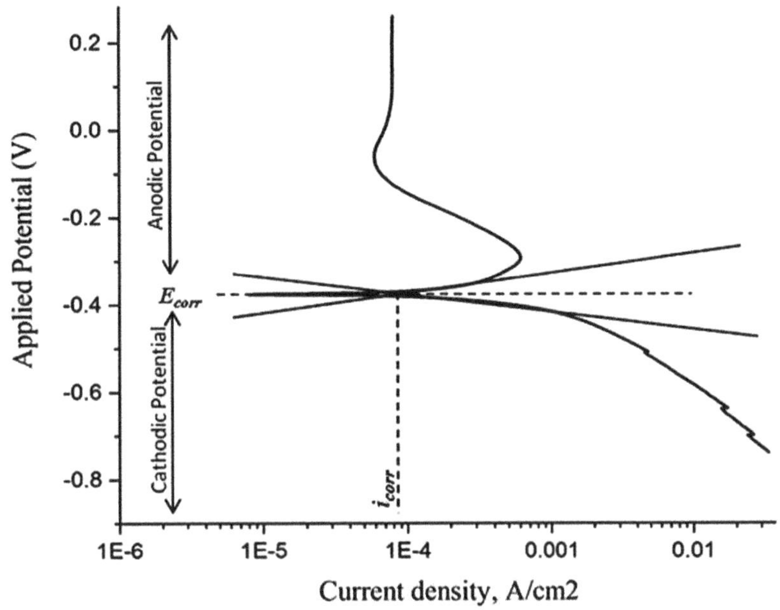

Table 1. Composition of a typical 304SS grade steel

Composition	C	Si	Mn	P	S	Ni	Cr	Fe
Weight (%)	0.1-0.03	0.26	1.76	0.034	0.025	8.01	18.64	Bal.

Table 2. Difference in properties between 304L and 304H variants of steel.

Serial No.	304SS	304L	304H
1	Carbon contents is 0.06%	Carbon contents are very low (0.035%)	Carbon contents are high (0.10%)
2	Good corrosion resistance against corrosive media, alkali, seashore environment etc.	Up to certain limits good corrosion resistance.	Low corrosion resistance
3	Applicable for industrial. infrastructure, bio-medical equipments, household equipments, food processing, universities, scientific research etc.	Applicable for welding and heavy gauge application.	Applicable for area where better mechanical properties (heat resistance, tensile yield strength, creep strength etc) needed e.g. oil refining, heat exchangers, clamp, hinges, pressure vessels etc.

Table 3a. Comparative study of corrosion protection methods used for 304SS

Source	Zuo, 2008	Subathira, 2012	Ren, 2010	Jing, 2019	Jiang, 2018
Anti-corrosion Material	Palladium	Polyaniline/Polypyrrole Composite	Polyaniline doped with dodecylbenzenesulfonate anions	Nickel coating	Ni (OH) $_2$ reinforced polyaniline
Method of Deposition	Electro-less plating	Electrodeposition			
Corrosion Parameters	Open circuit potential(OCP) for Uncoated sample – (0.345V) and Coated sample- (0.365V)	Coated sample E_{corr} values of over 200 mV obtained at deposition time 2 and 3 minutes respectively.	i_{corr} - uncoated (10^{-7} A cm^{-2}),Coated (10^{-9} A cm^{-2}) E_{corr} - Uncoated- (-130mV) Coated (325mV)	i_{Corr} – Uncoated(3.74×10^{-5} A/ cm^2) Coated (6.54×10^{-7} A/ cm^2) E_{corr} -Uncoated (-1.0883 V)Coated (-0.4464V)	OCP- Uncoated-(-201 mV),Coated- (274mV) E_{corr} -Uncoated- (-421mV), Coated- (-22mV) R_p - Coated (6.05 MΩ cm2)
Sl. No.	1	2	3	4	5

Table 3b. Comparative study of corrosion protection methods used for 304SS

Source	Huda, 2018	Eljaouhari,2019	Qinyi, 2018	Tiancheng,2001	Weining,2018	Xing, 2015
Anti-corrosion Material	HAP/rGO/Pd nanocomposite	Zinc Phosphate-Polypyrrole	WSe_2/TiO_2 nanocomposite	Yttrium alloyed	Yttrium addition	CeO_2- Cr modified titanium coatings
Method of Deposition	Electrodeposition			Alloying	Addition	Pack cementation
Corrosion Parameters	i_{corr} - Uncoated- (3.83x10^{-06} A/ cm^2),Coated- (1.05x10^{-07} A/cm^2) E_{corr} - Uncoated-(1.29 x 10^{-01} V),Coated- (4.54 x10^{-02} V) R_p - Uncoated - (308.160 kΩ cm^2), Coated- (603.754 kΩ cm^2)	i_{corr} - Uncoated-(30.19 A/cm^2),Coated-(1.99 A/cm^2) E_{corr} - Uncoated- (-313.30 mV),Coated- (-403.20 mV)	OCP - uncoated (−267 mV), coated(- 800mV)	i_{corr} -Uncoated-(58.4 μA/cm^2),Coated- (1.7 μA/cm^2) E_{corr} -Uncoated-(-100 mV), Coated-(-80mV)	Highest pitting corrosion resistance potential – (0.11V) obtained for yttrium (0.019%) by different contents added to the surface of 304SS	i_{corr} - Uncoated-(0.1363 μA/cm^2), Coated- (0.03086 μA/cm^2) E_{corr} - Uncoated- (-0.32V),
Sl. No.	6	7	8	9	10	11

Table 3c. Comparative study of corrosion protection methods used for 304SS

Source	Yangzhi,2016	Pan, 2016	Miranda,2016	Sultan,2018	Mihaela,2018
Anti-corrosion Material	Polyaniline doped electropolymerized phosphomolybdic acid	Niobium	Silica oxide(SiO_2)	Silane Coatings	Chromium nitride (Cr–N) and oxynitride (Cr(N,O))
Method of Deposition	Electro polymerization	High energy micro arc alloying	sol-gel and dip-coating		Reactive cathodic arc
Corrosion Parameters	OCP- Uncoated- (-388mV),Coated-(-138 mV) i_{corr} - Uncoated- (0.9188 mA/cm^2),Coated- (0.0264 mA/cm^2) E_{corr} - Uncoated- (-331.38 mV), Coated- (-119.13 mV)	$i_{corr\text{-}}$ Uncoated- (247.53 Acm^{-2}), Coated- (1.25 Acm^{-2}) E_{corr} -Uncoated-(-420 mV), Coated- (-310 mV) R_p -Uncoated- (155.57 Ωcm^2), Coated- 28050.79Ω cm^2	I_{corr} - uncoated (- 5.5μAcm$^{-2)}$ than coated one (0.234μAcm^{-2})	i_{corr} - Uncoated- (608.34 μA/cm^2), Coated- (3.21 × 10^{-3} μA/cm^2)	$i_{corr\text{-}}$ Uncoated- (14.689) μA/cm^2), Coated- (0.168 μA/ cm^2) E_{corr} - Uncoated- (34 mV), Coated- (250 mV) R_p - Uncoated- (2.109 kΩ), Coated- (223.047 kΩ)
Sl. No.	12	13	14	15	16

Chapter 10
Corrosion Assessment and Control Techniques for Reinforced Concrete Structures

Md Daniyal
Lingaya's Vidyapeeth, India

Sabih Akhtar
Aligarh Muslim University, India

ABSTRACT

The steel reinforced concrete structures perform well in various environmental conditions, but structures may undergo premature damage in aggressive environments such as marine or acidic, primarily due to steel corrosion, and substantial reduction in service life occurs. This also causes huge economical loss and create safety and environmental problems. The repair and maintenance of steel reinforced concrete structures for their safety needs effective monitoring and inspection systems for evaluating the corrosion condition of steel. Since the corrosion of steel reinforcement occurs through electrochemical reactions, electrochemical methods are suitable to study the corrosion processes. In this chapter, some commonly used electrochemical techniques have been comprehensively explained. In addition, there is a critical requirement to develop effective and long-lasting techniques to control the corrosion of steel. Hence, some of the commonly used corrosion control methods have been comprehensively described in this chapter.

1. INTRODUCTION

Prior to the invention of cement, mortars were made by mixing water, sand and slaked lime (Lazell, 1915). John Smeaton discovered hydraulic lime mortars in 1754 (Searle, 1935). A British mason named Joseph discovered the cement in 1824 and filed the first patent on Portland cement (Gagg, 2014). However, the objects prepared with the use of Portland cement were extremely brittle and incapable to tolerate shocks. Joseph Monier, a French gardener in 1867 made flowerpots with embedded iron nails and consequently observed a remarkable enhancement in the durability of his pots. Afterward, efficient scientific develop-

DOI: 10.4018/978-1-7998-2775-7.ch010

ments in concrete manufacture took place. In 1911, American Society for Testing and Materials (ASTM) formulated a detailed specification for the utilization of steel in concrete (Singh & Venugopalan, 2013). Subsequently, various developments ensued in designing the steel reinforcement bars in order to produce stronger and durable concrete structures. Consequently, the steel-reinforced concrete became the most widely used structural material in the world because of its economical, strength and durability properties. Steel-reinforced concrete structures were viewed as maintenance-free and unlimited service life until the mid-1970s. Conversely, since then, several durability related problems have emerged, such as alkali-silica reactions, sulphate attacks and corrosion of steel reinforcement. Among all durability related problems in steel-reinforced concrete structures, corrosion of steel reinforcement has been recognized as the main source of deterioration (Zhao & Jin, 2016).

Generally, the corrosion affects our daily lives directly as well as indirectly. In direct, it shortens the useful service life of our goods. In indirect, the manufacturer and provider of goods and services incur costs of corrosion from the clients. In particular, the corrosion of reinforcing steel bar in concrete results the collapse of bridges, failure of a part of highways, damage to buildings, and parking structures, etc. consequently endangers public safety and requires considerable repair costs. For instance, the unexpected collapse of the Silver Bridge over the Ohio River at Point Pleasant due to corrosion fatigue in 1967 resulted in deaths of 46 people and cost millions of dollars ("The Effects and Economic Impact of Corrosion," 2000). Therefore, in order to estimate the cost of corrosion in the United States (US), a study entitled "Corrosion Costs and Preventive Strategies in the United States" was conducted with the help of Federal Highway Administration (FHWA) and National Association of Corrosion Engineers (NACE) International, from 1999 to 2001 by CC Technologies Laboratories. This investigation estimated the average direct cost of corrosion of $8.3 billion per year for highway bridges alone and the total corrosion cost of US industries was estimated to $276 billion annually, which is approximately equal to 3.1% of the US Gross Domestic Product (GDP) (G. H. Koch, Brongers, Thompson, Virmani, & Payer, 2002). In 2014, a study entitled "International Measures of Prevention, Application, and Economic of Corrosion Technologies (IMPACT)" was initiated by NACE International and conducted by Det Norske Veritas (DNV), Germanischer Lloyd (GL), and American Productivity and Quality Center (APQC) and its industry and technology partners worldwide. From this study, the cost of corrosion at the global level is estimated to be $2.5 trillion per year, which is about 3.4% of the global GDP in 2013 (G. Koch et al., 2016). Hence, from the safety and economical perspective, the corrosion of steel reinforcement is very serious problem that can affect the sustainability of the steel-reinforced concrete structures directly. Numerous studies have been conducted worldwide in order to extend service life of steel reinforced concrete structures; regardless of this, several aspects are still not well known, and there is the requirement to incorporate present facts into practical field.

This chapter is organized in two major thematic sections. In the first section, an overview of various corrosion monitoring techniques for steel reinforced concrete has been undertaken. The second section includes studies related to the different corrosion control methods for steel reinforced concrete.

2. CORROSION ASSESSMENT TECHNIQUES

The maintenance and repair of steel reinforced concrete structures for their safety needs effective monitoring and inspection methods for evaluating the corrosion of steel reinforcement. These methods need to be able to identify any probable durability problems within structures before they become severe.

Since the corrosion of steel reinforcement occurs through electrochemical reactions involving charge (electrons) transfer via concrete pore solution (electrolyte), electrochemical methods are appropriate to study the corrosion processes. In this section, some electrochemical and non-destructive techniques commonly used for monitoring the corrosion of steel in concrete structures have been discussed. Moreover, a destructive technique viz. the gravimetric weight loss method has also been explained.

2.1 Open Circuit Potential Measurement

The basic principle involved in this method is the measurement of corrosion potential (also called half-cell potential or open circuit potential) of steel reinforcement with respect to a standard reference electrode (RE) such as Copper/Copper Sulphate Electrode (CSE), Silver/Silver Chloride Electrode (SSCE), Standard Hydrogen Electrode (SHE) and Saturated Calomel Electrode (SCE). The reference electrode has a predetermined potential. For example, the SCE has a potential of +242 mV vs. the SHE (assumed potential of 0.0 V) at room temperature. The schematic diagram for open circuit potential (E_{OC}) measurement is shown in Fig. 1 (H. Song & Saraswathy, 2007). In accordance with ASTM C 876, the probable corrosion conditions of steel reinforcement related with E_{OC} values is presented in Table 1 (ASTM C876 - 15, 2015).

Figure 1. Illustration of E_{OC} measurement technique

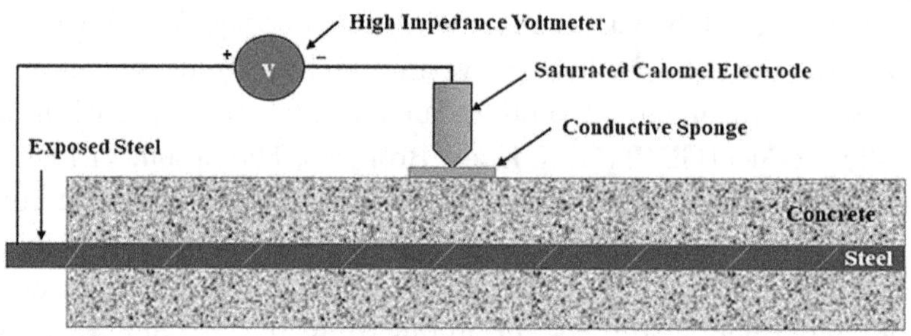

The open circuit potential (E_{OC}) of a steel is a measure of its tendency to corrode. However, these E_{OC} values are not sufficient criterion, as they are affected by many factors, such as polarization by partial diffusion of O_2, porosity of concrete and the existence of resistive layers. Therefore, it is commonly believed that E_{OC} measurement must be complemented by other techniques (Perveen, Bridges, Bhadra,

Table 1. Corrosion risk of steel reinforcement associated with E_{OC} values

E_{OC} Values				Corrosion Risk
(mV vs. CSE)	(mV vs. SSCE)	(mV vs. SHE)	(mV vs. SCE)	
< -500	< -406	< -184	< -426	Severe
< -350	< -256	< -34	< -276	High
-350 to -200	-106 to -256	+116 to -34	-126 to -276	Intermediate
> -200	> -106	> +116	> -126	Low

& Thomson, 2014). This technique is very useful in identifying the anodic and cathodic locations in steel-concrete structures by drawing potential map. Numerous investigators have studied the effectiveness of this method and got advantageous outcomes (Choi, Kim, & Lee, 2006; Erdoğdu, Kondratova, & Bremner, 2004; Razak & Choi, 2001). However, E_{OC} values can provide information for degree of corrosion risk only and cannot specify the corrosion rate (W. Yeih & Huang, 1998).

2.2 Surface Potential Measurement

This is another useful non-destructive method to identify the condition of steel embedded in concrete. In this method, two reference electrodes are used as shown in Fig. 2 (H. Song & Saraswathy, 2007). One electrode is kept fixed (called fixed electrode), the other electrode (called movable electrode) is moved along the structure on the nodal points. When movable electrode is placed at the nodal points, the potential is measured against the fixed electrode by means of a high impedance voltmeter. A more positive potential value signifies anodic region where corrosion is possible. The higher the potential difference between anodic and cathodic zones higher is the possibility of corrosion. Therefore, surface potential measurement is used for detecting anodic and cathodic areas in steel-concrete structures and ultimately identifying the possibility of corrosion of steel in concrete (H. Song & Saraswathy, 2007).

Figure 2. Schematic diagram of surface potential measurement

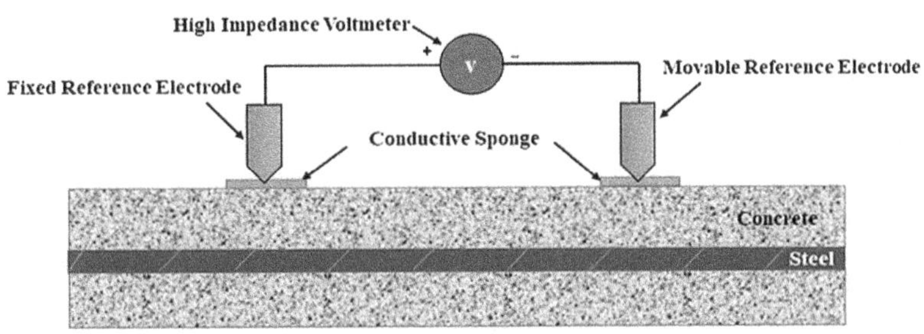

2.3 Potentiodynamic and Cyclic Polarization

The recent theory of electrochemical corrosion is based on electrode kinetics. The corroding system wherein a single anodic reaction and a single cathodic reaction take place on a single electrode surface, the relationship of the current as a function of potential can be written as Eq. (1). This equation is derived by the application of the mixed potential theory and Butler-Volmer equation (Butler, 1924; Erdey-Gruz & Volmer, 1930; Frankel, 2016):

$$i = i_{corr}\left[\exp\left\{\frac{2.303\left(E - E_{corr}\right)}{\beta_a}\right\} - \exp\left\{\frac{-2.303\left(E - E_{corr}\right)}{\beta_c}\right\}\right] \tag{1}$$

Where,

i = External current density (A/cm^2)
E = Potential applied to polarize the corroding system (V)
E_{corr} = Corrosion potential (V)
i_{corr} = Corrosion current density (A/cm^2)
β_a = Anodic Tafel constant (V/decade)
β_c = Cathodic Tafel constant (V/decade)

However, when E is far away from E_{corr}, then the Eq. (2) gives the Tafel equation (Tafel, 1905):

$$E = a \pm b \log |i| \quad (2)$$

Where, a = a constant and b = β_a or β_c

The Eq. (2) shows that the variation of the logarithm of the external current density with the potential is linear at high overpotential. The i_{corr} can be calculated by extrapolating the anodic and cathodic straight lines of E vs. log |i| plot at the $E_{corr.}$ Moreover, the E vs. log |i| plot and various corrosion kinetic parameters can be determined easily by conducting potentiodynamic and cyclic polarization test using inbuilt software provided by the potentiostat manufacturer. A potentiostat is a specifically designed instrument that measures the potential / current characteristics of an electrochemical (electrode/solution) interface (Zhang, Jiang, Yao, Song, & Wu, 2009).

In order to perform the potentiodynamic or the cyclic polarization scan, three electrodes, electrolyte (testing solution) and a potentiostat are required. The electrodes include a working electrode (the test sample itself), a counter or auxiliary electrode (Used to transport the current to the working electrode and to close the electrical circuit) and a reference electrode (Used to measure the potential difference). In order to conduct the scan, these electrodes are to be immersed in the electrolyte and attached to the potentiostat. Moreover, the potentiostat is a device that is used to apply potential and record the induced current or vice versa. More explicitly, both controlled-current (Galvanostatic) and controlled-potential (Potentiostatic) polarization can be applied. The term 'polarization' means the perturbation of the potential of a sample in electrolyte from its open circuit potential (E_{oc}). This can be done by two ways, i.e. when the polarization is accomplished galvanostatically, potential is measured and when it is achieved potentiostatically, current is measured. However, the potentiostatic method is much more common than galvanostatic method. The response (i.e., resulting current) of the sample is measured as it is polarized. The response is used to develop a model of the sample's corrosion behaviour (H. Song & Saraswathy, 2007).

The potentiodynamic polarization is used to study the corrosion behaviour and to calculate the different corrosion parameters such as corrosion current density (i_{corr}) of a corroding system in a certain environments. In a usual potentiodynamic scan, the potential (E) is swept over a fixed range by potentiostat and the induced current is recorded as a function of potential. The graphical output of a potentiodynamic scan is referred to as Tafel plot (named after Swiss chemist Julius Tafel), which is a plot of the potential (E) versus the logarithm of the current density (log |i|). In order to perform several tests on one sample (i.e., non-destructive test) the scan range should be within the Tafel region. Applying excessive anodic potential will force the system to corrode (i.e. destructive testing).

The analysis of Tafel plots is performed by corrosion test software (such as Gamry Instruments, ACM Instruments etc.). The corrosion parameters (such as E_{corr}, i_{corr}, β_a, and β_c) are determined by intersecting the open circuit potential (E_{OC}) and the extrapolation of the linear portions of logarithmic current plot as

Figure 3. Typical potentiodynamic polarization curve

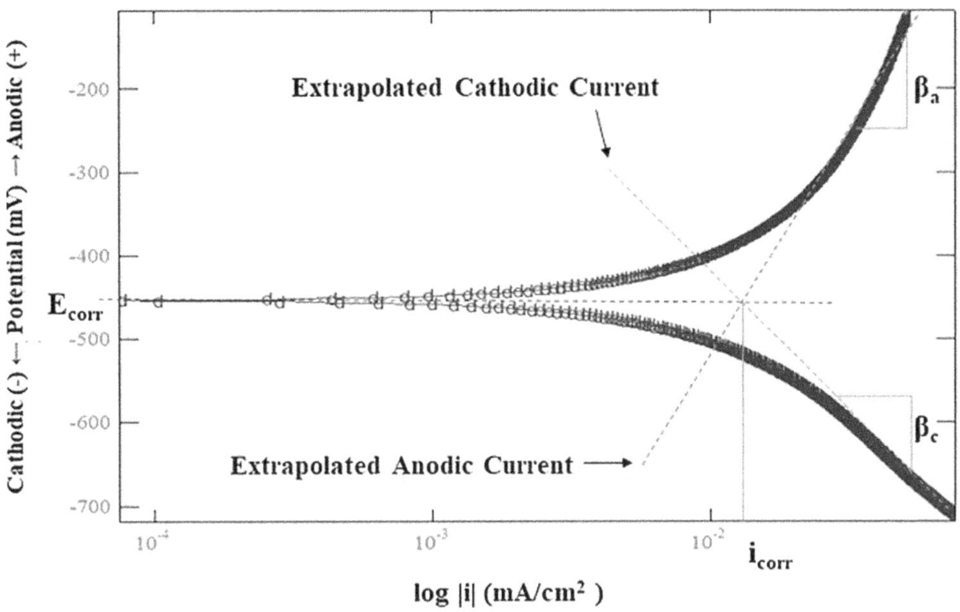

shown in Fig. 3. Knowing the corrosion current density (i_{corr}) the corrosion rate (CR) can be calculated using Eq. (3) (Aldabbagh & Alshimary, 2017). Though, this can also be estimated by corrosion test software provided by potentiostat manufacturer.

$$CR = \frac{i_{corr} \times K \times (EW)}{d \times A} \tag{3}$$

Where,

EW (Equivalent weight) = Theoretical mass of metal that will be lost from the sample after one Faraday of anodic charge has been passed (EW = atomic weight/valence)
i_{corr} = Corrosion current density (mA/cm^2)
K = A constant that defines the units for the corrosion rate (for CR in mm/year (mmpy), K= 3272; and CR in milli-inches/year (mpy), K= 1.288×10^5)
d = Density (g/cm^3)
A = specimen area (cm^2)

The cyclic polarization test is used primarily to study the pitting corrosion behaviour of metal in a certain environmental condition. Therefore, in this test, sufficiently high anodic potential is to be applied to ensure initiation of pitting corrosion. As soon as the scan is reached at anodic potential limit E_{rev} (reverse potential), the path of the scan is reversed on the way to the cathodic direction until the ultimate predetermined potential is reached. The potential at the scan plot in which the induced current sharply increases is called the pitting potential (E_{pit}). The potential on the reverse scan where the loop closes is called the protection potential (E_{prot}), below which no pitting corrosion is likely to take place. If the Eprot is more positive than the Epit, then pitting will not occur. Though, pitting could occur if the E_{prot} is more negative than the E_{pit}. In general, the extent of the hysteresis loop indicates the degree of pitting tendency. The greater the loop size, the higher the pitting tendency and vice versa. Fig. 4 shows a detailed cyclic polarization scan curve (Ghods, Isgor, Mcrae, & Gu, 2010; Mahallati & Saremi, 2006; Saremi & Mahallati, 2002).

Figure 4. Typical cyclic polarization curve

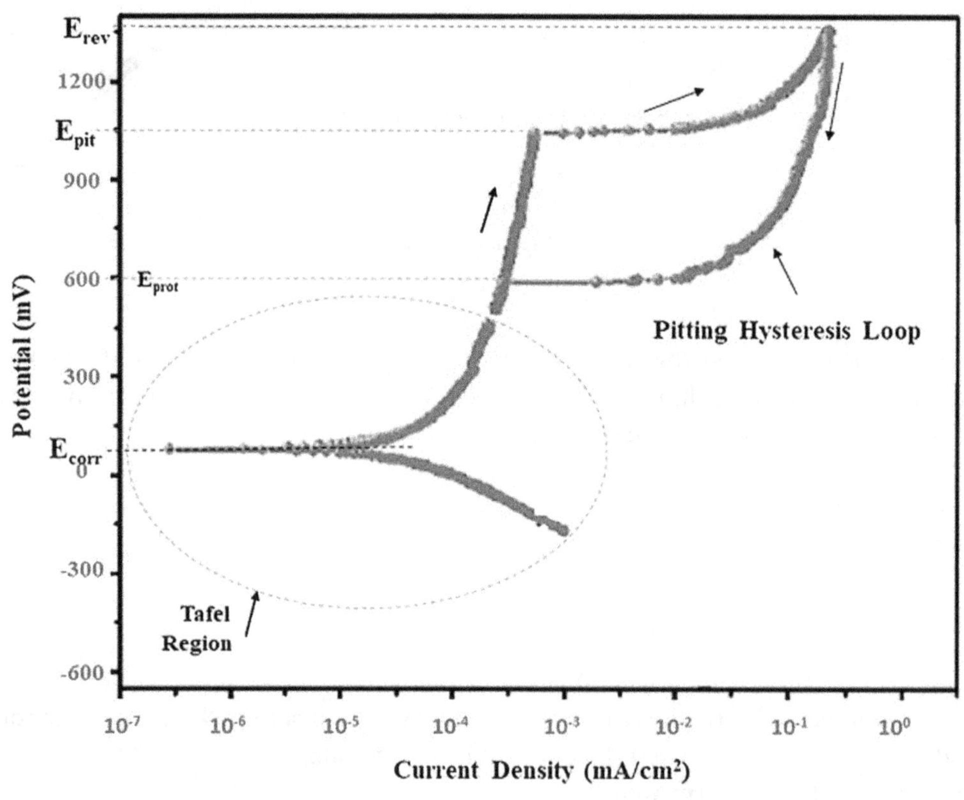

2.4 Linear Polarization Resistance

The linear polarization resistance (LPR) technique can be applied in both field and lab measurements. It is a direct current (DC) technique, first introduced by Stern and Geary in 1957 (Stern, 1957a, 1957b; Stern & Geary, 1957). In this technique, a small polarization perturbation (between 10 and 30 mV) is applied about E_{OC} of metal using a potentiostat and the resulting induced current is recorded. Although, the corrosion is an electrochemical process and does not comply with the Ohm's law (i.e. linear relationship between current and potential). However, it has been found that Ohm's law will be nearly true if polarization potential is very small (Millard, Law, Bungey, & Cairns, 2001). Since in this method the applied potential is small, therefore the current response will be linear. A typical potential and current

Figure 5. Typical linear polarization plot

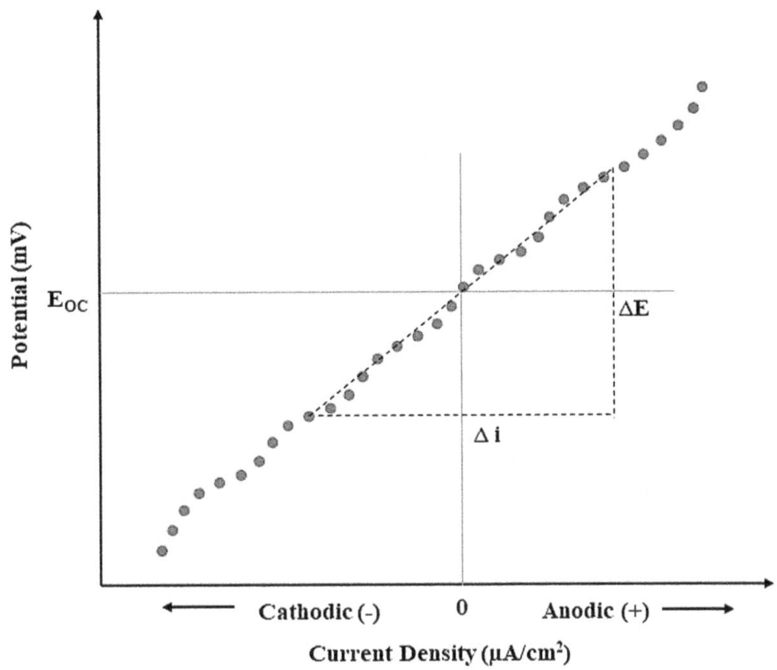

density (obtained by dividing the current response by known surface area of working electrode) plot is shown in Fig. 5.

The slope of the potential–current density plot near E_{OC} is defined as linear polarization resistance (R_p) ("ASTM G59 - 97: Standard test method for conducting potentiodynamic polarization resistance measurements," 2009). In case of steel reinforced concrete system, the R_p includes the concrete resistance (R_C) and charge transfer resistance (R_{ct}). Thus, linear polarization can be estimated as follows:

$$R_p = \frac{\Delta E}{\Delta i} = R_C + R_{ct} \tag{4}$$

The corrosion current density is then determined using the Stern-Geary formula:

$$i_{corr} = \frac{B}{R_p} \tag{5}$$

Where i_{corr} is the corrosion current density (μA/cm²), Rp is the polarization resistance (kΩ cm²) and B can be calculated by the following relationship:

$$B = \frac{\beta_a \times \beta_c}{2.3\left(\beta_a + \beta_c\right)} \tag{6}$$

Where β_a and β_c are the anodic and cathodic Tafel constants in mV/decade, respectively.

The Tafel constants are usually obtained from Tafel plot (by polarizing the steel to ±250 mV of the E_{OC}). However, in the absence of necessary data on β_a and β_c, B = 26 mV (for actively corroding steel in concrete) and B = 52 mV (for steel in passive situation) is used to compute the corrosion current density (Dehwah, Maslehuddin, & Austin, 2002). Moreover, the corrosion current density is inversely proportional to the linear polarization resistance. Hence, high polarization resistance value indicates low corrosion rate, and vice versa.

In order to carry out the linear polarization resistance investigation, similar to the potentiodynamic and cyclic polarization test, three electrodes are to be connected to a potentiostat. These are the working electrode (corroding metal), the auxiliary (counter) electrode and the reference electrode (Millard et al., 2001).

Several investigation has used a corrosion current density limit of 0.1- 0.2 μA/cm^2 to distinguish between passive and active corrosion state of steel embedded in concrete. In addition, many study reported different limits (in terms of corrosion current density, linear polarization resistance and corrosion penetration rate) in order to separate the different degree of corrosion risk associated with steel reinforcement, as listed in Table 2.

Table 2. Typical corrosion criteria of steel embedded in concrete

Corrosion Condition	Polarization Resistance (kΩ cm^2)	Current Density (μA/cm^2)	Corrosion Penetration Rate (μm/year)
Very high	2.5–0.25	10–100	100–1000
High	25–2.5	1–10	10–100
Low/moderate	250–25	0.1–1	1–10
Passive	>250	<0.1	<1

2.5 Electrochemical Impedance Spectroscopy

This technique is also known as AC (Alternating Current) impedance spectroscopy, because in this method, an AC excitation potential with variable frequencies is applied to the specimen (working electrode) and the induced AC current response is measured. The AC excitation potential should be very low (usually between 5 and 20 mV) in order to retain in the linear zone and to perform non-destructive testing. In the linear zone, the AC current (induced) will have the different amplitude and identical frequency as the applied AC excitation potential, but with a phase difference. The impedance (that includes real and imaginary parts) is the ratio of AC potential to AC current, and can be calculated from the following relationships:

$$Z = \frac{E(t)}{I(t)} = \frac{E_0 \operatorname{Cos}(\omega t)}{I_0 \operatorname{Cos}(\omega t - \Phi)} = Z_0 \frac{\operatorname{Cos}(\omega t)}{\operatorname{Cos}(\omega t - \Phi)} \quad (7)$$

$$Z = Z_0 [\exp (j.\ \phi)] = Z_0 (\mathrm{Cos}\phi + j.\ \mathrm{Sin}\phi) = Z' + jZ'' \quad (8)$$

Where,

Z = Complex impedance
$E(t)$ = Potential at time t
$I(t)$ = Current at time t
E_0 = Potential amplitude
I_0 = Current amplitude
ω = Radial frequency
ϕ = Phase difference between potential and current
Z_0 = Magnitude of complex impedance
$j = \sqrt{-1}$

The real part (Z') and imaginary part (Z") of the impedance (Z) represent resistive and capacitive/inductive terms respectively. The graph of Z' on the abscissa and Z" on the ordinate, measured at various frequencies (usually between 100 KHz and 10 mHz) is called as impedance spectrum or impedance graph or Nyquist plot. In order to determine various parameters (such as solution resistance, charge transfer resistance and double layer capacitance), the Nyquist plot has to be fitted and modelled with an equivalent electrical circuit that indicate the electrochemical processes at the specimen. Particularly in the steel reinforced concrete specimens, various equivalent circuits have been suggested due to complex nature of the steel-concrete interface. The simplest electrochemical equivalent circuit known as Randels circuit shown in Fig. 6 can be used for modelling the typical Nyquist plot shown in Fig. 7 ("ASTM-STP1065-Study of the corrosion of concrete reinforcement by electrochemical impedance measurement," n.d.; Martínez & Andrade, 2008; Ribeiro & Abrantes, 2016).

Figure 6. Rendles electrical equivalent circuit for steel reinforcement

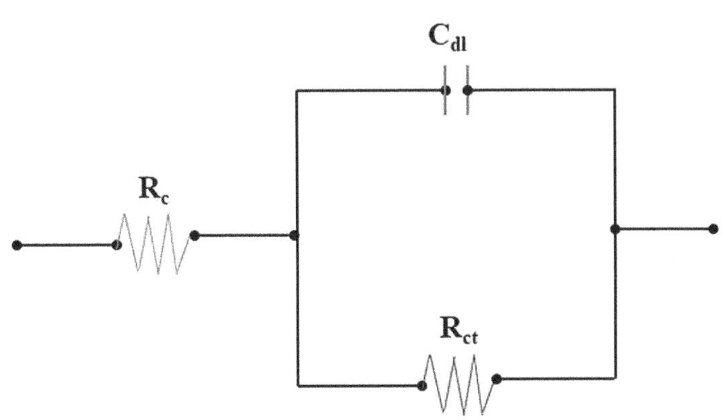

The equivalent circuit comprises of a resistance R_c linked in series to a loop that contains a different resistance R_{ct} and a capacitance C_{dl} joined in parallel. The R_c indicates the concrete pore electrolyte resistance or concrete cover resistance. The R_{ct} is the charge transfer resistance at the surface of steel that

Figure 7. Typical Nyquist plot for steel reinforcement

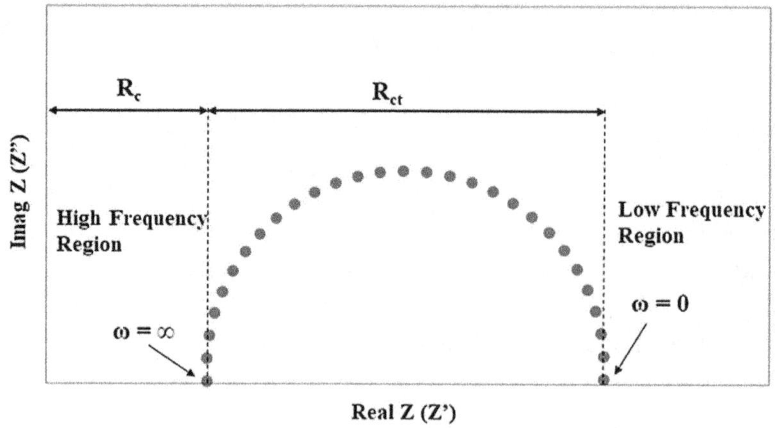

Figure 8. Impedance components in steel concrete interface

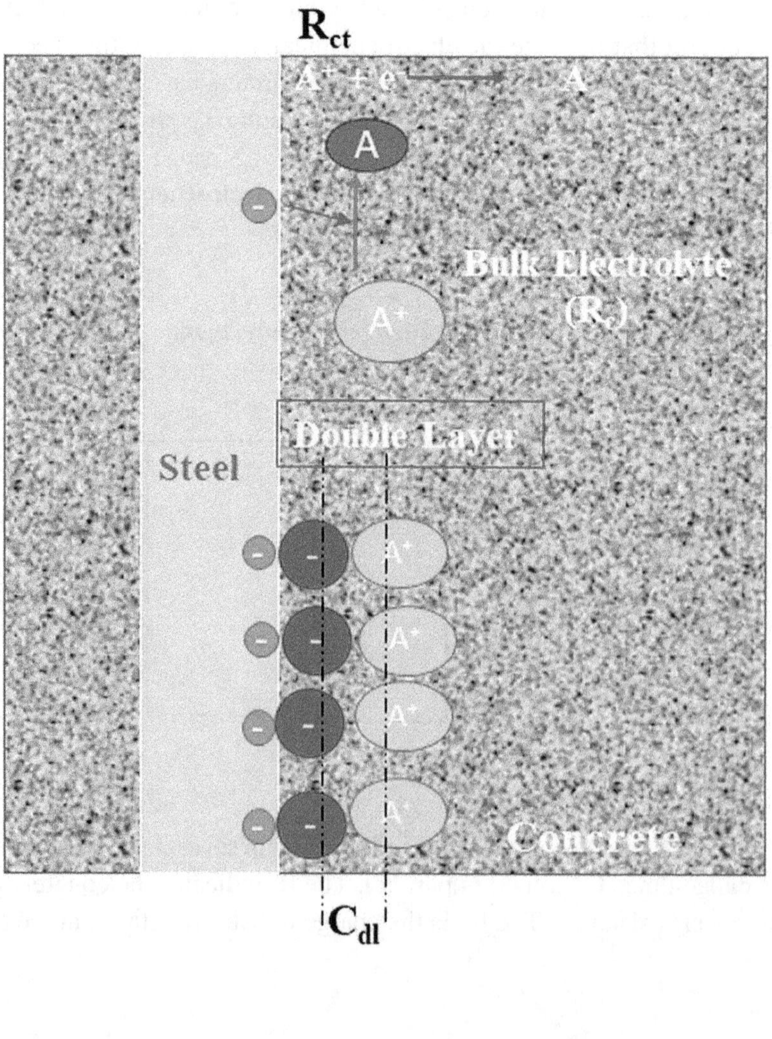

represents the ease with which the charged ions can leave the steel surface and reach into the electrolyte (i.e. metal dissolution resistance). The C_{dl} is a double layer capacitor, which refers to the development of two layers on the steel/electrolyte interface. The first layer is formed due to the attraction of molecules towards the steel surface anions (free electrons). The second layer is formed due to the attraction of molecules towards cations in the pore solution. Moreover, R_{ct} and C_{dl} both resembles to the ionic resistance of the corrosion products film developed on the steel surface. These terms are illustrated in Fig. 8.

From the Rendles equivalent circuit, the equivalent impedance (Z) of the Nyquist plot is represented as follows (Natarajan & Ramakrishnan, 2007):

$$Z = R_C + \frac{R_{ct}}{1 + j\omega C_{dl} R_{ct}} \tag{9}$$

Where,

R_c = Concrete resistance
R_{ct} = Charge transfer resistance
$j = \sqrt{-1}$
$\omega = 2\pi f$
C_{dl} = Double layer capacitance

When, $\omega \rightarrow 0$ (i.e. at low frequency), the circuit turn into DC circuit and therefore

$$Z = R_c + R_{ct} = R_p$$

When, $\omega \rightarrow \infty$ (i.e. at high frequency, AC circuit), $Z = R_c$

The difference of DC and AC impedance gives the true polarization resistance as:

True polarization resistance $= (R_c + R_{ct}) - R_c = R_{ct}$

Accordingly, the diameter of the extrapolated semicircle in the Nyquist plot indicates the charge transfer resistance (R_{ct}), which is equivalent to the true polarization resistance. Thus, if the diameter of the semicircle will be large, then polarization resistance will be high and corrosion rate will be low, and vice versa (Ribeiro et al., 2015).

2.6 Galvanostatic Pulse

This is a rapid and non-destructive polarisation method. It was introduced in 1988 for field application (Newton & Sykes, 1988). In this technique, galvanostatically a short-time anodic current pulse (usually 10 to 100 μA with pulse duration of 5 to 30 seconds) is applied between the steel reinforcement and auxiliary (counter) electrode placed on the surface of concrete. The reference electrode is commonly placed in the centre of the auxiliary electrode. The steel is polarised anodically and the resulting electrochemical potential response of the steel is recorded using a reference electrode as a function of

Figure 9. Electrochemical potential response of corroding steel

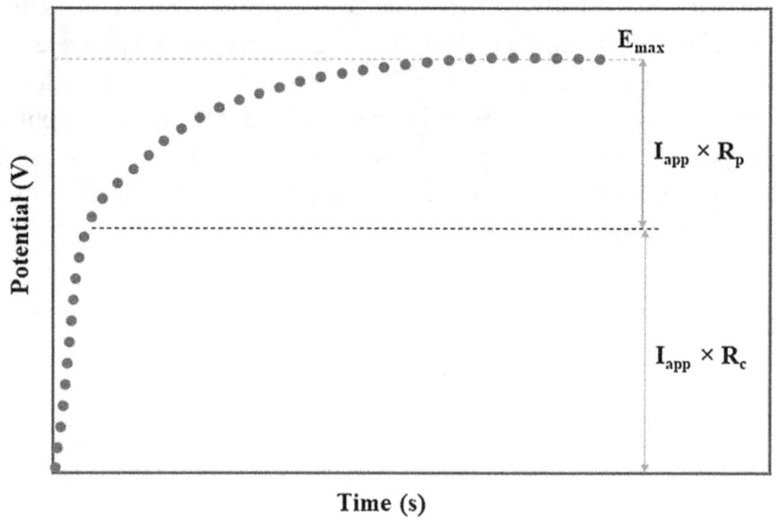

polarisation time. The characteristic electrochemical potential response of corroding steel is shown in Fig. 9 (Elsener et al., 1997).

When the current pulse, I_{app}, is applied to the steel reinforcement, then the polarisation of the steel at time t, E_t, can be written as follows (Elsener, 2005; Klinghoffer, 1995; Law, Millard, & Bungey, 1999):

$$E_t = I_{app}\left[R_p\left\{1 - e^{\left(\frac{-t}{R_p C_{dl}}\right)}\right\} + R_c\right] \tag{10}$$

Where,

R_p = Polarisation resistance
C_{dl} = Double layer capacitance
R_c = Concrete cover resistance

R_p and C_{dl} can be determined by transforming the Eq. (10) into linear form as (Klinghoffer, 1995; Law et al., 1999):

$$\log\left(E_{max} - E_t\right) = \log\left(I_{app} \times R_p\right) - \left(\frac{t}{R_p C_{dl}}\right) \tag{11}$$

Where E_{max} = Final steady state potential

A typical plot of Eq. (11) is shown in Fig. 10 (Klinghoffer, 1995). If the straight line passing through the data points is extrapolated to time (t) = 0, then the intercept at ordinate will give $\log(I_{app} \times R_p)$ and slope of the straight line will yield $1/(R_p \times C_{dl})$. The residual overpotential analogous to $(I_{app} \times R_c)$ i.e.

Figure 10. A typical plot of Eq. (11)

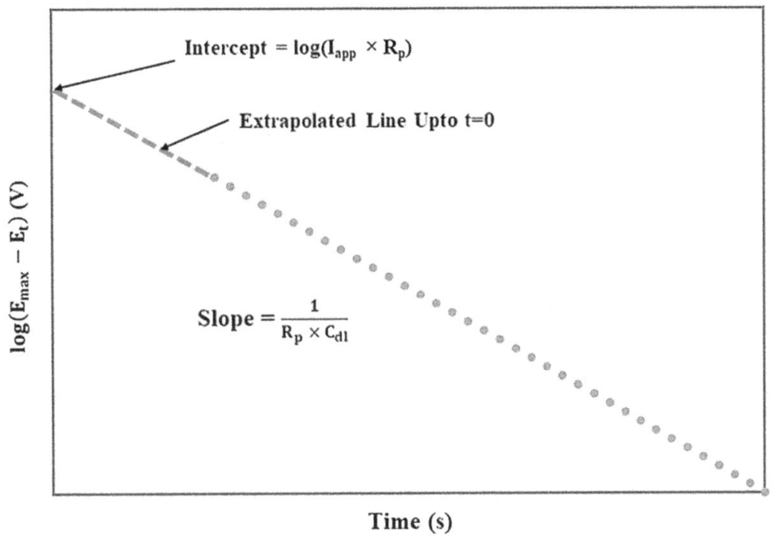

potential drop across the cover concrete. After finding the Rp by applying above technique, the corrosion current density (i_{corr}) can be determined from Sterne Geary formula (Eq. (5)).

2.7 Electrical Resistivity Measurement

The durability characteristics of concrete can be assessed by determining its electrical resistivity. Various investigations have given the correlation between electrical resistivity of concrete and corrosion of steel reinforcement (B.Hope, K. Ip, & G.Manning, 1985; Cavalier & Vassie, 1981; López & González, 1993). Several researchers have tried to made the relationship between the concrete resistivity and the different concrete properties such as water/cement ratio, cement content, coarse aggregate content, age, diffusion coefficient of chloride and porosity (Buenfeld, Newman, & Page, 1986; I. L. H. Hansson & Hansson, 1983; Hughes, Soleit, & Brierly, 1985).

The electrical resistivity of concrete is commonly measured using four-probe technique. F. Wenner firstly introduced this technique in order to evaluate the soil resistivity (Wenner, 1916). After that, this method was widely applied for the concrete resistivity measurement in the lab as well as field. The Fig. 11 illustrates principle involved in the Wenner four-probe electrical resistivity measurement technique (Chen, Chang, & Yeih, 2014). In this method, four equally spaced electrodes (probes) are used to measure the concrete resistivity. A small AC current (I) is applied between the outermost probes whereas potential difference (V) between the inner probes is recorded. Then, the surface resistivity of concrete is determined from the Eq. (12) (Gowers & Millard, 1999).

$$\rho = ka\frac{V}{I} \quad (12)$$

Where;

Figure 11. Illustrations of the Wenner four-probe resistivity measurement

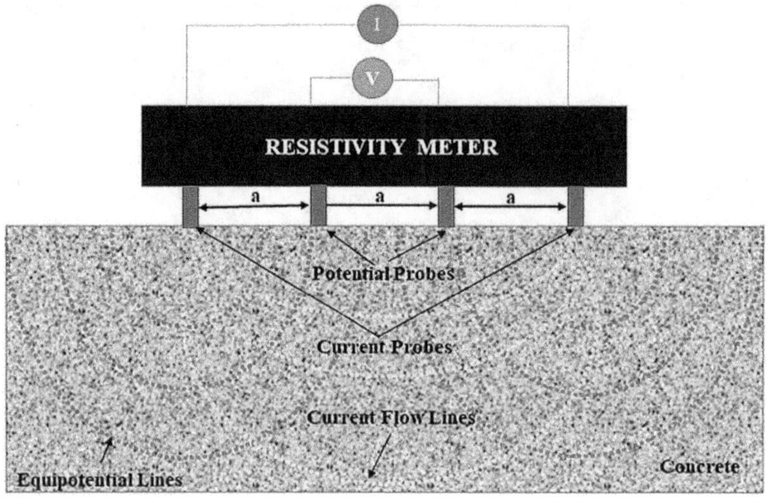

a = Distance between the probes
k = Geometry factor = 2π (for semi-infinite body such as concrete slabs)

However, the value of k is different for small cubic or cylindrical specimens used in laboratory testing condition. The AASTHO TP 95-11 Standard specified the typical value of 'a = 38 mm' and 'AC frequency = 13 Hz' in order to conduct the concrete surface electrical resistivity measurement (Azarsa & Gupta, 2017).

Moreover, it is extensively recognized that resistivity of concrete can be easily determined, particularly in the field, compared to other corrosion parameters such as corrosion rate. Various investigators as presented in Table 3 have developed the correlation between the resistivity of concrete and steel reinforcement corrosion risk (Azarsa & Gupta, 2017). The variations in the results show that the concrete resistivity can not be used with full confidence as an indicator for corrosion activity in steel reinforcement and need more investigations. Furthermore, various factors that might affect the results of resistivity measurement of concrete include water/cement ratio, microstructural properties of concrete, aging, existence of steel reinforcement, geometrical constraints, surface contact of probes, spacing between probes, temperature, and moisture content (Chen et al., 2014).

2.8 Gravimetric Weight Loss Measurement

This is the most commonly used destructive technique for estimating the corrosion rate of steel reinforcement. The detailed guideline for the preparation of test specimens and evaluation of corrosion rate by means of gravimetric weight loss measurement has been specified in ASTM G1-03 ("ASTM G1-03 - Standard practice for preparing, cleaning, and evaluating corrosion test specimens," 2017). In order to carry out this test, firstly, the steel is to be immersed in the Clark solution (1L of HCl + 50g of $SnCl_2$ + 20g of Sb_2O_3) for 40 seconds, then, degreased with acetone, rinsed with distilled water and dried in air. Thereafter, the initial weight of the steel is taken using an electronic weighing balance with 0.1mg precision before embedding in concrete. After a specified period of exposure in the corrosive medium,

Table 3. Relationship between concrete resistivity and corrosion risk

Corrosion Risk in Terms of Concrete Resistivity (kΩ.cm)			Reference
High	Moderate	Low	
<5	5–12	>12	(Cavalier & Vassie, 1981)
<6.5	6.5–8.5	>8.5	(B.Hope et al., 1985)
<7	7–30	>30–40	(López & González, 1993)
<10	10–30	>30	(Morris, Vico, Vazquez, & DeSanchez, 2002)
<20	20–100	>100	(González, Miranda, & Feliu, 2004)
<10	10–100	>100–200	(Andrade & Alonso, 1996)
<5	5–20	>20	(Broomfield & Millard, 2002)
<8	8–12	>12	(Smith, Schokker, & Tikalsky, 2004)

the steel is taken out by splitting the concrete specimens. After that, the corroded steel is to be cleaned by the above procedure and finally, weighed again. The obtained weight loss value is to be used to estimate the corrosion rate (CR) by the following relationship (Chinwko, Odio, Chukwuneke, & Sinebe, 2014):

$$CR = \frac{K \times \Delta W}{A \times T \times \rho} \tag{13}$$

Where,

ΔW = Weight loss (g)
A = Exposed surface area (cm^2)
T = Time of exposure (hours)
K = Unit conversion constant = 8.76×10^4 (for mm/year) = 534 (for mils/year)
ρ = Density (g/cm^3)

3. CORROSION CONTROL TECHNIQUES

The corrosion control methods for steel reinforced cementitious composite include cathodic protection, electrochemical chloride extraction, surface treatments of the steel reinforcement, surface treatment of concrete, utilization of mineral admixtures and chemical corrosion inhibitors. In the subsequent sections, these methods have been discussed comprehensively.

3.1 Cathodic Protection

The first investigation on corrosion control of steel in a reinforced concrete bridge through cathodic protection was reported by Stratfull in 1957 (Stratfull, 1957), and the application of this technique began in 1972 in order to protect Sky Park bridge decks contaminated by de-icing salts in Placerville, California (Stratfull, 1973). Thereafter, research has been carried out by various government and private industries for the development of cathodic protection systems in steel reinforced concrete structures. For example, on the basis of research investigations, Federal Highway Administration has identified that the cathodic protection is the only rehabilitation system that has proven to stop corrosion in chloride contaminated bridge decks irrespective of the chloride content in concrete (Barnhart, 1982).

The cathodic protection system can be established by two ways: galvanic (sacrificial) anode and impressed current systems. The galvanic anode system is based on the principles of dissimilar metal corrosion and the relative position of specific metals in the electro-chemical series. Hence, this system employs reactive metals as auxiliary anodes, which are connected to the steel that is to be protected. The difference in potentials between the reactive metal and the steel, as specified by their relative positions in the electro-chemical series, causes a positive current to flow in the electrolyte, from the reactive metal to the steel. Thus, the whole surface of the steel converted to negatively charge and becomes the cathode. The metals commonly used, as sacrificial anodes are aluminium, zinc and magnesium. No external power source and low maintenance is required in this system. Patch-repair and plug-type anodes are examples of galvanic anodes (Bennett & Schue, 1996).

In contrast, an impressed current system requires a current source (generally less than 50 volts) and inert (zero or low dissolution) anode material, fixed either on the surface, or embedded within the concrete. To achieve an electrical circuit, the positive terminal is connected to anode material and the negative terminal is connected to steel (cathode) to be protected, the anode and cathode are separated by an electrolyte (concrete) as shown in Fig. 12. This system works well until power supply is not interrupted. Therefore, this system consumes substantial electrical energy and also requires specialized services to design and verify the system's functioning (Ahmad, Basavaraja, & Bhattacharjee, 2000).

Figure 12. Impressed current cathodic protection system

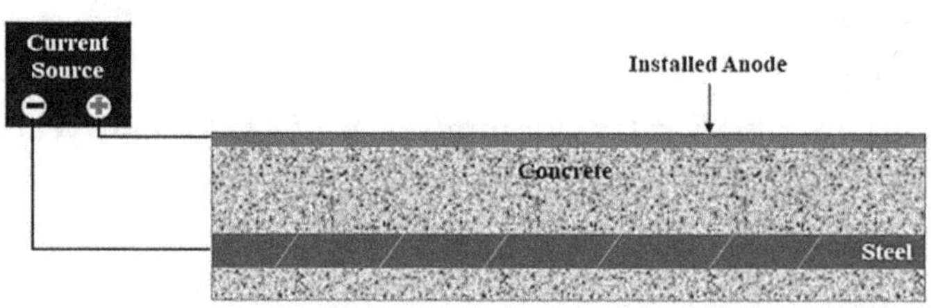

3.2 Electrochemical Chloride Extraction

The aim of the electrochemical chloride extraction (ECE) technique is to remove the chloride ions from the steel rebar and out of the concrete body. This method is similar to cathodic protection, but there are two main differences. First, the current density used for impressed current cathodic protection (ICCP) is much lower than that for ECE. Second, the external anode for ECE is temporary and installed for few weeks only (during the process) (Liu & Shi, 2009). However, both protection techniques have been recognized to increase the service lifespan of the structures. However, ECE technique is more advantageous because of no long-term regular maintenance requirement.

In this technique, an electric field is applied between the steel reinforcement and an external anode metal fixed to the surface of the concrete body. The commonly used anodes are titanium-activated

Figure 13. Illustration of the chloride extraction technique

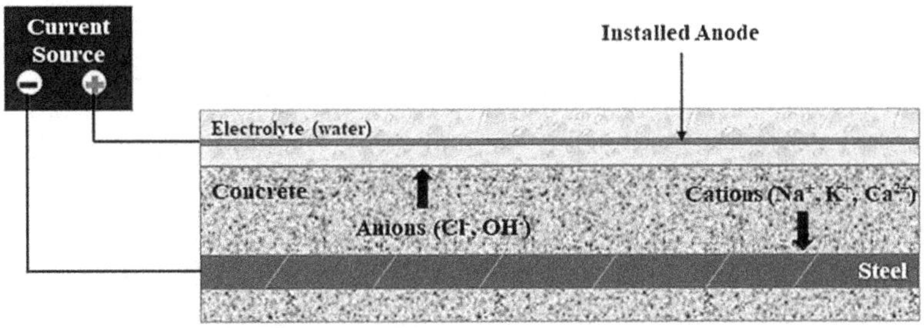

meshes, steel meshes, and mortar pastes prepared with graphite powder (Baldenebro Lopez et al., 2010). The positive terminal of current source is connected to anode, while negative terminal is connected to steel rebar embedded in concrete. When the electric current is applied, anions (such as Cl^-) are attracted towards the external anode placed on the surface of the concrete, while cations (such as Na^+, K^+) are migrated towards steel reinforcement and thereby produced hydroxyl ions (OH^-) on the surface steel, as a result of cathodic reactions. The schematic diagram of the chloride extraction technique is shown in Fig. 13. The following reactions occur during electrochemical extraction treatment due to the development of electric potential difference between the external anode and the steel (cathode) (Marcotte, Hansson, & Hope, 1999).

At anode:

$$2H_2O_{(l)} \rightarrow 4\,H^+_{(aq)} + O_{2(g)} + 4e^- \quad (14)$$

$$2Cl^-_{(aq)} \rightarrow Cl_{2(g)} + 2e^- \quad (15)$$

The pH of the electrolyte should be sufficiently high in order to suppress the reaction represented by Eq. (15). The highly alkaline electrolytes also neutralize the hydrogen ions produced by the reaction in Eq. (14) and protect the surface of concrete from acidification. Therefore, solution of calcium hydroxide, sodium hydroxide and treated water are commonly used as external electrolytes (Saraswathy, Lee, Karthick, & Kwon, 2018). However, water is the most widely used owing to the deficiency of salts, thus Cl^- ions travel more easily in the concrete and are extracted effortlessly (Elsener & Angst, 2007).

At cathode (steel):

$$O_{2(g)} + 2H_2O_{(l)} + 4e^- \rightarrow 4OH^-_{(aq)} \quad (16)$$

$$H_2O_{(l)} + 2e^- \rightarrow 2OH^-_{(aq)} + H_{2(g)} \quad (17)$$

The Eq. (17) shows the evolution of hydrogen gas at the level of steel reinforcement due to the low cathodic potential that is induced by the high current density. Also, it is believed that higher current densities can produce cracking in the concrete as a function of the chloride extraction velocity. Therefore, the current density has been taken between 1 A/m^2 and 5 A/m^2 in various studies (Liu & Shi, 2009). Despite this, some authors determine that the current density should not be more than 2 A/m^2 (Page, 2000). However, some of the studies focused on densities up to 1 A/m^2 (Siegwart, Lyness, & McFarland, 2003).

The period of application of this system varies from 6 to 10 weeks (Weichung Yeih, Chang, & Hung, 2006), in this period the chloride concentration is reduced below 0.4% of total chlorides, which is considered as critical value (Angst, Elsener, Larsen, & Vennesland, 2009). According to results of an investigation, the chloride content was found to be reduced by around 40% within 7 weeks of treatment and simultaneously, substantial quantities of alkali ions were detected nearby the steel (Orellan, Escadeillas, & Arliguie, 2004). The reduction percentage of chloride content can reach up to 70% (Sánchez, Garcés, & Climent, 2006). The reduction of chloride ions content below the critical level and simultaneous increment of hydroxyl concentration on the surface of the steel reinforcement creates favourable environment for the steel repassivation (de Almeida Souza, de Medeiros, Pereira, & Capraro, 2017).

3.3 Surface Treatment of Steel

The reinforcing bars commonly used are mild steel, mainly due to its significant low cost. Alternatively, stainless steel shows excellent mechanical and durability properties, but impractical for use in reinforced concrete because of its high cost (Callaghan, 1993). The poor corrosion resistance of mild steel is a common cause of deterioration in reinforced concrete structure. Therefore, various treatments have been given to steel surface in order to increase the corrosion resistance. One such treatment is coating with epoxy, which acts as a barrier to aggressive corrosion agents, and thus significantly decrease the corrosion rate of steel rebar. But, the poor bonding between epoxy coated steel and concrete, and the damaged areas of epoxy coated steel rebar (at which pitting corrosion may occur) are the common problems with epoxy coating (Pyc, Weyers, & Sprinkel, 1998). Another similar treatment is coating with zinc (*galvanization*), which improve the corrosion resistance of steel by acting as a sacrificial anode. It has been observed in many studies that the galvanized steel develops better bond to concrete than the epoxy coated steel, and also the tendency of the zinc coating to debond is very less as compared to epoxy coating (Rasheeduzzafar, Dakhil, Bader, & Khan, 1992). Subsequently, techniques for surface treatment of steel that increases the corrosion resistance as well as bond strength have been developed. These techniques include inhibited cement slurry coating, surface oxidation and sand blasting.

The Central Electrochemical Research Institute (CECRI), Karaikudi, India, has developed inhibited cement slurry coating for corrosion control in steel reinforcement. This coating system involves of four stages i.e. pickling, phosphating, two times application of inhibited cement slurry coating and sealing. This coating is an in-situ process that is carried out after all bending and shaping operations are completed at the construction site. As this coating is applied at the site of working, the damage caused on account of transporting and lifting are significantly minimised. Moreover, in situ patch repairing is easy to execute in this coating system. Though, strict quality control measures are required to ensure the performance of this coating (Vedalakshmi, Kumar, Raju, & Rengaswamy, 2000). In an investigation, it was concluded that the inhibited cement slurry coating is economical and efficient to control the corrosion of steel reinforcement (Rengaswamy, N.S. Srinivasan, S. Balasubramanian, 1988). In an another study, it was observed that inhibited cement slurry coated steel show higher bond strength as compared to galvanized and epoxy coated steel (Thangavel, Rengaswamy, & Balakrishnan, 1995).

Sand blasting technique involves the blasting of Al_2O_3 particles (size about 250 μm) on the steel rebar under pressure of around 0.6 MPa. It is used for cleaning of the surface of the steel. The cleaning includes the removal of rust and/or contaminants from the surface of steel. This cleaning improves the corrosion resistance by making the surface of the steel more uniform in composition. Moreover, the cleaning increases the roughness of surface and thus improves the bond strength. However, the disadvantage of this method is that it is uneconomical and difficult to use in the construction site (Hou, Fu, & Chung, 1997).

Water immersion method involves the immersion of the steel rebar under water at ambient temperature for 2 days. This treatment forms a black coating (oxide layer) on the steel surface, hence the surface of steel turns out to be more uniform in composition and thereby, enhancing the corrosion resistance. Furthermore, the oxide layer improves the bond between concrete and rebar, thus increasing the bond strength. The immersion periods less than or more than 2 days provide less beneficial effects on both corrosion resistance and bond strength. This method is found to be very simple and economical as compared to other methods (Fu & Chung, 1996).

3.4 Surface Treatment of Concrete

Since numerous hostile agents (e.g. chloride, CO_2 etc.) ingress into the concrete through air or water and creates durability problems (e.g. carbonation, chloride attack and sulphate attack etc.), therefore, the permeation property of the surface concrete is considered as a main factor that impact the durability of entire concrete structure (H.-W. Song, Lee, & Ann, 2008).

Concrete surface treatment has gained sufficient attention since 1980s, for example, American and German transportation agencies applied some hydrophobic agents on the surface concrete of bridges in order to prevent chloride penetration. Also, a chemical named isobutiltrimetoxisilane (100% pure and hydrophobic agent) has been used on bridges in the United Kingdom in 1986 to protect from chloride attack (de Vries & Polder, 1997). However, many progresses has been occurred in terms of understanding the mechanism of already known surface treatment agents; and various new treatment agents has emerged in last two decades. Moreover, the surface treatment agents can be classified into two categories based on chemical composition i.e. organic and inorganic. Organic agents are widely used for surface treatment of concrete owing to their good barrier properties. But, there are some problems associated with organic treatments viz. poor fire resistance, crack and detach easily, inadequate service life, and after losing protecting effects their removal is difficult (Delucchi, Barbucci, & Cerisola, 1997).In contrast, the inorganic surface treatment agents are more stable and have better durability performance. The most commonly used inorganic surface treatment agent is sodium silicate. Also, it has been reported that lithium silicate, potassium silicate and fluosilicates can be used as inorganic surface treatment agents for concrete (Franzoni, Pigino, & Pistolesi, 2013).

According to the functions of surface treatments, they were grouped into four types: surface coating, hydrophobic impregnation, pore blocking treatment and multifunctional surface treatments (Pan, Shi, Shi, Ling, & Li, 2017). Surface coating forms a continuous layer and produce a physical barrier to prevent the ingress of the corrosive substances into cementitious substrate (Diamanti et al., 2013). There are many types of surface coatings which have been used in the construction industry, for example, acrylic, chlorinated rubber, butadiene copolymer, polyethylene copolymer, epoxy resin, polyester resin, vinyl, polyurethane, coal tar and polymer modified cementitious coatings (Bertolini, Elsener, Pedeferri, Redaelli, & Polder, 2013).

Hydrophobic impregnation works via penetrating concrete pores, expanding the contact angle. The surface turns out to be hydrophobic when the contact angle is greater than 90°. Thus, hydrophobic impregnation can prevent the ingress of water and water-born ions; but permits penetration of water vapour (Fu & Chung, 1996). The most commonly used hydrophobic impregnation are silane, siloxane and a combination of these two chemicals. The molecular structures of silane (1.0 to 1.5μm diameter) and siloxane (1.5 to 7.5μm diameter) are extremely small; and therefore, efficiently penetrate into dense concrete substrate. Both silane and siloxane comprise an alkyl group and various alkoxy groups (Batis, Pantazopoulou, & Routoulas, 2003; Johnson, Schultz, & French, 2009). The alkyl group can decrease the surface tension of the substrate, whereas the alkoxy group is related to the bonds between silane/siloxane and concrete. For example, the chemical reaction between silane and concrete substrate can be proceeded as: firstly, silane hydrolysis reacts with pore water, and silanol groups are formed due to hydrolysis alkoxy groups (McGettigan, 1995). Then, the unstable silanol molecules lose water molecules and transformed into silicone resin. Subsequently, the silanol groups react with hydroxyl groups in the cementitious substrate via hydrogen bonds. Lastly, the silicon resin bonds to cementitious substrate during drying, and creates water repellent pore surfaces. The alkalinity of the cementitious substrate serves as a catalyst in these reactions (Woo, Zhu, Chow, Leung, & Kim, 2008).

Pore blocking surface treatment agents have been used in construction industries for many years (Dai, Akira, Wittmann, Yokota, & Zhang, 2010). They are able to partially or completely block the existing capillary pores in concrete surface and thus increase impermeability of surface layer. Silicate and fluosilicate based pore blockers (e.g. calcium silicate, lithium silicate, sodium silicate, magnesium fluosilicate and sodium fluosilicate) have been demonstrated to be effective in blocking capillary pores in concrete surfaces. Sodium silicate is considered as one of the most common pore blocking agents (Jiang et al., 2015).

Multifunctional surface treatments should have at least two functions, for example, ethyl silicate (or tetraethylorthosilicate), an alkoxysilane compound, which itself does not have binding property, and nevertheless it can produce silica gels and block the pores through hydrolysing process. Also, it can prevent the entry of corrosive substances because of its hydrophobic effect (Sandrolini, Franzoni, & Pigino, 2012). Similarly, Silane-clay nanocomposites are not only show a hydrophobic effect, but also improve the microstructure of concrete cover (Russo, Simon, & Incarnato, 2006).

Surface treatment of concrete may be an effective and economical technique in order to enhance the quality of surface concrete and improve the durability of concrete structures. However, this technique of corrosion control suffers from the poor durability of the coating and the loss of corrosion protection in the parts where the coating is damaged.

3.5 Admixtures in Concrete

Admixtures are solid or liquid substances that are added to the concrete mixture in order to enhance the properties of the resulting concrete. Admixtures that are primarily used for improving the corrosion resistance of steel reinforcement, commonly known as corrosion inhibitors. These include mainly organic and inorganic chemicals. However, admixtures that are mainly used for enhancing the structural performance of concrete are more attractive due to their multi-functionality (such as, improvement in workability, mechanical strengths, bond strength, impermeability, corrosion control, etc.). They are usually mineral particles, for example, fly ash, silica fume and slag.

Fly ash (FA) is a by-product of coal combustion and mainly composed of SiO_2 and CaO. The recent advancement in concrete technology exhibits that the fly ash is the most common pozzolan and used worldwide as an essential component for making high-performance concrete. This is mainly because of pozzolanic reaction between fly ash and hydrated cement paste resulting a denser microstructure over a period of time. However, studies have revealed that fly ash does not reduce the diffusion coefficient of concrete at early ages, but at the later ages, significantly lowers the diffusion coefficient as compared to control concrete (Basheer, Kropp, & Cleland, 2001). Moreover, Class F fly ash (low CaO content) leads to a slightly lower diffusion coefficient than Class C fly ash (high CaO content) admixed concrete (Papadakis, 2000). As far as steel reinforcement corrosion is concerned, it was observed that the corrosion resistance of steel in concrete is improved with the addition of fly ash as a replacement of Portland cement (PC) upto the 30% (Ha et al., 2007). In another study, it was concluded that the incorporation of 15% fly ash as replacement of cement can produced quite durable concretes against corrosion (Boa & Topu, 2012). Also, it was found that the chloride penetration resistance of concrete increases with increasing the fineness of fly ash (Chindaprasirt, Chotithanorm, Cao, & Sirivivatnanon, 2007). In general, it is recognised that the addition of fly ash improves the corrosion resistance properties of steel reinforcement by reducing porosity and increasing the resistivity of cementitious composite (Güneyisi, Özturan, & Gesoğlu, 2005). Moreover, fly ash is typically used at replacement levels between 15% and 35% by weight of Portland cement so as to increase the mechanical strengths as well as the resistance to deterioration mechanisms (such as chloride attack, sulphate attack, carbonation etc.) of cementitious composites (Holland, Kurtis, & Kahn, 2016).

Silica fume (SF) is a by-product from silicon alloy production in electric arc furnaces. It is almost pure SiO_2 and its particles size are approximately 100 times finer than cement grains. When added in cementitious composite, silica fume reacts quickly with hydrated cement due to small particle size and high surface area to volume ratio and consequently improving the microstructural and mechanical properties and decreasing the diffusion coefficient of cementitious composite at all ages with respect to plain cementitious composite. Though, it was observed that chloride diffusion coefficient of concrete in a marine environment reduces with lowering the water binder ratio and increasing silica fume content upto 10%. Also, the chloride diffusion coefficient is found to be high for early ages and drops over time (Farahani, Taghaddos, & Shekarchi, 2015). The reduction in diffusivity leads to decrease the chloride threshold level required to initiate corrosion, and this may increase the service life of the exposed concrete structures by 15-fold in certain circumstances (Bentz, Jensen, Coats, & Glasser, 2000). From the sensitivity analysis, it is calculated that with the addition of 10% silica fume in concrete, the chloride transport in terms of apparent diffusion coefficient is reduced by approximately 74% and corrosion-free life is increased by 270% as compared to OPC concrete (Jung, Kim, Lee, & Ann, 2018). Moreover, the incorporation of silica fume in appropriate quantity, significantly enhances the corrosion resistance properties of steel reinforced cementitious composite. Furthermore, the typical replacement levels of Portland cement with silica fume range from 3% to 10% in order to enhance the mechanical strengths and durability properties of concrete structures (Güneyisi et al., 2005).

Ground granulated blast-furnace slag (GGBFS), commonly known as slag, is a by-product of steel production. It mainly contains CaO, SiO_2, Al_2O_3 and MgO (Muhmood, Vitta, & Venkateswaran, 2009). When slag is used as part of Portland cement in concrete, it reacts with both the hydrated cement (*pozzolanic reaction*) and the water (*latent hydraulic reaction*), and consequently refined the pore structure and enhanced interfacial transition zone (ITZ) of concrete (Taylor, Richardson, & Brydson, 2010). Also, the addition of slag, improves the consistency and workability of fresh concrete. The mechanical strengths and durability properties of concrete is found to be substantially improved with the incorporation of GGBFS (Teng, Lim, & Sabet Divsholi, 2013). The diffusion coefficient of GGBFS admixed cementitious composite is found to be significantly lower as compared to the plain cementitious composite, particularly at the later exposure ages. Therefore, when incorporated appropriate amount of GGBFS in concrete, the transportation of corrosive agents is greatly reduced and consequently improved the corrosion resistance of steel reinforcement, and thereby, increasing the predicted service life of concrete structures. From most of the investigational studies, it may be concluded that the use of 40% to 80% of GGBS as a replacement of Portland cement does not cause any negative impact on the performance of cementitious composite.

Metakaolin (MK) is obtained by the calcination of clay mineral kaolinite ($Al_2Si_2O_7$) at the temperature range of 500°C to 800°C. It is primarily composed of SiO_2 and Al_2O_3. The size of 99.9% metakaolin particles is below 16 μm and the average particle size is approximately 3 μm. It is a pozzolanic material and commonly used in the manufacture of ceramics (Siddique & Klaus, 2009). The small particle size, high surface area to volume ratio and pozzolanic property of metakaolin, make it possible to be befittingly used as partial replacement of cement upto 10% and fine aggregate upto 20% in concrete (Batis, Pantazopoulou, Tsivilis, & Badogiannis, 2005). The studies revealed that when metakaolin is used in concrete, it reacts quickly due to small particle size and high surface area, and undergoes pozzolanic reaction, and consequently improves the microstructure of the hydrated cement paste (Barbhuiya, Chow, & Memon, 2015). This improvement leads to further enhancement in the mechanical and durability properties and reduction in the diffusion coefficient of concrete (Gruber, Ramlochan, Boddy, Hooton, & Thomas, 2001). In particular, the corrosion resistance of steel reinforcement in aggressive environment is improved substantially with the inclusion of suitable amount of metakaolin. Besides, the appropriate replacement level of cement by metakaolin ranges from 5% to 10% (Keleştemur & Demirel, 2015).

Moreover, the use of binary blends (Portland cement plus one admixture, such as PC + SF, PC + SF, PC + GGBFS and so on.) and ternary blends (Portland cement plus two admixtures, such as PC + FA + SF, PC + MK + FA, PC + FA + GGBFS etc.) can significantly enhance the performance of cementitious composite exposed to corrosive environments. Research has revealed that the use of admixtures can reduce the diffusion coefficient of concrete and consequently improve the corrosion resistance of steel reinforcement. However, it has been observed that ternary blends result in superior resistance to chloride ingress compared with binary blends at both early and later ages. This is because the ternary blends contain a slow-reacting admixture (e.g. fly ash or slag) and a fast-reacting admixture (e.g. silica fume or metakaolin) (Kannan & Ganesan, 2014). In addition, such replacement of cement will contribute to the reduction in CO_2 production (carbon footprint) and therefore encourage the use of such sustainable and green concrete. Furthermore, recently some new generation nano-admixtures (such as carbon nanotubes, SiO_2, TiO_2, Fe_2O_3, CuO, ZrO_2, ZnO_2, Al_2O_3, $CaCO_3$, Cr_2O_3) emerges, which can be used to improve the microstructure and corrosion resistance properties of reinforced concrete (Daniyal, Azam, & Akhtar, 2018).

3.6 Chemical Corrosion Inhibitors

International Organization for Standardization (ISO 8044:2015) defines corrosion inhibitor as a chemical substance that decreases the corrosion rate when present in the corrosion system at suitable concentration, without significantly changing the concentration of any other corrosion agent ("ISO 8044:2015, Corrosion of metals and alloys — Basic terms and definitions," 2015). The National Association of Corrosion Engineers (NACE) defines corrosion inhibitors as substances that, when added to an environment, decrease or slow down the rate of attack of the metal (Nathan, 1973). ACI 116R-85 defines a corrosion inhibitor as a chemical compound, either liquid or powder, that effectively decreases or slows down reinforcement corrosion in hardened concrete if introduced, usually in very small concentrations, as an admixture (ACI Committee 116, 1985). Therefore, an ideal corrosion inhibitor for cementitious composite may be defined as a chemical compound, which, when incorporated in suitable quantities to cementitious composite, can control corrosion of embedded steel reinforcement and has no adverse effect on the fresh and hardened properties of cementitious composite (Hope & Ip, 1987). These definitions exclude other corrosion control methods such as pore blockers, coatings etc., which modify the oxygen, water and chloride concentrations. Nevertheless, some chemical corrosion inhibitors may also act as pore blockers, which is a secondary property. The inhibitors may be the better than the other methods of protection owing to lower cost and easy to use.

The inhibitors could be classify in many ways: according to their application methods, according to their mechanism of protection, and their chemical nature. The main application methods for corrosion inhibitors are:

1. Added to fresh concrete as an admixture
2. Added to repair mortars
3. Used as a surface treatment on the steel reinforcement bars before concreting
4. Applied on the hardened concrete surface, as a penetrating corrosion inhibitor (also surface-applied corrosion inhibitor and migrating corrosion inhibitor).

According to the mechanisms of protection, corrosion inhibitors could be divide into three types:

1. **Anodic inhibitors**: These inhibitors passivate the metal by forming a film adsorbed or insoluble layer on anodic surfaces of the metal. They mainly act by affecting the anodic reaction and consequently reducing the corrosion rate by increasing the corrosion potential and decreasing the corrosion current density of the metal. Though, the sufficient concentration of anodic inhibitors must be used for covering the entire surface of metal. The inappropriate concentration of inhibitors affects the formation of film and leaves some exposed sites, and thus causes pitting corrosion. The commonly used anodic inhibitor are nitrites, nitrates, benzoates, chromates, molybdates, phosphates, hydroxides, silicates and carbonates.
2. **Cathodic inhibitors**: These inhibitors act by forming a barrier of insoluble precipitates over the metal surface and particularly affecting the cathodic reaction. Accordingly, reduced the corrosion rate by decreasing the corrosion potential as well as corrosion current density of metal. These inhibitors are usually less effective but more secure than anodic inhibitors. Examples of cathodic inhibitors are silicates, phosphates polyphosphates, tannins and ions of the zinc, nickel and magnesium that react with the hydroxyl (OH^-) of the water forming the insoluble hydroxides as $Zn(OH)_2$, $Ni(OH)_2$, $Mg(OH)_2$, which are deposited on the cathodic site of the metal surface, protecting it.
3. **Mixed inhibitors**: These inhibitors act on both anodic and cathodic sites by adsorption on overall surface of the metal. They reduce the corrosion rate by decreasing the corrosion current density without substantial changing the corrosion potential of metal. This type of inhibitor includes amines, amino alcohols, sulfonates, esters and compound with the hydrophobic group that have polar groups such as N, S and OH.

According to chemical functionality, the inhibitors can be classified as follows:

1. **Inorganic Inhibitors**: These inhibitors are generally crystalline salts such as nitrite salts, chromate salts, molybdate compounds, zinc salts, phosphates, silicate compounds etc. Usually, the anions of

these salts are involved in decreasing the corrosion of metal. In general, these inhibitors have anodic or cathodic actions. However, when zinc salt is used, the zinc cation can add some advantageous effect. These zinc-added salts are termed as mixed-charge inhibitors.

2. **Organic Inhibitors**: The inhibitors are characterized by high molecular weight structures, incorporating $-NH_2$, $-SH$, $-COOH$, $-OH$ or $-SO_3H$ functional groups. In their concentrated forms, these are either liquids or wax like solids. Their active portions are generally large aliphatic or aromatic compounds with positively charged amine groups. These inhibitors are said to be chelating agents, which can form five or six-membered chelate rings. These rings are formed as a result of the bonding between two or more functional groups from the inhibitor and the cation metal, thus, completely cover the metal surface. However, occasionally, they act as cathodic, anodic or mixed inhibitors. Some examples are amines, mercaptobenzothiazole, aldehydes, sulphur-containing compounds, heterocyclic nitrogen compounds and acetylenic compounds.

As compared to the other corrosion control techniques, chemical corrosion inhibitors have some benefits such as economical and versatility. Application of inhibitors in concrete can aid to delay the corrosion initiation in the embedded steel exposed to aggressive environments. However, after the corrosion initiation, their efficiency was less significant. In general, the side effects of the inhibitors on fresh and hardened concrete were reported to be insignificant (Söylev & Richardson, 2008). Very limited studies on the long-term performance of the inhibitors in real structures are available. Hence, more researches are needed to identify such inhibitors that are effective throughout the service life of structures and are non-hazardous.

4. CONCLUSION

Presently numerous electrochemical techniques for corrosion assessment of steel reinforcement are available. Each method possesses certain advantages and limitations. A combination of evaluating methods is recommended to use in order to find maximum information about the corrosion condition of steel embedded in concrete. Although, the electrochemical methods are generally qualitative and/or semi quantitative, and hence various corrosion kinetic parameters can be derived from them.

A number of corrosion control techniques have been identified. Each method has certain benefits and constraints. Thus, more researches are needed to develop such methods of corrosion protection of embedded steel that are economical, durable, environment-friendly and do not cause any adverse effect on the structural performance of concrete and steel.

REFERENCES

ACI Committee 116. (1985). *ACI 116R-85, Cement and Concrete Terminology.* Author.

Ahmad, S., Basavaraja, L. R., & Bhattacharjee, B. (2000). Design procedures for cathodic protection systems for RC members. *Indian Concrete Journal*, *74*(4), 208–213.

Aldabbagh, B. M., & Alshimary, H. J. (2017). Polyamide nanofibers coating by electrospinning technique for anti corrosion behavior. *Engineering & Technology Journal*, *35*(10), 987–991.

Andrade, C., & Alonso, C. (1996). Corrosion rate monitoring in the laboratory and on-site. *Construction & Building Materials*, *10*(5), 315–328. doi:10.1016/0950-0618(95)00044-5

Angst, U., Elsener, B., Larsen, C. K., & Vennesland, Ø. (2009). Critical chloride content in reinforced concrete - A review. *Cement and Concrete Research*, *39*(12), 1122–1138. doi:10.1016/j.cemconres.2009.08.006

ASTM G1-03 - Standard practice for preparing, cleaning, and evaluating corrosion test specimens. (2017).

ASTM G59 - 97: Standard test method for conducting potentiodynamic polarization resistance measurements. (2009).

ASTM C876 - 15. (2015). Standard Test Method for Corrosion Potentials of Uncoated Reinforcing Steel in Concrete. doi:10.1520/C0876-15

ASTM-STP1065-Study of the corrosion of concrete reinforcement by electrochemical impedance measurement. (n.d.). doi:10.1520/STP25019S

Azarsa, P., & Gupta, R. (2017). Electrical resistivity of concrete for durability evaluation: A review. *Advances in Materials Science and Engineering*, *2017*, 1–30. Advance online publication. doi:10.1155/2017/8453095

Baldenebro Lopez, F. J., Barrios Durstewitz, C. P., Fajardo San Miguel, G., Nuñez Jaquez, R. E., Almeraya Calderon, F., Almaral-Sánchez, J. L., & Castorena Gonzalez, J. H. (2010). Electrochemical realkalisation of reinforced concrete using a conductive mortar anode. *ECS Transactions*, *29*(1), 125–131. doi:10.1149/1.3532310

Barbhuiya, S., Chow, P., & Memon, S. (2015). Microstructure, hydration and nanomechanical properties of concrete containing metakaolin. *Construction & Building Materials*, *95*, 696–702. doi:10.1016/j.conbuildmat.2015.07.101

Barnhart, R. A. (1982). FHWA position on cathodic protection. Academic Press.

Basheer, L., Kropp, J., & Cleland, D. J. (2001). Assessment of the durability of concrete from its permeation properties: A review. *Construction & Building Materials*, *15*(2–3), 93–103. doi:10.1016/S0950-0618(00)00058-1

Batis, G., Pantazopoulou, P., & Routoulas, A. (2003). Corrosion protection investigation of reinforcement by inorganic coating in the presence of alkanolamine-based inhibitor. *Cement and Concrete Composites*, *25*(3), 371–377. doi:10.1016/S0958-9465(02)00061-6

Batis, G., Pantazopoulou, P., Tsivilis, S., & Badogiannis, E. (2005). The effect of metakaolin on the corrosion behavior of cement mortars. *Cement and Concrete Composites*, *27*(1), 125–130. doi:10.1016/j.cemconcomp.2004.02.041

Bennett, J. E., & Schue, T. J. (1996). *Galvanic cathodic protection of reinforced concrete bridge members using sacrificial anodes attached by conductive adhesives*. Academic Press.

Bentz, D. P. (2000). Influence of silica fume on diffusivity in cement-based materials II. Multi-scale modeling of concrete diffusivity. *Cement and Concrete Research*, *30*(7), 1121–1129. doi:10.1016/S0008-8846(00)00263-5

Bentz, D. P., Jensen, O. M., Coats, A. M., & Glasser, F. P. (2000). Influence of silica fume on diffusivity in cement-based materials I. Experimental and computer modeling studies on cement pastes. *Cement and Concrete Research, 30*(6), 953–962. doi:10.1016/S0008-8846(00)00264-7

Bertolini, L., Carsana, M., & Pedeferri, P. (2007). Corrosion behaviour of steel in concrete in the presence of stray current. *Corrosion Science, 49*(3), 1056–1068. doi:10.1016/j.corsci.2006.05.048

Bertolini, L., Elsener, B., Pedeferri, P., Redaelli, E., & Polder, R. (2013). *Corrosion of Steel in Concrete : Prevention, Diagnosis, Repair*. John Wiley & Sons. doi:10.1002/9783527651696

Boa, A. R., & Topu, L. B. (2012). Influence of fly ash on corrosion resistance and chloride ion permeability of concrete. *Construction & Building Materials, 31*, 258–264. doi:10.1016/j.conbuildmat.2011.12.106

Broomfield, J., & Millard, S. (2002). Measuring concrete resistivity to assess corrosion rates. *Concrete (London), 128*, 37–39.

Buenfeld, N. R., Newman, J. B., & Page, C. L. (1986). The resistivity of mortars immersed in sea-water. *Cement and Concrete Research, 16*(4), 511–524. doi:10.1016/0008-8846(86)90089-X

Butler, J. A. V. (1924). Studies in heterogeneous equilibria. Part III. A kinetic theory of reversible oxidation potentials at inert electrodes. *Transactions of the Faraday Society, 19*(March), 734–739. doi:10.1039/tf9241900734

Cabrera, J. G., & Claisse, P. A. (1999). Oxygen and water vapour transport in cement-silica fume pastes. *Construction & Building Materials, 13*(7), 405–414. doi:10.1016/S0950-0618(99)00039-2

Cabrera, J. G., Claisse, P. A., & Hunt, D. N. (1995). A statistical analysis of the factors which contribute to the corrosion of steel in Portland cement and silica fume concrete. *Construction & Building Materials, 9*(2), 105–113. doi:10.1016/0950-0618(95)00003-X

Callaghan, B. G. (1993). The performance of a 12% chromium steel in concrete in severe marine environments. *Corrosion Science, 35*(5–8), 1535–1541. doi:10.1016/0010-938X(93)90381-P

Cavalier, P. G., & Vassie, P. R. (1981). Investigation and repair of reinforcement corrosion in a bridge deck. *PI Civil Engineering Part-1., 70*, 461–480.

Chen, C. T., Chang, J. J., & Yeih, W. C. (2014). The effects of specimen parameters on the resistivity of concrete. *Construction & Building Materials, 71*, 35–43. doi:10.1016/j.conbuildmat.2014.08.009

Chindaprasirt, P., Chotithanorm, C., Cao, H. T., & Sirivivatnanon, V. (2007). Influence of fly ash fineness on the chloride penetration of concrete. *Construction & Building Materials, 21*(2), 356–361. doi:10.1016/j.conbuildmat.2005.08.010

Chinwko, E. C., Odio, B. O., Chukwuneke, J. L., & Sinebe, J. E. (2014). Investigation of the effect of corrosion on mild steel in five different environments. *International Journal of Scientific & Technology Research, 3*(7), 306–310.

Choi, Y. S., Kim, J. G., & Lee, K. M. (2006). Corrosion behavior of steel bar embedded in fly ash concrete. *Corrosion Science, 48*(7), 1733–1745. doi:10.1016/j.corsci.2005.05.019

Dai, J.-G., Akira, Y., Wittmann, F. H., Yokota, H., & Zhang, P. (2010). Water repellent surface impregnation for extension of service life of reinforced concrete structures in marine environments: The role of cracks. *Cement and Concrete Composites*, *32*(2), 101–109. doi:10.1016/j.cemconcomp.2009.11.001

Daniyal, M., Azam, A., & Akhtar, S. (2018). Application of Nanomaterials in Civil Engineering. In *Advanced Structured Materials* (Vol. 84, pp. 169–189). Springer Nature Singapore Pte Ltd., doi:10.1007/978-981-10-6214-8_6

de Almeida Souza, L. R., de Medeiros, M. H. F., Pereira, E., & Capraro, A. P. B. (2017). Electrochemical chloride extraction: Efficiency and impact on concrete containing 1% of NaCl. *Construction & Building Materials*, *145*, 435–444. doi:10.1016/j.conbuildmat.2017.04.010

de Vries, I. J., & Polder, R. B. (1997). Hydrophobic treatment of concrete. *Construction & Building Materials*, *11*(4), 259–265. doi:10.1016/S0950-0618(97)00046-9

Dehwah, H. A. F., Maslehuddin, M., & Austin, S. A. (2002). Long-term effect of sulfate ions and associated cation type on chloride-induced reinforcement corrosion in Portland cement concretes. *Cement and Concrete Composites*, *24*(1), 17–25. doi:10.1016/S0958-9465(01)00023-3

Delucchi, M., Barbucci, A., & Cerisola, G. (1997). Study of the physico-chemical properties of organic coatings for concrete degradation control. *Construction & Building Materials*, *11*(7–8), 365–371. doi:10.1016/S0950-0618(97)00060-3

Diamanti, M. V., Brenna, A., Bolzoni, F., Berra, M., Pastore, T., & Ormellese, M. (2013). Effect of polymer modified cementitious coatings on water and chloride permeability in concrete. *Construction & Building Materials*, *49*, 720–728. doi:10.1016/j.conbuildmat.2013.08.050

Elsener, B. (2005). Corrosion rate of steel in concrete-Measurements beyond the Tafel law. *Corrosion Science*, *47*(12), 3019–3033. doi:10.1016/j.corsci.2005.06.021

Elsener, B., & Angst, U. (2007). Mechanism of electrochemical chloride removal. *Corrosion Science*, *49*(12), 4504–4522. doi:10.1016/j.corsci.2007.05.019

Elsener, B., Klinghoffer, O., Frolund, T., Rislund, E., Schiegg, Y., & Böhni, H. (1997). Assessment of reinforcement corrosion by means of galvanostatic pulse technique. In *Int.Conf. Repair of Concrete Structures* (pp. 1–10). Academic Press.

Erdey-Gruz, T., & Volmer, M. (1930). The theory of hydrogen overvoltage. *Zeitschrift für Physikalische Chemie*, *150*, 203–213.

Erdoğdu, Ş., Kondratova, I. L., & Bremner, T. W. (2004). Determination of chloride diffusion coefficient of concrete using open-circuit potential measurements. *Cement and Concrete Research*, *34*(4), 603–609. doi:10.1016/j.cemconres.2003.09.024

Farahani, A., Taghaddos, H., & Shekarchi, M. (2015). Prediction of long-term chloride diffusion in silica fume concrete in a marine environment. *Cement and Concrete Composites*, *59*, 10–17. doi:10.1016/j.cemconcomp.2015.03.006

Frankel, G. S. (2016). *Active Protective Coatings* (A. E. Hughes, J. M. C. Mol, M. L. Zheludkevich, & R. G. Buchheit, Eds.; Vol. 233). Springer Netherlands. doi:10.1007/978-94-017-7540-3_2

Franzoni, E., Pigino, B., & Pistolesi, C. (2013). Ethyl silicate for surface protection of concrete: Performance in comparison with other inorganic surface treatments. *Cement and Concrete Composites*, *44*, 69–76. doi:10.1016/j.cemconcomp.2013.05.008

Fu, X., & Chung, D. D. L. (1996). Improving the bond strength between steel rebar and concrete by oxidation treatments of the rebar. *Cement and Concrete Research*, *26*(10), 1499–1503. doi:10.1016/0008-8846(96)00133-0

Gagg, C. R. (2014). Cement and concrete as an engineering material: An historic appraisal and case study analysis. *Engineering Failure Analysis*, *40*, 114–140. doi:10.1016/j.engfailanal.2014.02.004

Ghods, P., Isgor, O. B., Mcrae, G. A., & Gu, G. P. (2010). Electrochemical investigation of chloride-induced depassivation of black steel rebar under simulated service conditions. *Corrosion Science*, *52*(5), 1649–1659. doi:10.1016/j.corsci.2010.02.016

González, J. A., Miranda, J. M., & Feliu, S. (2004). Considerations on reproducibility of potential and corrosion rate measurements in reinforced concrete. *Corrosion Science*, *46*(10), 2467–2485. doi:10.1016/j.corsci.2004.02.003

Gowers, K. R., & Millard, S. G. (1999). Measurement of concrete resistivity for assessment of corrosion severity of steel using Wenner technique. *ACI Materials Journal*, *96*(5), 536–541. doi:10.14359/655

Gruber, K. A., Ramlochan, T., Boddy, A., Hooton, R. D., & Thomas, M. D. A. (2001). Increasing concrete durability with high-reactivity metakaolin. *Cement and Concrete Composites*, *23*(6), 479–484. doi:10.1016/S0958-9465(00)00097-4

Güneyisi, E., Özturan, T., & Gesoğlu, M. (2005). A study on reinforcement corrosion and related properties of plain and blended cement concretes under different curing conditions. *Cement and Concrete Composites*, *27*(4), 449–461. doi:10.1016/j.cemconcomp.2004.05.006

Ha, T. H., Muralidharan, S., Bae, J. H., Ha, Y. C., Lee, H. G., Park, K. W., & Kim, D. K. (2007). Accelerated short-term techniques to evaluate the corrosion performance of steel in fly ash blended concrete. *Building and Environment*, *42*(1), 78–85. doi:10.1016/j.buildenv.2005.08.019

Hansson, C. M., Mammoliti, L., & Hope, B. B. (1998). Corrosion inhibitors in concrete - Part I: The principles. *Cement and Concrete Research*, *28*(12), 1775–1781. doi:10.1016/S0008-8846(98)00142-2

Hansson, I. L. H., & Hansson, C. M. (1983). Electrical resistivity measurements of Portland cement based materials. *Cement and Concrete Research*, *13*(5), 675–683. doi:10.1016/0008-8846(83)90057-1

Heniegal, A. M., Amin, M., & Youssef, H. (2017). Effect of silica fume and steel slag coarse aggregate on the corrosion resistance of steel bars. *Construction & Building Materials*, *155*(September), 846–851. doi:10.1016/j.conbuildmat.2017.08.111

Holland, R. B., Kurtis, K. E., & Kahn, L. F. (2016). Effect of different concrete materials on the corrosion of the embedded reinforcing steel. In *Corrosion of Steel in Concrete Structures* (pp. 131–147). Elsevier. doi:10.1016/B978-1-78242-381-2.00007-9

Hope, B., Ip, A. K., & Manning, D. G. (1985). Corrosion and electrical impedance in concrete. *Cement and Concrete Research*, *15*(3), 525–534. doi:10.1016/0008-8846(85)90127-9

Hope, B. B., & Ip, A. K. C. (1987). *The Research and Development Branch.* Report No. ME-87-09, Ontario Ministry of Transportation.

Hou, J., Fu, X., & Chung, D. D. L. (1997). Improving both bond strength and corrosion resistance of steel rebar in concrete by water immersion or sand blasting of rebar. *Cement and Concrete Research, 27*(5), 679–684. doi:10.1016/S0008-8846(97)00043-4

Huang, Q. A., Hui, R., Wang, B., & Zhang, J. (2007). A review of AC impedance modeling and validation in SOFC diagnosis. *Electrochimica Acta, 52*(28), 8144–8164. doi:10.1016/j.electacta.2007.05.071

Hughes, B. P., Soleit, A. K. O., & Brierly, R. W. (1985). New technique for determining the electrical resistivity of concrete. *Magazine of Concrete Research, 37*(133), 243–248. doi:10.1680/macr.1985.37.133.243

ISO 8044:2015, Corrosion of metals and alloys — Basic terms and definitions. (2015). Retrieved from https://www.iso.org/obp/ui/#iso:std:iso:8044:ed-4:v1:en

Jiang, L., Xue, X., Zhang, W., Yang, J., Zhang, H., Li, Y., Zhang, R., Zhang, Z., Xu, L., Qu, J., Song, J., & Qin, J. (2015). The investigation of factors affecting the water impermeability of inorganic sodium silicate-based concrete sealers. *Construction & Building Materials, 93*, 729–736. doi:10.1016/j.conbuildmat.2015.06.001

Johnson, K., Schultz, A. E., & French, C. (2009). *Crack and concrete deck sealant performance* (Vol. 89261). Academic Press.

Jung, M. S., Kim, K. B., Lee, S. A., & Ann, K. Y. (2018). Risk of chloride-induced corrosion of steel in SF concrete exposed to a chloride-bearing environment. *Construction & Building Materials, 166*, 413–422. doi:10.1016/j.conbuildmat.2018.01.168

Kannan, V., & Ganesan, K. (2014). Chloride and chemical resistance of self compacting concrete containing rice husk ash and metakaolin. *Construction & Building Materials, 51*, 225–234. doi:10.1016/j.conbuildmat.2013.10.050

Keleştemur, O., & Demirel, B. (2015). Effect of metakaolin on the corrosion resistance of structural lightweight concrete. *Construction & Building Materials, 81*, 172–178. doi:10.1016/j.conbuildmat.2015.02.049

Klinghoffer, O. (1995). In situ monitoring of reinforcement corrosion by means of electrochemical methods. *Nordic Concrete Research-Publications, 95*(1), 1–13.

Koch, G., Varney, J., Thompson, N., Moghissi, O., Gould, M., & Payer, J. (2016). International Measures of Prevention, Application, and Economics of Corrosion Technologies Study. NACE International.

Koch, G. H., Brongers, M. P. H., Thompson, N. G., Virmani, Y. P., & Payer, J. H. (2002). Corrosion costs and preventive strategies in the United States. *NACE International, FHWA-RD, 01*, 1–12.

Law, D. W., Millard, S. G., & Bungey, J. H. (1999). The use of galvanostatic pulse measurements to determine corrosion parameters. *Durability of Building Materials and Components, 8*, 310–319.

Lazell, E. W. (1915). *Hydrated lime: history, manufacture and uses in plaster, mortar, concrete; a manual for the architect, engineer, contractor and builder*. Jackson-Remlinger Printing Co.

Liu, Y. J., & Shi, X. M. (2009). Electrochemical chloride extraction and electrochemical injection of corrosion inhibitor in concrete: State of the knowledge. *Corrosion Reviews*, *27*(1), 53–81. doi:10.1515/CORRREV.2009.27.1-2.53

López, W., & González, J. A. (1993). Influence of the degree of pore saturation on the resistivity of concrete and the corrosion rate of steel reinforcement. *Cement and Concrete Research*, *23*(2), 368–376. doi:10.1016/0008-8846(93)90102-F

Mahallati, E., & Saremi, M. (2006). An assessment on the mill scale effects on the electrochemical characteristics of steel bars in concrete under DC-polarization. *Cement and Concrete Research*, *36*(7), 1324–1329. doi:10.1016/j.cemconres.2006.03.015

Marcotte, T. D., Hansson, C. M., & Hope, B. B. (1999). The effect of the electrochemical chloride extraction treatment on steel-reinforced mortar Part I: Electrochemical measurements. *Cement and Concrete Research*, *29*(10), 1555–1560. doi:10.1016/S0008-8846(99)00118-0

Martínez, I., & Andrade, C. (2008). Application of EIS to cathodically protected steel: Tests in sodium chloride solution and in chloride contaminated concrete. *Corrosion Science*, *50*(10), 2948–2958. doi:10.1016/j.corsci.2008.07.012

McGettigan, E. (1995). Factors affecting the selection of water-repellent treatments. *APT Bulletin*, *26*(4), 22–26. doi:10.2307/1504446

Millard, S. G., Law, D., Bungey, J. H., & Cairns, J. (2001). Environmental influences on linear polarisation corrosion rate measurement in reinforced concrete. *NDT & E International*, *34*(6), 409–417. doi:10.1016/S0963-8695(01)00008-1

Morris, W., Vico, A., Vazquez, M., & DeSanchez, S. R. (2002). Corrosion of reinforcing steel evaluated by means of concrete resistivity measurements. *Corrosion Science*, *44*(1), 81–99. doi:10.1016/S0010-938X(01)00033-6

Muhmood, L., Vitta, S., & Venkateswaran, D. (2009). Cementitious and pozzolanic behavior of electric arc furnace steel slags. *Cement and Concrete Research*, *39*(2), 102–109. doi:10.1016/j.cemconres.2008.11.002

Natarajan, P., & Ramakrishnan, K. R. (2007). LPR meter for concrete corrosion studies. *Journal Instrument Society of India*, *37*(1), 1–6.

Nathan, C. C. (1973). Corrosion inhibitors. National Association of Corrosion Engineers.

Newton, C. J., & Sykes, J. (1988). A galvanostatic pulse technique for investigation of steel corrosion in concrete. *Corrosion Science*, *28*(11), 1051–1074. doi:10.1016/0010-938X(88)90101-1

Orellan, J. C., Escadeillas, G., & Arliguie, G. (2004). Electrochemical chloride extraction: Efficiency and side effects. *Cement and Concrete Research*, *34*(2), 227–234. doi:10.1016/j.cemconres.2003.07.001

Page, C. L. (2000). *Repair of corroded reinforcement in concrete using sacrificial anodes.* US Patent 6, 022, 469.

Pan, X., Shi, Z., Shi, C., Ling, T. C., & Li, N. (2017). A review on concrete surface treatment Part I: Types and mechanisms. *Construction & Building Materials*, *132*, 578–590. doi:10.1016/j.conbuildmat.2016.12.025

Papadakis, V. G. (2000). Effect of supplementary cementing materials on concrete resistance against carbonation and chloride ingress. *Cement and Concrete Research*, *30*(2), 291–299. doi:10.1016/S0008-8846(99)00249-5

Pech-Canul, M. A., & Castro, P. (2002). Corrosion measurements of steel reinforcement in concrete exposed to a tropical marine atmosphere. *Cement and Concrete Research*, *32*(3), 491–498. doi:10.1016/S0008-8846(01)00713-X

Perveen, K., Bridges, G. E., Bhadra, S., & Thomson, D. J. (2014). Corrosion Potential Sensor for Remote Monitoring of Civil Structure Based on Printed Circuit Board Sensor. *IEEE Transactions on Instrumentation and Measurement*, *63*(10), 2422–2431. doi:10.1109/TIM.2014.2310092

Pyc, W. A., Weyers, R. E., & Sprinkel, M. M. (1998). Corrosion protection performance of corrosion inhibitors and epoxy-coated reinforcing steel in a simulated concrete pore water solution. Academic Press.

Rasheeduzzafar, D. F. H., Bader, M. A., & Khan, M. M. (1992). Performance of corrosion-resisting steels in chloride-bearing concrete. *ACI Materials Journal*, *89*(5), 439–448. doi:10.14359/2379

Razak, H. A., & Choi, F. C. (2001). Damage assessment of corroded reinforced concrete beams using modal testing. In Structural Engineering, Mechanics and Computation (pp. 1203–1215). Elsevier. doi:10.1016/B978-008043948-8/50134-X

Rengaswamy, N. S., Srinivasan, S., & Balasubramanian, T. M. (1988). Inhibited and sealed cement slurry coating of steel rebar-A state of art report. *Transactions of the SAEST*, *23*(2–3), 163–173.

Ribeiro, D. V., & Abrantes, J. C. C. (2016). Application of electrochemical impedance spectroscopy (EIS) to monitor the corrosion of reinforced concrete: A new approach. *Construction & Building Materials*, *111*, 98–104. doi:10.1016/j.conbuildmat.2016.02.047

Ribeiro, D. V., Souza, C. A. C., & Abrantes, J. C. C. (2015). Use of Electrochemical Impedance Spectroscopy (EIS) to monitoring the corrosion of reinforced concrete. *Revista IBRACON de Estruturas e Materiais*, *8*(4), 529–546. doi:10.1590/S1983-41952015000400007

Roberge, P. (1995). Corrosion Inhibitors. In Handbook of Corrosion Engineering (pp. 833–862). Academic Press.

Russo, G. M., Simon, G. P., & Incarnato, L. (2006). Correlation between rheological, mechanical, and barrier properties in new copolyamide-based nanocomposite films. *Macromolecules*, *39*(11), 3855–3864. doi:10.1021/ma052178h

Sánchez, M. J., Garcés, P., & Climent, M. A. (2006). Electrochemical extraction of chlorides from reinforced concrete: Variables affecting treatment efficiency. *Materiales de Construcción*, *56*(284), 17–26. doi:10.3989/mc.2006.v56.i284.15

Sandrolini, F., Franzoni, E., & Pigino, B. (2012). Ethyl silicate for surface treatment of concrete – Part I: Pozzolanic effect of ethyl silicate. *Cement and Concrete Composites*, *34*(3), 306–312. doi:10.1016/j.cemconcomp.2011.12.003

Saraswathy, V., Lee, H. S., Karthick, S., & Kwon, S. J. (2018). Extraction of chloride from chloride contaminated concrete through electrochemical method using different anodes. *Construction & Building Materials*, *158*, 549–562. doi:10.1016/j.conbuildmat.2017.10.052

Saremi, M., & Mahallati, E. (2002). A study on chloride-induced depassivation of mild steel in simulated concrete pore solution. *Cement and Concrete Research*, *32*(12), 1915–1921. doi:10.1016/S0008-8846(02)00895-5

Searle, A. B. (1935). *Limestone and Its Products: Their Nature, Production, and Uses*. Ernest and Benn Ltd.

Siddique, R., & Klaus, J. (2009). Influence of metakaolin on the properties of mortar and concrete: A review. *Applied Clay Science*, *43*(3–4), 392–400. doi:10.1016/j.clay.2008.11.007

Siegwart, M., Lyness, J. F., & McFarland, B. J. (2003). Change of pore size in concrete due to electrochemical chloride extraction and possible implications for the migration of ions. *Cement and Concrete Research*, *33*(8), 1211–1221. doi:10.1016/S0008-8846(03)00047-4

Singh, D. D. N., & Venugopalan, T. (2013). Understanding the corrosion and remedial measures to control the deterioration of reinforcement steel bars by modification in their chemistry and application of surface coatings. *Transactions of the Indian Institute of Metals*, *66*(5–6), 677–687. doi:10.100712666-013-0335-x

Smith, K., Schokker, A., & Tikalsky, P. (2004). Performance of supplementary cementitious materials in concrete resistivity and corrosion monitoring evaluations. *ACI Materials Journal*, *101*(5), 385–389.

Song, H., & Saraswathy, V. (2007). Corrosion monitoring of reinforced concrete structures - A review. *International Journal of Electrochemical Science*, *2*, 1–28.

Song, H.-W., Lee, C.-H., & Ann, K. Y. (2008). Factors influencing chloride transport in concrete structures exposed to marine environments. *Cement and Concrete Composites*, *30*(2), 113–121. doi:10.1016/j.cemconcomp.2007.09.005

Söylev, T. A., & Richardson, M. G. (2008). Corrosion inhibitors for steel in concrete: State-of-the-art report. *Construction & Building Materials*, *22*(4), 609–622. doi:10.1016/j.conbuildmat.2006.10.013

Stern, M. (1957a). Electrochemical polarization-II. Ferrous-Ferric electrode kinetics on stainless steel. *Journal of the Electrochemical Society*, *104*(9), 559–563. doi:10.1149/1.2428653

Stern, M. (1957b). Electrochemical polarization-III. Further aspects of the shape of polarization curves. *Journal of the Electrochemical Society*, *104*(9), 645–650. doi:10.1149/1.2428438

Stern, M., & Geary, A. L. (1957). Electrochemical polarization-I. A theoretical analysis of the shape of polarization curves. *Journal of the Electrochemical Society*, *104*(9), 56–63. doi:10.1149/1.2428496

Stratfull, R. F. (1957). The corrosion of steel in a reinforced concrete bridge. *Corrosion*, *13*(3), 43–48. doi:10.5006/0010-9312-13.3.43

Stratfull, R. F. (1973). *Preliminary investigation of cathodic protection of a bridge deck.* California Department of Transportation.

Tafel, J. (1905). Über die Polarisation bei kathodischer Wasserstoffentwicklung. *Zeitschrift für Physikalische Chemie*, *50*, 641–712. doi:10.1515/zpch-1905-5043

Taylor, R., Richardson, I. G., & Brydson, R. M. D. (2010). Composition and microstructure of 20-year-old ordinary Portland cement-ground granulated blast-furnace slag blends containing 0 to 100% slag. *Cement and Concrete Research*, *40*(7), 971–983. doi:10.1016/j.cemconres.2010.02.012

Teng, S., Lim, T. Y. D., & Sabet Divsholi, B. (2013). Durability and mechanical properties of high strength concrete incorporating ultra fine ground granulated blast-furnace slag. *Construction & Building Materials*, *40*, 875–881. doi:10.1016/j.conbuildmat.2012.11.052

Thangavel, K., Rengaswamy, N. S., & Balakrishnan, K. (1995). Influence of protective coatings on steel-concrete bond. *Indian Concrete Journal*, *69*(4), 191–196.

The Effects and Economic Impact of Corrosion. (2000). In Corrosion: Understanding the Basics (pp. 1–20). ASM International.

Thomas, M. D. A., & Bamforth, P. B. (1999). Modeling chloride diffusion in concrete effect of fly ash and slag. *Cement and Concrete Research*, *29*, 487–495. doi:10.1016/S0008-8846(98)00192-6

Topcu, I. B., & Boga, A. R. (2010). Effect of ground granulate blast-furnace slag on corrosion performance of steel embedded in concrete. *Materials & Design*, *31*(7), 3358–3365. doi:10.1016/j.matdes.2010.01.057

Vedalakshmi, R., Kumar, K., Raju, V., & Rengaswamy, N. S. (2000). Effect of prior damage on the performance of cement based coatings on rebar: Macrocell corrosion studies. *Cement and Concrete Composites*, *22*(6), 417–421. doi:10.1016/S0958-9465(00)00041-X

Wenner, F. (1916). A method of measuring earth resistivity. *Bulletin of the Bureau of Standards*, 469–478.

Woo, R. S. C., Zhu, H., Chow, M. M. K., Leung, C. K. Y., & Kim, J.-K. (2008). Barrier performance of silane–clay nanocomposite coatings on concrete structure. *Composites Science and Technology*, *68*(14), 2828–2836. doi:10.1016/j.compscitech.2007.10.028

Yeih, W., Chang, J. J., & Hung, C. C. (2006). Selecting an adequate procedure for the electrochemical chloride removal. *Cement and Concrete Research*, *36*(3), 562–570. doi:10.1016/j.cemconres.2005.12.008

Yeih, W., & Huang, R. (1998). Detection of the corrosion damage in reinforced concrete members by ultrasonic testing. *Cement and Concrete Research*, *28*(7), 1071–1083. doi:10.1016/S0008-8846(98)00060-X

Zhang, X. L., Jiang, Z. H., Yao, Z. P., Song, Y., & Wu, Z. D. (2009). Effects of scan rate on the potentiodynamic polarization curve obtained to determine the Tafel slopes and corrosion current density. *Corrosion Science*, *51*(3), 581–587. doi:10.1016/j.corsci.2008.12.005

Zhao, Y., & Jin, W. (2016). Chapter-1. In Steel Corrosion-Induced Concrete Cracking (pp. 1–17). China Science Publishing & Media Ltd. doi:10.1016/B978-0-12-809197-5.00001-3

Compilation of References

Abbouda, Y., Abourriche, A., Saffaj, T., Berrada, M., Charrouf, M., Bennamara, A., Al Himidi, N., & Hannache, H. (2007). 2,3-Quinoxalinedione as a novel corrosion inhibitor for mild steel in 1 M HCl. *Materials Chemistry and Physics*, *105*(1), 1–5. doi:10.1016/j.matchemphys.2007.03.037

Abd El-Raouf, M., Khamis, E. A., Abou Kana, M. T. H., & Negm, N. A. (2018). Electrochemical and quantum chemical evaluation of new bis (coumarins) derivatives as corrosion inhibitors for carbon steel corrosion in 0.5 M H2SO4. *Journal of Molecular Liquids*, *255*, 341–353. doi:10.1016/j.molliq.2018.01.148

Abdallah, M. (2002). Rhodanine azosulpha drugs as corrosion inhibitors for corrosion of 304 stainless steel in hydrochloric acid solution. *Corrosion Science*, *44*(4), 717–728. doi:10.1016/S0010-938X(01)00100-7

Abdelaal, M. S., Radwan, S., & El Saied, A. (1983). Phenothiazines as corrosion inhibitors for zinc in NHAC Solution. *British Corrosion Journal*, *18*(2), 102–106. doi:10.1179/bcj.1983.18.2.102

Abe, Y., Kokubo, T., & Yamamuro, T. (1990). Apatite coating on ceramics, metals and polymers utilizing a biological process. *Journal of Materials Science. Materials in Medicine*, *1*(4), 233–238. doi:10.1007/BF00701082

Abiola, O. K., & Oforka, N. C. (2004). Adsorption of (4-amino-2-methyl-5-pyrimidinyl methylthio) acetic acid on mild steel from hydrochloric acid solution (HCl)—Part 1. *Materials Chemistry and Physics*, *83*, 315.

Abouchane, M., El Bakri, M., Touir, R., Rochdi, A., Elkhattabi, O., Touhami, M. E., ... Mernari, B. (2015). Corrosion inhibition and adsorption behavior of triazoles derivatives on mild steel in 1 MH3PO4 and synergistic effect of iodide ions. *Research on Chemical Intermediates*, *41*(4), 1907–1923. doi:10.100711164-013-1319-5

Abousalem, A. S., Ismail, M. A., & Fouda, A. S. (2019). A complementary experimental and in silico studies on the action of fluorophenyl-2, 2′-bichalcophenes as ecofriendly corrosion inhibitors and biocide agents. *Journal of Molecular Liquids*, *276*, 255–274. doi:10.1016/j.molliq.2018.11.125

ACI Committee 116. (1985). *ACI 116R-85, Cement and Concrete Terminology.* Author.

Adardour, K., Touir, R., Elbakri, M., Ramli, Y., Ebn Touhami, M., El Kafsaoui, H., Kalonji Mubengayi, C., & Essassi, E. M. (2013). Thermodynamic properties and comparative studies of quinoxaline derivatives as a corrosion inhibitor for mild steel in 1 M H_2SO_4. *Research on Chemical Intermediates*, *39*, 4175–4188. doi:10.100711164-012-0934-x

Agrawal, R., & Namboodhiri, T. K. G. (1990). The inhibition of sulphuric acid corrosion of 410 stainless steel by thioureas. *Corrosion Science*, *30*, 37.

Ahamad, I., Prasad, R., & Quraishi, M. A. (2010). Inhibition of mild steel corrosion in acid solution by Pheniramine drug: Experimental and theoretical study. *Corrosion Science*, *52*(9), 3033–3041. doi:10.1016/j.corsci.2010.05.022

Aherwar, A., Singh, A. K., & Patnaik, A. (2015). Current and future biocompatibility aspects of biomaterials for hip prosthesis. *AIMS Bioengineering*, *3*(1), 23–43. doi:10.3934/bioeng.2016.1.23

Ahmad, S., Basavaraja, L. R., & Bhattacharjee, B. (2000). Design procedures for cathodic protection systems for RC members. *Indian Concrete Journal*, *74*(4), 208–213.

Alaa, S. A., & Ayman, A. (2006). El-Fetouh. Gouda., Ragaa, El-Sheikh., Faten, Zahran. *Spectrochimica Acta. Part A: Molecular and Biomolecular Spectroscopy*, *14*, 171.

Alaoui, K., Touir, R., Galai, M., Serrar, H., Ouakki, M., Kaya, S., Tüzün, B., Boukhris, S., Ebn Touhami, M., & El Kacimi, Y. (2018), Electrochemical and Computational Studies of Some Triazepine Carboxylate Compounds as Acid Corrosion Inhibitors for Mild Steel. *Journal of Bio- and Tribo-Corrosion, 4*, 37.

Alaoui, K., Touir, R., Galai, M., Serrar, H., Ouakki, M., Kaya, S., Tüzün, B., Boukhris, S., Ebn Touhami, M., & El Kacimi, Y. (2018). Electrochemical and Computational Studies of Some Triazepine Carboxylate Compounds as Acid Corrosion Inhibitors for Mild Steel. *Journal of Bio- and Tribo-Corrosion*, *4*, 37.

Alaoui,, K., El Kacimi,, Y., Galai,, M., Dahmani,, K., Touir,, R., & El Harfi,, A., & EbnTouhami, M. (2016). Poly(1-phenylethene): As a novel corrosion inhibitor for carbon steel/hydrochloric acid interface. *Anal Bioanal Electrochem.*, *8*, 830.

Alaoui, K., El Kacimi, Y., Galai, M., Dahmani, K., Touir, R., El Harfi, A., & Ebn Touhami, M. (2016). Poly (1-phenylethene): As a Novel Corrosion Inhibitor for Carbon Steel / Hydrochloric Acid Interface. *Anal. Bioanal. Electrochem*, *7*, 830.

Alaoui, K., El Kacimi, Y., Galai, M., Serrar, H., Touir, R., Kaya, S., Kaya, C., & Ebn Touhami, M. (2020). New triazepine carboxylate derivatives: Correlation between corrosion inhibition property and chemical structure. *Int J Ind Chem*, *11*(1), 23–42. doi:10.100740090-019-00199-5

Alaoui, K., El Kacimi, Y., Galai, M., Touir, R., Dahmani, K., Harfi, A., & Ebn Touhami, M. (2016). Anti-corrosive properties of Polyvinyl-Alcohol for carbon steel in Hydrochloric acid media: Electrochemical and Thermodynamic investigation. *J. Mater. Environ. Sci.*, *7*, 2389–2403.

Alaoui, K., Ouakki, M., Abousalem, A. S., Serrar, H., Galai, M., Derbali, S., Nouneh, K., Boukhris, S., Touhami, M. E., & El Kacimi, Y. (2019). Molecular Dynamics, Monte-Carlo Simulations and Atomic Force Microscopy to Study the Interfacial Adsorption Behaviour of Some Triazepine Carboxylate Compounds as Corrosion Inhibitors in Acid Medium. *J Bio Tribo Corros*, *5*(1), 1. doi:10.100740735-018-0196-2

Alaoui, K., Ouakki, M., Abousalem, A. S., Serrar, H., Touhami, M. E., & Kacimi, Y. E. (2020). Chemical, Electrochemical, Theoretical and Surface Morphology Studies of Triazepine Carboxylate Compounds as Corrosion Inhibitors for Mild Steel in Hydrochloric Acid Medium. *J Bio Tribo Corros*, *6*(2), 54. doi:10.100740735-020-00352-z

Alaoui, K., Touir, R., Galai, M., Serrar, H., Ouakki, M., Kaya, S., Tüzün, B., Boukhris, S., Ebn Touhami, M., & El Kacimi, Y. (2018). Electrochemical and Computational Studies of Some Triazepine Carboxylate Compounds as Acid Corrosion Inhibitors for Mild Steel. *J Bio Tribo Corros*, *4*(3), 37. doi:10.100740735-018-0154-z

Aldabbagh, B. M., & Alshimary, H. J. (2017). Polyamide nanofibers coating by electrospinning technique for anti corrosion behavior. *Engineering & Technology Journal*, *35*(10), 987–991.

Al-Jabbari, Y. S., Ntasi, A., Gaintatzopoulou, M., Mueller, W. D., Eliades, G., Sherif, E. S. M., & Zinelis, S. (2016). Elemental, Morphological, and Corrosion Characterization of Different Surface States of Co-Cr Alloy for Prosthodontic Applications. Academic Press.

Al-Ketan, O., Rowshan, R., & Al-Rub, R. K. A. (2018). Topology-mechanical property relationship of 3D printed strut, skeletal, and sheet based periodic metallic cellular materials. *Additive Manufacturing*, *19*, 167–183. doi:10.1016/j.addma.2017.12.006

Allo, B. A., Costa, D. O., Dixon, S. J., Mequanint, K., & Rizkalla, A. S. (2012). Bioactive and biodegradable nanocomposites and hybrid biomaterials for bone regeneration. *Journal of Functional Biomaterials*, *3*(2), 432–463. doi:10.3390/jfb3020432 PMID:24955542

Amin, M. A., Abd-El-Rehim, S. S., El-Sherbini, E. E. F., & Hazzazi, O. A. (2009). Polyacrylic acid as a corrosion inhibitor for aluminium in weakly alkaline solutions. Part I: Weight loss, polarization, impedance EFM and EDX studies. *Corrosion Science*, *51*(3), 658–667. doi:10.1016/j.corsci.2008.12.008

Amjad, Z. (1995). *Mineral Scale Formation and inhibition*. Plenum Press. doi:10.1007/978-1-4899-1400-2

Anacona, J. R., Martell, T., & Sanchez, I. (2005). Metal complexes of a new ligand derived from 2, 3-quinoxalinedithiol and 2, 6-bis (bromomethyl) pyridine. *Journal of the Chilean Chemical Society*, *50*(1), 375–378. doi:10.4067/S0717-97072005000100001

Andrade, C., & Alonso, C. (1996). Corrosion rate monitoring in the laboratory and on-site. *Construction & Building Materials*, *10*(5), 315–328. doi:10.1016/0950-0618(95)00044-5

Anejjar, A., Zarrouk, A., Salghi, R., Zarrok, H., Ben Hmamou, D., Hammouti, B., Elmahi, B., & Al-Deyab, S. S. (2013). Studies on the Inhibitive Effect of the Ammonium Iron (II) Sulphate on the Corrosion of Carbon Steel in HCl Solution. *Journal of Materials and Environmental Science*, *4*(5), 583–592.

Angst, U., Elsener, B., Larsen, C. K., & Vennesland, Ø. (2009). Critical chloride content in reinforced concrete - A review. *Cement and Concrete Research*, *39*(12), 1122–1138. doi:10.1016/j.cemconres.2009.08.006

Ansari, F. A., Verma, C., Siddiqui, Y. S., Ebenso, E. E., & Quraishi, M. A. (2018). Volatile corrosion inhibitors for ferrous and non-ferrous metals and alloys: A review. *Int J Corros Scale Inhib.*, *7*, 126.

Arabnejad, S., Johnston, R. B., Pura, J. A., Singh, B., Tanzer, M., & Pasini, D. (2016). High-strength porous biomaterials for bone replacement: A strategy to assess the interplay between cell morphology, mechanical properties, bone ingrowth and manufacturing constraints. *Acta Biomaterialia*, *30*, 345–356. doi:10.1016/j.actbio.2015.10.048 PMID:26523335

Arellanes-Lozada, P., Olivares-Xometl, O., Likhanova, N. V., Lijanova, I. V., Vargas-García, J. R., & Hernández-Ramírez, R. E. (2018). Adsorption and performance of ammonium-based ionic liquids as corrosion inhibitors of steel. *Journal of Molecular Liquids*, *265*, 151–163. doi:10.1016/j.molliq.2018.04.153

Arrieta-González, C. D., Porcayo-Calderon, J., Salinas-Bravo, V. M., Chacon-Nava, J. G., Martinez-Villafañe, A., & Gonzalez-Rodriguez, J. G. (2011). Corrosion behavior of Ni-Cr based coatings in simulated human body fluid environment. *International Journal of Electrochemical Science*, *6*, 3644–3655.

Arrieta-Gonzalez, C. D., Porcayo-Calderon, J., Salinas-Bravo, V. M., Gonzalez-Rodriguez, J. G., & Chacon-Nava, J. G. (2011). Electrochemical behavior of Fe3Al modified with Ni in Hank's solution. *International Journal of Electrochemical Science*, *6*(9), 4016–4031.

Asan, A., Soylu, S., Kıyak, T., Yıldırım, F., Oztas, S. G., Ancın, N., & Kabasakaloğlu, M. (2006). Investigation on some Schiff bases as corrosion inhibitors for mild steel. *Corrosion Science*, *48*(12), 3933–3944. doi:10.1016/j.corsci.2006.04.011

Ashita, J., Tolwani, M. D., & Ruth, C. (2001). Simplified citrate anticoagulation for continuous renal replacement therapy. *Kidney International*, *60*(1), 370–374. doi:10.1046/j.1523-1755.2001.00809.x PMID:11422774

ASM International Handbook Committee. (1990). *Metals Handbook. ASM International.*

ASTM C876 - 15. (2015). Standard Test Method for Corrosion Potentials of Uncoated Reinforcing Steel in Concrete. doi:10.1520/C0876-15

ASTM G1-03 - Standard practice for preparing, cleaning, and evaluating corrosion test specimens. (2017).

ASTM G1-03. (2011). *Standard practice for preparing, cleaning, and evaluating corrosion test specimens*. http://www.astm.org/Standards/G1

ASTM G59 - 97: Standard test method for conducting potentiodynamic polarization resistance measurements. (2009).

ASTM-STP1065-Study of the corrosion of concrete reinforcement by electrochemical impedance measurement. (n.d.). doi:10.1520/STP25019S

Atapour, M., Pilchak, A. L., Frankel, G. S., & Williams, J. C. (2011). Corrosion behavior of β titanium alloys for biomedical applications. *Materials Science and Engineering: C, 31*(5), 885-891.

Atkinson, R.J., Posner, A.M., & Quirk, J.P. (1972). Kinetics of isotopic exchange of phosphate at the L-FeOOH-aqueous solution interface. *Journal of Organic and Nuclear Chemistry, 34*, 2201-2211.

Atwell, G. J., Baguley, B. C., & Denny, W. A. (1989). Potential antitumor agents. 57. 2-Phenylquinoline-8-carboxamides as minimal DNA-intercalating antitumor agents with in vivo solid tumor activity. *Journal of Medicinal Chemistry*, *32*(2), 396–401. doi:10.1021/jm00122a018 PMID:2913299

Au, A. G., Raso, V. J., Liggins, A. B., & Amirfazli, A. (2007). Contribution of loading conditions and material properties to stress shielding near the tibial component of total knee replacements. *Journal of Biomechanics*, *40*(6), 1410–1416. doi:10.1016/j.jbiomech.2006.05.020 PMID:16846605

Avci, G. (2008). Inhibitor effect of N, N¢-methylenediacrylamide on corrosion behavior of mild steel in 0.5 M HCl. *Materials Chemistry and Physics*, *112*, 234.

Avci, G. (2008). Inhibitor effect of N,N¢-methylenediacrylamide on corrosion behavior of mild steel in 0.5 M HCl. *Materials Chemistry and Physics*, *112*(1), 234–238. doi:10.1016/j.matchemphys.2008.05.036

Azarsa, P., & Gupta, R. (2017). Electrical resistivity of concrete for durability evaluation: A review. *Advances in Materials Science and Engineering*, *2017*, 1–30. Advance online publication. doi:10.1155/2017/8453095

Balakrishnan, A., Lee, B. C., Kim, T. N., & Panigrahi, B. B. (2008). Corrosion behaviour of ultra fine grained titanium in simulated body fluid for implant application. *Trends in Biomaterials & Artificial Organs*, *22*(1), 58–64.

Balani, K., Verma, V., Agarwal, A., & Narayan, R. (2015). *A Materials Science and Engineering Perspective*. John Wiley & Sons, The American Ceramic Society.

Baldenebro Lopez, F. J., Barrios Durstewitz, C. P., Fajardo San Miguel, G., Nuñez Jaquez, R. E., Almeraya Calderon, F., Almaral-Sánchez, J. L., & Castorena Gonzalez, J. H. (2010). Electrochemical realkalisation of reinforced concrete using a conductive mortar anode. *ECS Transactions*, *29*(1), 125–131. doi:10.1149/1.3532310

Bammou, L., Belkhaouda, M., Salghi, R., Benali, O., Zarrouk, A., Zarrok, H., & Hammouti, B. (2014). Corrosion inhibition of steel in sulfuric acidic solution by the chenopodium ambrosioides extracts. *J Assoc Arab Univ Basic Appl Sci.*, *16*(1), 83–90. doi:10.1016/j.jaubas.2013.11.001

Bandyopadhyay, A., Espana, F., Balla, V. K., Bose, S., Ohgami, Y., & Davies, N. M. (2010). Influence of porosity on mechanical properties and in vivo response of Ti6Al4V implants. *Acta Biomaterialia*, *6*(4), 1640–1648. doi:10.1016/j.actbio.2009.11.011 PMID:19913643

Bandyopadhyay, A., Mitra, I., Shivaram, A., Dasgupta, N., & Bose, S. (2019). Direct comparison of additively manufactured porous titanium and tantalum implants towards in vivo osseointegration. *Additive Manufacturing*, *28*, 259–266. doi:10.1016/j.addma.2019.04.025 PMID:31406683

Banks, J. (2013). Adding value in additive manufacturing: Researchers in the United Kingdom and Europe look to 3D printing for customization. *IEEE Pulse*, *4*(6), 22–26. doi:10.1109/MPUL.2013.2279617 PMID:24233187

Ban, S., Iwaya, Y., Kono, H., & Sato, H. (2006). Surface modification of titanium by etching in concentrated sulfuric acid. *Dental Materials*, *22*(12), 1115–1120.

Barbhuiya, S., Chow, P., & Memon, S. (2015). Microstructure, hydration and nanomechanical properties of concrete containing metakaolin. *Construction & Building Materials*, *95*, 696–702. doi:10.1016/j.conbuildmat.2015.07.101

Barcia, O. E., D'Elia, E., Frateur, I., Mattos, O. R., Pébère, N., & Tribollet, B. (2002). Application of the impedance model of de Levie for the characterization of porous electrodes. *Electrochimica Acta*, *47*(13-14), 2109–2116. doi:10.1016/S0013-4686(02)00081-6

Barnhart, R. A. (1982). FHWA position on cathodic protection. Academic Press.

Barrere, F. V., Van Blitterswijk, C. A., De Groot, K., & Layrolle, P. (2002). Nucleation of biomimetic Ca–P coatings on Ti6Al4V from a SBF× 5 solution: Influence of magnesium. *Biomaterials*, *23*(10), 2211–2220. doi:10.1016/S0142-9612(01)00354-4 PMID:11962662

Barrere, F., van Blitterswijk, C. D., De Groot, K., & Layrolle, P. (2002). Influence of ionic strength and carbonate on the Ca-P coating formation from SBF× 5 solution. *Biomaterials*, *23*(9), 1921–1930. doi:10.1016/S0142-9612(01)00318-0 PMID:11996032

Basheer, L., Kropp, J., & Cleland, D. J. (2001). Assessment of the durability of concrete from its permeation properties: A review. *Construction & Building Materials*, *15*(2–3), 93–103. doi:10.1016/S0950-0618(00)00058-1

Bashir, S., Sharma, V., Lgaz, H., Chung, M., Singh, A., & Kumar, A. (2018). The inhibition action of analgin on the corrosion of mild steel in acidic medium: A combined theoretical and experimental approach. *Journal of Molecular Liquids*, *263*, 454–462. doi:10.1016/j.molliq.2018.04.143

Bastian, M., & Frankel, G. S. (2010). Pitting Corrosion of Bare Stainless Steel 304 under Chloride Solution Droplets. *Journal of the Electrochemical Society*, *157*(10), C302–C312. doi:10.1149/1.3467850

Bastidas, J. M., Pinilla, P., Cano, E., Polo, J. L., & Miguel, S. (2003). Copper corrosion inhibition by triphenylmethane derivatives in sulphuric acid media. *Corrosion Science*, *45*(2), 427–449. doi:10.1016/S0010-938X(02)00123-3

Batis, G., Pantazopoulou, P., & Routoulas, A. (2003). Corrosion protection investigation of reinforcement by inorganic coating in the presence of alkanolamine-based inhibitor. *Cement and Concrete Composites*, *25*(3), 371–377. doi:10.1016/S0958-9465(02)00061-6

Batis, G., Pantazopoulou, P., Tsivilis, S., & Badogiannis, E. (2005). The effect of metakaolin on the corrosion behavior of cement mortars. *Cement and Concrete Composites*, *27*(1), 125–130. doi:10.1016/j.cemconcomp.2004.02.041

Bayol, E., Gurten, A. A., Dursun, M., & Kayakırılmaz, K. (2008). Adsorption Behavior and Inhibition Corrosion Effect of Sodium Carboxymethyl Cellulose on Mild Steel in Acidic Medium. *Wuli Huaxue Xuebao*, *24*, 2236.

Be Bellman, C., & Stamm, M. (2008). *Polymer Surfaces and Interfaces*. Springer.

Bedair, M. A., El-Sabbah, M. M. B., Fouda, A. S., & Elaryian, H. M. (2017). Synthesis,electrochemical and quantum chemical studies of some prepared surfactants based on azodye and Schiff base as corrosion inhibitors for steel in acid medium. *Corrosion Science*, *128*, 54–72. doi:10.1016/j.corsci.2017.09.016

Behpour, M., Ghoreishi, S. M., Soltani, N., Salavati, N. M., Hamadanian, M., & Gandomi, A. (2008). Electrochemical and Theoretical Investigation on the Corrosion Inhibition of Mild Steel by Thiosalicylaldehyde Derivatives in Hydrochloric Acid Solution. *Corrosion Science*, *50*(8), 2172–2181. doi:10.1016/j.corsci.2008.06.020

Belakhmima, R. A., Dkhireche, N., Touir, R., & Ebn Touhami, M. (2015). Development of a multi-component SG with CTAB as corrosion, scale, and microorganism inhibitor for cooling water systems. *Materials Chemistry and Physics*, *152*, 85–94. doi:10.1016/j.matchemphys.2014.12.018

Benali, O., Larabi, L., Tabti, B., & Harek, Y. (2005). Influence of 1-methyl 2-mercapto imidazole on corrosion inhibition of carbon steel in 0.5 M H2SO4. *Anti-Corrosion Methods and Materials*, *52*(5), 280–285. doi:10.1108/00035590510615776

Benarioua, M., Mihi, A., Bouzeghaia, N., & Naoun, M. (2019). Mild steel corrosion inhibition by Parsley (Petroselium Sativum) extract in acidic media. *Egyptian Journal of Petroleum*, *28*(2), 155–159. doi:10.1016/j.ejpe.2019.01.001

Benhiba, F., Serrar, H., Hsissou, R., Guenbour, A., Bellaouchou, A., Tabyaoui, M., Boukhris, S., Oudda, H., Warad, I., & Zarrouk, A. (2020). Tetrahydropyrimido-Triazepine derivatives as anti-corrosion additives for acid corrosion: Chemical, electrochemical, surface and theoretical studies. *Chemical Physics Letters*, *743*, 137181. doi:10.1016/j.cplett.2020.137181

Bennett, J. E., & Schue, T. J. (1996). *Galvanic cathodic protection of reinforced concrete bridge members using sacrificial anodes attached by conductive adhesives*. Academic Press.

Bentiss, F., Lebrini, M., & Lagrenee, M. (2005). Thermodynamic characterization of metal dissolution and inhibitor adsorption processes in mild steel/2,5-bis(n-thienyl)-1,3,4-thiadiazoles/hydrochloric acid system. *Corrosion Science*, *47*(12), 2915–2931. doi:10.1016/j.corsci.2005.05.034

Bentz, D. P. (2000). Influence of silica fume on diffusivity in cement-based materials II. Multi-scale modeling of concrete diffusivity. *Cement and Concrete Research*, *30*(7), 1121–1129. doi:10.1016/S0008-8846(00)00263-5

Bentz, D. P., Jensen, O. M., Coats, A. M., & Glasser, F. P. (2000). Influence of silica fume on diffusivity in cement-based materials I. Experimental and computer modeling studies on cement pastes. *Cement and Concrete Research*, *30*(6), 953–962. doi:10.1016/S0008-8846(00)00264-7

Berndt, C. C., Hasan, F., Tietz, U., & Schmitz, K. P. (2014). A review of HAP coatings manufactured by thermal spray. In *Advances in calcium phosphate biomaterials* (pp. 267–329). Springer. doi:10.1007/978-3-642-53980-0_9

Bertolini, L., Carsana, M., & Pedeferri, P. (2007). Corrosion behaviour of steel in concrete in the presence of stray current. *Corrosion Science*, *49*(3), 1056–1068. doi:10.1016/j.corsci.2006.05.048

Bertolini, L., Elsener, B., Pedeferri, P., Redaelli, E., & Polder, R. (2013). *Corrosion of Steel in Concrete : Prevention, Diagnosis, Repair*. John Wiley & Sons. doi:10.1002/9783527651696

Biswas, B., Purandare, Y., Sugumaran, A. A., Loch, D. A., Creasey, S., Khan, I., Ehiasarian, A. P., & Hovsepian, P. E. (2017). Defect growth in multilayer chromium nitride/niobium nitride coatings produced by combined high power impulse magnetron sputtering and unbalance magnetron sputtering technique. *Thin Solid Films*, *636*, 558–566. doi:10.1016/j.tsf.2017.06.027

Boa, A. R., & Topu, L. B. (2012). Influence of fly ash on corrosion resistance and chloride ion permeability of concrete. *Construction & Building Materials*, *31*, 258–264. doi:10.1016/j.conbuildmat.2011.12.106

Bobbio, A. (1972). The first endosseous alloplastic implant in the history of man. *Bulletin of the History of Dentistry*, *20*, 1–6. PMID:4505221

Bochris, J. O'M., & Reddy, A.K.N. (1977). Modern Electrochemistry (vol. 2). Plenum Press.

Boffardi, B. (1993). The chemistry of polyphosphate. *Materials Performance*, *32*, 50–53.

Bommersbach, P., Alemany-Dumont, C., Millet, J. P., & Normand, B. (2005). Formation and behaviour study of an environment-friendly corrosion inhibitor by electrochemical methods. *Electrochimica Acta*, *51*, 1076–1084.

Bosch, R., Hubrecht, J., Syrett, B., & Bogaerts, W. (2001). Electrochemical Frequency Modulation: A New Electrochemical Technique for Online Corrosion Monitoring. *Corrosion*, *57*(1), 57. doi:10.5006/1.3290331

Bose, S., Banerjee, D., & Bandyopadhyay, A. (2017). Introduction to biomaterials and devices for bone disorders. In *Materials for Bone Disorders* (pp. 1–27). Academic Press. doi:10.1016/B978-0-12-802792-9.00001-X

Bouckamp, A. (1993). Faculty of Chem. Tech., Universidal of Twente, The Netherlands. User's Manual Equivalent Circuit, Ver, 4, 51.

Bouckamp, A. (1993). Users manual equivalent circuit. Faculty of Chemical Technology, 4.

Bousskri, A., Anejjar, A., Messali, M., Salghi, R., Benali, O., Karzazi, Y., Jodeh, S., Zougagh, M., Ebenso, E. E., & Hammouti, B. (2015). Corrosion inhibition of carbon steel in aggressive acidic media with 1-(2-(4-chlorophenyl)-2-oxoethyl) pyridazinium bromide. *Journal of Molecular Liquids*, *211*, 1000–1008. doi:10.1016/j.molliq.2015.08.038

Broomfield, J., & Millard, S. (2002). Measuring concrete resistivity to assess corrosion rates. *Concrete (London)*, *128*, 37–39.

Brown, P. W., & Constantz, B. (2017). *HAP and related materials*. CRC press.

Brug, G. J., Van Den Eeden, A. L. G., Sluyters-Rehbach, M., & Sluyters, J. H. (1984). The analysis of electrode impedances complicated by the presence of a constant phase element. *Journal of Electroanalytical Chemistry and Interfacial Electrochemistry*, *176*(1-2), 275–295. doi:10.1016/S0022-0728(84)80324-1

Buenfeld, N. R., Newman, J. B., & Page, C. L. (1986). The resistivity of mortars immersed in sea-water. *Cement and Concrete Research*, *16*(4), 511–524. doi:10.1016/0008-8846(86)90089-X

Butler, J. A. V. (1924). Studies in heterogeneous equilibria. Part III. A kinetic theory of reversible oxidation potentials at inert electrodes. *Transactions of the Faraday Society*, *19*(March), 734–739. doi:10.1039/tf9241900734

Cabrera, J. G., & Claisse, P. A. (1999). Oxygen and water vapour transport in cement-silica fume pastes. *Construction & Building Materials*, *13*(7), 405–414. doi:10.1016/S0950-0618(99)00039-2

Cabrera, J. G., Claisse, P. A., & Hunt, D. N. (1995). A statistical analysis of the factors which contribute to the corrosion of steel in Portland cement and silica fume concrete. *Construction & Building Materials*, *9*(2), 105–113. doi:10.1016/0950-0618(95)00003-X

Cachet, C., & Wiart, R. (1994). Zinc Electrowinning in Acidic Sulfate Electrolytes: Impedance Analysis and Modeling of the Influence on Nickel Impurities. *Journal of the Electrochemical Society*, *141*(1), 131–140. doi:10.1149/1.2054672

Calignano, F., Galati, M., Iuliano, L., & Minetola, P. (2019). Design of Additively Manufactured Structures for Biomedical Applications: A Review of the Additive Manufacturing Processes Applied to the Biomedical Sector. *Journal of Healthcare Engineering*, *2019*, 2019. doi:10.1155/2019/9748212 PMID:30992744

Callaghan, B. G. (1993). The performance of a 12% chromium steel in concrete in severe marine environments. *Corrosion Science*, *35*(5–8), 1535–1541. doi:10.1016/0010-938X(93)90381-P

Cavalier, P. G., & Vassie, P. R. (1981). Investigation and repair of reinforcement corrosion in a bridge deck. *PI Civil Engineering Part-1.*, *70*, 461–480.

Cenoui, M., Dkhireche, N., Kassou, O., Ebn Touhami, M., Touir, R., Dermaj, A., & Hajjaji, N. (2010). Synergistic influence of molybdate ions with TDMTAA on corrosion inhibition of ordinary steel in cooling water system. *Journal of Materials Environmental Science. Journal Mater. Environ. Sci.*, *1*(2), 84–95.

Chen, C. T., Chang, J. J., & Yeih, W. C. (2014). The effects of specimen parameters on the resistivity of concrete. *Construction & Building Materials*, *71*, 35–43. doi:10.1016/j.conbuildmat.2014.08.009

Chen, Y., Hong, T., Gopal, M., & Jepson, W. P. (2000). EIS studies of a corrosion inhibitor behavior under multiphase flow conditions. *Corrosion Science*, *42*(6), 979–990. doi:10.1016/S0010-938X(99)00127-4

Chindaprasirt, P., Chotithanorm, C., Cao, H. T., & Sirivivatnanon, V. (2007). Influence of fly ash fineness on the chloride penetration of concrete. *Construction & Building Materials*, *21*(2), 356–361. doi:10.1016/j.conbuildmat.2005.08.010

Chinwko, E. C., Odio, B. O., Chukwuneke, J. L., & Sinebe, J. E. (2014). Investigation of the effect of corrosion on mild steel in five different environments. *International Journal of Scientific & Technology Research*, *3*(7), 306–310.

Choi, D. J., You, S.J., & Kim, J. G. (2002). Development of an environmentally safe corrosion, scale, and microorganism inhibitor for open recirculating cooling systems. *Journal of Materials Science and Engineering: A*, *335*(1-2), 228–236.

Choi, Y. S., Kim, J. G., & Lee, K. M. (2006). Corrosion behavior of steel bar embedded in fly ash concrete. *Corrosion Science*, *48*(7), 1733–1745. doi:10.1016/j.corsci.2005.05.019

Choubey, A., Basu, B., & Balasubramaniam, R. (2005). Electrochemical behavior of Ti-based alloys in simulated human body fluid environment. *Trends in Biomaterials & Artificial Organs*, *18*(2), 64–72.

Chung, K. S., Shin, D. A., Kim, K. N., Ha, Y., Yoon, H., & Yi, S. (2019). Vertebral reconstruction with a customized 3D-printed spine implant replacing large vertebral defect with 3-year follow-up: A case report. *World Neurosurgery*.

Cimpoeşu, N., Săndulache, F., Istrate, B., Cimpoeşu, R., & Zegan, G. (2018). Electrochemical Behavior of Biodegradable FeMnSi–MgCa Alloy. *Metals*, *8*(7), 541. doi:10.3390/met8070541

Conner, K. A., Sabatini, R., Mealey, B. L., Takacs, V. J., Mills, M. P., & Cochrann, D. L. (2003). Guided bone regeneration around titanium plasma-sprayed, acid-etched, and HAP-coated implants in the canine model. *Journal of Periodontology*, *74*(5), 658–668. doi:10.1902/jop.2003.74.5.658 PMID:12816298

Contu, F., Elsener, B., & Böhni, H. (2004). Serum effect on the electrochemical behaviour of titanium, Ti6Al4V and Ti6Al7Nb alloys in sulphuric acid and sodium hydroxide. *Corrosion Science*, *46*(9), 2241–2254. doi:10.1016/j.corsci.2004.01.005

Cruz-Borbolla, J., Garcia-Ochoa, E., Narayanan, J., Maldonado-Rivas, P., Pandiyan, T., & Vásquez-Pérez, J. M. (2017). Electrochemical and theoretical studies of the interactions of a pyridyl-based corrosion inhibitor with iron clusters (Fe 15, Fe 30, Fe 45, and Fe 60). *Journal of Molecular Modeling*, *23*(12), 342. doi:10.100700894-017-3510-x PMID:29143152

Cruz, P. V., Nishikata, A., & Tsuru, T. (1998). Pitting corrosion Mechanism of Stainless steel under wet-dry exposure in Chloride-Containing Environments. *Corrosion Science*, *40*(1), 125–139. doi:10.1016/S0010-938X(97)00124-8

Cubillos, M., Sancy, M., Pavez, J., Vargas, E., Urzua, R., Henríquez-Roman, J., Tribollet, B., Zagal, J. H., & Páez, M. A. (2010). Influence of 8-aminoquinoline on the corrosion behaviour of copper in 0.1 M NaCl. *Electrochimica Acta, 55*(8), 2782–2792. doi:10.1016/j.electacta.2009.12.064

Cui, X., Kim, H. M., Kawashita, M., Wang, L., Xiong, T., Kokubo, T., & Nakamura, T. (2009). Preparation of bioactive titania films on titanium metal via anodic oxidation. *Dental Materials, 25*(1), 80–86. doi:10.1016/j.dental.2008.04.012 PMID:18599115

Dahmani, K., Galai, M., & Cherkaoui, M., El hasnaoui, A., & El Hessni, A. (2017). Cinnamon essential oil as a novel eco-friendly corrosion inhibitor of copper in 0.5 M Sulfuric Acid medium. *J. Mater. Environ. Sci, 8*, 1676.

Dai, J.-G., Akira, Y., Wittmann, F. H., Yokota, H., & Zhang, P. (2010). Water repellent surface impregnation for extension of service life of reinforced concrete structures in marine environments: The role of cracks. *Cement and Concrete Composites, 32*(2), 101–109. doi:10.1016/j.cemconcomp.2009.11.001

Daniyal, M., Azam, A., & Akhtar, S. (2018). Application of Nanomaterials in Civil Engineering. In *Advanced Structured Materials* (Vol. 84, pp. 169–189). Springer Nature Singapore Pte Ltd., doi:10.1007/978-981-10-6214-8_6

Daoud, D., Douadi, T., Issaadi, S., & Chafaa, S. (2014). Adsorption and corrosion inhibition of new synthesized thiophene Schiff base on mild steel X52 in HCl and H2SO4 solutions. *Corrosion Science, 79*, 50–58. doi:10.1016/j.corsci.2013.10.025

Davis, J. R. (Ed.). (2004). *Handbook of thermal spray technology*. ASM International.

de Almeida Souza, L. R., de Medeiros, M. H. F., Pereira, E., & Capraro, A. P. B. (2017). Electrochemical chloride extraction: Efficiency and impact on concrete containing 1% of NaCl. *Construction & Building Materials, 145*, 435–444. doi:10.1016/j.conbuildmat.2017.04.010

De Souza, F. S., & Spinelli, A. (2009). Caffeic acid as a green corrosion inhibitor for mild steel. *Corrosion Science, 51*(3), 642–649. doi:10.1016/j.corsci.2008.12.013

de Vries, I. J., & Polder, R. B. (1997). Hydrophobic treatment of concrete. *Construction & Building Materials, 11*(4), 259–265. doi:10.1016/S0950-0618(97)00046-9

Dehghanghadikolaei, A., Ibrahim, H., Amerinatanzi, A., & Elahinia, M. (2019). Biodegradable magnesium alloys. In *Metals for Biomedical Devices* (pp. 265–289). Woodhead Publishing. doi:10.1016/B978-0-08-102666-3.00009-2

Dehwah, H. A. F., Maslehuddin, M., & Austin, S. A. (2002). Long-term effect of sulfate ions and associated cation type on chloride-induced reinforcement corrosion in Portland cement concretes. *Cement and Concrete Composites, 24*(1), 17–25. doi:10.1016/S0958-9465(01)00023-3

Delucchi, M., Barbucci, A., & Cerisola, G. (1997). Study of the physico-chemical properties of organic coatings for concrete degradation control. *Construction & Building Materials, 11*(7–8), 365–371. doi:10.1016/S0950-0618(97)00060-3

Deng, S., Li, X., & Fu, H. (2011). Two pyrazine derivatives as inhibitors of the cold rolled steel corrosion in hydrochloric acid solution. *Corrosion Science, 53*(2), 822–828. doi:10.1016/j.corsci.2010.11.019

Denkhaus, E., & Salnikow, K. (2002). Nickel essentiality, toxicity, and carcinogenicity. *Critical Reviews in Oncology/Hematology, 42*(1), 35–56. doi:10.1016/S1040-8428(01)00214-1 PMID:11923067

Diamanti, M. V., Brenna, A., Bolzoni, F., Berra, M., Pastore, T., & Ormellese, M. (2013). Effect of polymer modified cementitious coatings on water and chloride permeability in concrete. *Construction & Building Materials, 49*, 720–728. doi:10.1016/j.conbuildmat.2013.08.050

Dkhireche, N., Dahami, A., Rochdi, A. J. Hmimou, J., Touir, R. M., Ebn Touhami. M., El Bakri, M., El Hallaoui, A., Anouar, A., & Takenouti, H. (2013). Corrosion and scale inhibition of low carbon steel in cooling water system by 2-propargyl-5-o-hydroxyphenyltetrazole. *Journal of Industrial and Engineering Chemistry*, *19*, 1996–2003.

Dkhireche, N., Abdelhadi, R., Ebn Touhami, M., Oudda, H., Touir, R., Elbakri, M., Saira, M., Hammouti, B., Senhaji, O., & Taouil, R. (2012). Elucidation of Dimethyldodecylphosphonate and CTAB Synergism on Corrosion and Scale Inhibition of Mild Steel in Simulated Cooling Water System. *International Journal of Electrochemical Science*, *7*, 5314–5330.

Ðlhan, S., Temel, H., & Sekerci, M. (2002). Synthesis and characterization of new N, N¢-bis (cinnemaldeydene)-1, 2-bis (p-aminophenoxy) ethane and its transition metal complexes. Synt. React. Inorg. Met-Org. *Chem*, *32*, 1625–1634.

Drelich, J. W. (2019). Characterization of Biodegradable Medical Materials. *JOM*, *71*(4), 1404–1405. doi:10.100711837-019-03381-3 PMID:31693017

Duffo, G. S., & Castillo, E. Q. (2004). Development of an artificial saliva solution for studying the corrosion behavior of dental alloys. *Corrosion*, *60*(6), 594–602. doi:10.5006/1.3287764

Ebenso, E. E. (2001). Inhibition of corrosion of mild steel hydrochloric acid by some azo dyes. *Niger J Chem Res*, *6*, 8–12.

Ech-chihbi, E., Nahlé, A., Salim, R., Oudda, H., El Hajjaji, F., El Kalai, F., El Aatiaoui, A., & Taleb, M. (2019). An Investigation into Quantum Chemistry and Experimental Evaluation of Imidazopyridine Derivatives as Corrosion Inhibitors for C-Steel in Acidic Media. *Journal of Bio- and Tribo-Corrosion, 5*(24).

Ech-chihbi, E., Belghiti, M. E., Salim, R., Oudda, H., Taleb, M., Benchat, N., Hammouti, B., & El-Hajjaji, F. (2017). Experimental and computational studies on the inhibition performance of the organic compound "2-phenylimidazo [1,2-a]pyrimidine-3-carbaldehyde" against the corrosion of carbon steel in 1.0M HCl solution. *J. Surf. Interface*, *9*, 206–217. doi:10.1016/j.surfin.2017.09.012

Ehsani, A., Babaei, F., & Nasrollahzadeh, M. (2013). Electrosynthesis and absorbance spectra of TiO2 nanoparticles dispersed in the conductive polymer. *Applied Surface Science*, *283*, 1060–1064. doi:10.1016/j.apsusc.2013.07.067

Ehtemam-Haghighi, S., Attar, H., Dargusch, M. S., & Kent, D. (2019). Microstructure, phase composition and mechanical properties of new, low cost Ti-Mn-Nb alloys for biomedical applications. *Journal of Alloys and Compounds*, *787*, 570–577. doi:10.1016/j.jallcom.2019.02.116

Ekanem, U. F., Umoren, S. A., Udousoro, I. I., & Udoh, A. P. (2010). Inhibition of mild steel corrosion in HCl using pineapple leaves (Ananas comosus L.) extract. *Journal of Materials Science*, *45*(20), 5558–5566. doi:10.100710853-010-4617-y

El Azhar, M., Mernari, B., Traisnel, M., Bentiss, F., & Lagrenée, M. (2001). Corrosion inhibition of mild steel by the new class of inhibitors [2,5-bis(*n*-pyridyl)-1,3,4-thiadiazoles] in acidic media. *Corrosion Science*, *43*, 2229.

El Bakri, M., Touir, R., Tazouti, A., Dkhireche, N., Touhami, M. E., Rochdi, A., & Zarrouk, A. (2016). Corrosion inhibition study of brass in simulated cooling water by triazole derivatives, cetyltrimethylammonium bromide and their mixture. *Arabian Journal for Science and Engineering*, *41*(1), 75–88. doi:10.100713369-014-1532-6

El Bakri, Y., El Aoufir, Y., Bourazmi, H., Harmaoui, A., Sebhaoui, J., Tabyaoui, M., Guenbour, A., Oudda, H., Lgaz, H., El Hajjaji, F., Ben Ali, A., Ramli, Y., & Essassi, E. M. (2017). Corrosion control of carbon steel in phosphoric acid by 6-methyl-7H-1,2,4-triazolo[4,3-b][1,2,4]-triazepine-8(9H)-thione: Electrochemical studies. *Journal of Materials and Environmental Science*, *8*(8), 2657–2666.

El Bakri, Y., Lei, G., El Hassane, A., & El Mokhtar, E. (2019). Electrochemical, DFT and MD simulation of newly synthesized triazolotriazepine derivatives as corrosion inhibitors for carbon steel in 1 M HCl. *Journal of Molecular Liquids*, *274*, 759–769. doi:10.1016/j.molliq.2018.11.048

El Faydy, M., Galai, M., Rbaa, M., Ouakki, M., Lakhrissi, B., Ebn Touhami, M., & El Kacimi, Y. (2018). Synthesis and Application of New Quinoline as Hydrochloric Acid Corrosion Inhibitor of Carbon Steel. *Anal. Bioanal. Electrochem*, *10*(7), 815–839.

El Guerraf, A., Titi, A., Cherrak, K., Mechbal, N., El Azzouzi, M., Touzani, R., Hammouti, B., & Lgaz, H. (2018). The synergistic effect of chloride ion and 1,5-diaminonaphthalene on the corrosion inhibition of mild steel in 0.5 M sulfuric acid: Experimental and theoretical insights. *Surfaces and Interfaces*, *13*, 168–177. doi:10.1016/j.surfin.2018.09.004

El Hezzat, M., Assouag, M., Zarrok, H., Benzekri, Z., El Assyry, A., Boukhris, S., Souizi, A., Galai, M., Touir, R., Ebn Touhami, M., Oudda, H., & Zarrouk, A. (2015). Correlated DFT and electrochemical study on inhibition behavior of ethyl 6-amino-5-cyano-2-methyl-4-(p-tolyl)-4H-pyran-3-carboxylate for the corrosion of mild steel in HCl. *Der Pharma Chemica*, *7*(10), 77–88.

El Hezzat, M., Assouag, M., Zarrok, H., Benzekri, Z., El Assyry, A., Boukhris, S., Souizi, A., Galai, M., Touir, R., Ebn Touhami, M., Oudda, H., & Zarrouk, A. (2015). Electrochemical and theoretical evaluation of ethyl 6-amino-5-cyano-2-methyl-4-phenyl-4H-pyran-3-carboxylate as corrosion inhibitor for low carbon steel in 1.0 M HCl. *Der Pharma Chemica*, *7*(10), 77–88.

El Kacimi, Y., Galai, M., Alaoui, K., Touir, R., & Ebn Touhami, M. (2018). Surface Morphology Studies and Kinetic-thermodynamic characterization of Steels treated in 5.0 M HCl medium: Hot-Dip Galvanizing application. *Anti-Corrosion Methods and Materials*, *65*(2), 176–189. doi:10.1108/ACMM-11-2016-1735

El Kacimi, Y., Touir, R., Galai, M., Belakhmima, R. A., Zarrouk, A., Alaoui, K., Harcharras, M., El Kafssaoui, H., & Ebn Touhami, M. (2016). Effect of silicon and phosphorus contents in steel on its corrosion inhibition in 5 M HCl solution in the presence of Cetyltrimethylammonium/KI. *J. Mater. Environ. Sci*, *7*, 371–381.

El Mehdi, B., Mernari, B., Traisnel, M., Bentiss, F., & Lagrenee, M. (2003). Synthesis and comparative study of the inhibitive effect of some new triazole derivatives towards corrosion of mild steel in hydrochloric acid solution. *Materials Chemistry and Physics*, *77*(2), 489–496. doi:10.1016/S0254-0584(02)00085-8

El Ouali, I., Hammouti, B., Aouniti, A., Ramli, Y., Azougagh, M., Essassi, E. M., & Bouachrine, M. (2010). Thermodynamic characterisation of steel corrosion in HCl in the presence of 2- phenylthieno (3, 2-b) quinoxaline. *J. Mater. Environ. Sci*, *1*(1), 1–8.

El Ouali, I., Hammouti, B., Aouniti, A., Ramli, Y., Azougagh, M., Essassi, E. M., & Bouachrine, M. (2010). Thermodynamic characterisation of steel corrosion in HCl in the presence of 2-phenylthieno (3, 2-b) quinoxaline. *J. Mater. Environ. Sci.*, *1*, 1–8.

El Ouali, I., Hammouti, B., Aouniti, A., Ramli, Y., Azougagh, M., & Essassi, E.M., Bouachrine. (2010). Thermodynamic characterisation of steel corrosion in HCl in the presence of 2-phenylthieno (3, 2-b) quinoxaline. *J. Mater. Envir. Sci*, *1*, 1–8.

Elaatiaoui, A., Rokni, Y., Mohammed, K., Asehraou, A., Chelfi, T., Saddik, R., Oussaid, A., Villalgordo, J. M., Abouricha, S., El Mahi, B., Oussaid, A., Zarrouk, A., & Benchat, N. (2015). Synthesis, spectroscopicand antimicrobial activityof some new 7-methyl-2-phenylimidazo[1,2-a]pyridin-3-amine derivatives. *J Mater Environ Sci.*, *6*, 2083.

Elattar, K. M., Abozeid, M. A., Mousab, I. A., & El-Mekabatya, A. (2015). Advances in 1,2,4-triazepines chemistry. *RSC Advances*, *5*(129), 106710–106753. doi:10.1039/C5RA21108E

El-Awady, A. R., Semreen, M. H., Saber, M. M., Cyprian, F., Menon, V., & Al-Tel, T. H. (2016). Modulation of DNA damage response and induction of apoptosis mediates synergism between doxorubicin and a new imidazopyridine derivative in breast and lung cancer cells. *DNA Repair*, *37*, 1–11. doi:10.1016/j.dnarep.2015.10.004 PMID:26590797

Elayyachy, M., El Idrissi, A., & Hammouti, B. (2006). New thio-compounds as corrosion inhibitor for steel in 1 M HCl. *Corrosion Science*, *48*, 2470.

ElBelghiti, M., Karzazi, Y., Dafali, A., Hammouti, B., Bentiss, F., Obot, I. B., Bahadur, I., & Ebenso, E. E. (2016). Experimental, quantum chemical and Monte Carlo simulation studies of 3,5-disubstituted-4-amino-1,2,4-triazoles as corrosion inhibitors on mild steel in acidic mediumJ. *Mol. Liq.*, *218*, 281–293. doi:10.1016/j.molliq.2016.01.076

El-Hajjaji,, F., Belkhmima,, R. A., Zerga,, B., Sfaira,, M., & Taleb,, M., EbnTouhami, M., Hammouti, B., Al-Deyab, S. S., & Ebenso, E. (2014). Temperature performance of a thionequinoxaline compound as mild steel corrosion inhibitor in hydrochloric acid medium. *International Journal of Electrochemical Science*, *9*, 4721.

Eliaz, N. (2019). Corrosion of metallic biomaterials: A review. *Materials (Basel)*, *12*(3), 407. doi:10.3390/ma12030407 PMID:30696087

Eljaouhari, A., Chennah, A., Jaddi, S. B., & Ahsaine, H. (2019). Electrosynthesis of Zinc Phosphate-Polypyrrole coatings for improved corrosion resistance of steel. *Surfaces and Interfaces*, *15*, 224–231. doi:10.1016/j.surfin.2019.02.011

Elsener, B., Klinghoffer, O., Frolund, T., Rislund, E., Schiegg, Y., & Böhni, H. (1997). Assessment of reinforcement corrosion by means of galvanostatic pulse technique. In *Int.Conf. Repair of Concrete Structures* (pp. 1–10). Academic Press.

Elsener, B. (2005). Corrosion rate of steel in concrete-Measurements beyond the Tafel law. *Corrosion Science*, *47*(12), 3019–3033. doi:10.1016/j.corsci.2005.06.021

Elsener, B., & Angst, U. (2007). Mechanism of electrochemical chloride removal. *Corrosion Science*, *49*(12), 4504–4522. doi:10.1016/j.corsci.2007.05.019

El-Taib Heakal, F., Deyab, M.A., Osman, M.M., Nessim, M.I., & Elkholy, A.E. (2017). Synthesis and assessment of new cationic gemini surfactants as inhibitors for carbon steel corrosion in oilfield water. *RSC Adv*, 47335–47352.

Erdey-Gruz, T., & Volmer, M. (1930). The theory of hydrogen overvoltage. *Zeitschrift für Physikalische Chemie*, *150*, 203–213.

Erdoğdu, Ş., Kondratova, I. L., & Bremner, T. W. (2004). Determination of chloride diffusion coefficient of concrete using open-circuit potential measurements. *Cement and Concrete Research*, *34*(4), 603–609. doi:10.1016/j.cemconres.2003.09.024

Essaadaoui, Y., Galai, M., Ouakki, M., Kadiri, L., Ouass, A., Cherkaoui, M., Rifi, E. H., & Lebkiri, A. (2019). Study of the anticorrosive action of eucalyptus Camaldulensis extract in case of mild steel in 1.0 m HCL. *J. Chem. Technol. Metall*, *54*(2), 431–442.

Fang, D., Jun-mao, Q., & Hao-hua, D. (2015). *Study on Thermal Physical Properties of 304 Stainless Steel*. The Minerals, Metals & Materials Society.

Farahani, A., Taghaddos, H., & Shekarchi, M. (2015). Prediction of long-term chloride diffusion in silica fume concrete in a marine environment. *Cement and Concrete Composites*, *59*, 10–17. doi:10.1016/j.cemconcomp.2015.03.006

Farr, J. P. G., & Saremi, M. (1982). Molybdate in aqueous corrosion inhibition part II: Effects of molybdate, citrate and phosphate on the anodic behaviour of steel. *Surface Technology*, *17*(1), 19–27. doi:10.1016/0376-4583(82)90057-7

Felhősi, I., Telegdi, J., Pálinkás, G., & Kálmán, E. (2002). Kinetics of self-assembled layer formation on iron. *Electrochimica Acta*, *47*(13-14), 2335–2340. doi:10.1016/S0013-4686(02)00084-1

Fergachi, O., Benhiba, F., Rbaa, M., Glai, M., Touir, R., & Ebn Touhami, M. (2019). Corrosion Inhibition of Ordinary Steel in 5.0 M HCl Medium by Benzimidazole Derivatives: Electrochemical, UV-Visible Spectrometry, and DFT Calculations. *J. Bio-and Tribo-Corros.*, *5*(1), 2–13. doi:10.100740735-018-0215-3

Fergachi, O., Benhiba, F., Rbaa, M., Ouakki, M., Galai, M., Touir, R., Lakhrissi, B., Oudda, H., & Ebn Touhami, M. (2019). Corrosion inhibition of ordinary steel in 5.0 M HCl medium by benzimidazole derivatives: electrochemical, UV–visible spectrometry, and DFT calculations. *Journal of Bio- and Tribo-Corrosion, 5*, 21.

Fergachi, O., Benhiba, F., Rbaa, M., Touir, R., Ouakki, M., Galai, M., Lakhrissi, B., Oudda, H., & Ebn Touhami, M. (2018). Experimental and Theoretical Study of Corrosion Inhibition of Mild Steel in 1.0 M HCl Medium by 2(-4(hloro phenyl-1H- benzo[d]imidazol)-1-yl)phenyl)methanone. *Materials Research*, *21*(6), e20171038. doi:10.1590/1980-5373-mr-2017-1038

Ficai, A., Andronescu, E., Voicu, G., & Ficai, D. (2011). Article. In B. Attaf (Ed.), Advances in Collagen/Hydroxyapatite Composite Materials (pp. 3–32). Advances in Composite Materials for Medicine and Nanotechnology.

Finley, H. F., & Hackerman, N. (1960). Effect of adsorption of polar organic compounds on the reactivity of steel. *Journal of the Electrochemical Society*, *107*(4), 259–263. doi:10.1149/1.2427675

Finley, H. F., & Hackerman, N. (1960). Effect of Adsorption of Polar Organic Compounds on the Reactivity of Steel. *Journal of the Electrochemical Society*, *107*, 259–263.

Flis, J., & Zakroczymski, T. (1996). Impedance study of reinforcing steel in simulated pore solution with tannin. *Journal of the Electrochemical Society*, *143*(8), 2458. doi:10.1149/1.1837031

Fouda, A. S., Ismail, M. A., EL-ewady, G. Y., & Abousalem, A. S. (2017). Evaluation of 4-amidinophenyl-2, 2¢-bithiophene and its aza-analogue as novel corrosion inhibitors for CS in acidic media: Experimental and theoretical study. *Journal of Molecular Liquids*, *240*, 372–388. doi:10.1016/j.molliq.2017.05.089

Fouda, A. S., Shalabi, K., & Elmogazy, H. (2014). Corrosion inhibition of α-brass in HNO3 by indole and 2-oxyindole. *J Mater Environ Sci.*, *5*, 1691.

Frangini, S., & De Cristofaro, N. (2003). Analysis of the galvanostatic polarization method for determining reliable pitting potentials on stainless steels in crevice-free conditions. *Corrosion Science*, *45*(12), 2769–2786. doi:10.1016/S0010-938X(03)00102-1

Frankel, G. S. (2016). *Active Protective Coatings* (A. E. Hughes, J. M. C. Mol, M. L. Zheludkevich, & R. G. Buchheit, Eds.; Vol. 233). Springer Netherlands. doi:10.1007/978-94-017-7540-3_2

Franzoni, E., Pigino, B., & Pistolesi, C. (2013). Ethyl silicate for surface protection of concrete: Performance in comparison with other inorganic surface treatments. *Cement and Concrete Composites*, *44*, 69–76. doi:10.1016/j.cemconcomp.2013.05.008

Fu, X., & Chung, D. D. L. (1996). Improving the bond strength between steel rebar and concrete by oxidation treatments of the rebar. *Cement and Concrete Research*, *26*(10), 1499–1503. doi:10.1016/0008-8846(96)00133-0

Gabrielli, C. (1980). *Identification of Electrochemical Processes by Frequency Response Analysis*. Solartron.

Gadow, R., Killinger, A., & Stiegler, N. (2010). HAP coatings for biomedical applications deposited by different thermal spray techniques. *Surface and Coatings Technology*, *205*(4), 1157–1164. doi:10.1016/j.surfcoat.2010.03.059

Gagg, C. R. (2014). Cement and concrete as an engineering material: An historic appraisal and case study analysis. *Engineering Failure Analysis*, *40*, 114–140. doi:10.1016/j.engfailanal.2014.02.004

Galai, M., El Gouri, M., Dagdag, O., El Kacimi, Y., El Harfi, A., & Ebn Touhami, M. (2015). The effect of temperature on the inhibition of acid hydrochloric corrosion of carbon steel by hexa propylene glycol cyclotriphosphazene. *Journal of Chemical and Pharmaceutical Research*, *7*, 712–719.

Galai, M., Rbaa, M., El Kacimi, Y., Ouakki, M., Dkhirech, N., Touir, R., & Lakhrissi, B., EbnTouhami, M. (2016). Anti-corrosion properties of some triphenylimidazole substituted compounds in corrosion inhibition of carbon steel in 1. 0 M hydrochloric acid solution. *Anal. Bioanal. Electrochem*, *9*, 80.

Galai, M., Rbaa, M., El Kacimi, Y., Ouakki, M., Dkhirech, N., Touir, R., Lakhrissi, B., & Ebn Touhami, M. (2017). Anti-corrosion Properties of some Triphenylimidazole Substituted Compounds in Corrosion Inhibition of Carbon Steel in 1. 0 M Hydrochloric Acid Solution. *Anal. Bioanal. Electrochem*, *9*, 80.

Galai, M., Rbaa, M., El Kacimi, Y., Ouakki, M., Dkhirech, N., Touir, R., Lakhrissi, B., & Ebn Touhami, M. (2017). Anti-corrosion Properties of some Triphenylimidazole Substituted Compounds in Corrosion Inhibition of Carbon Steel in 1.0 M Hydrochloric Acid Solution. *Anal. Bioanal. Electrochem.*, *9*, 80–101.

Galati, M., & Iuliano, L. (2018). A literature review of powder-based electron beam melting focusing on numerical simulations. *Additive Manufacturing*, *19*, 1–20. doi:10.1016/j.addma.2017.11.001

Gao, G., & Liang, C. H. (2007). 1, 3-Bis-diethylamino-propan-2-ol as volatile corrosion inhibitor for brass. *Corrosion Science*, *49*(9), 3479–3493. doi:10.1016/j.corsci.2007.03.030

Gao, M., Wang, M., & Zheng, Q. H. (2016). Synthesis of carbon-11-labeled imidazopyridine- and purinethioacetamide derivatives as new potential PET tracers for imaging of nucleotide pyrophosphatase/phosphodiesterase. *Bioorganic & Medicinal Chemistry Letters*, *26*(5), 1371–1375. doi:10.1016/j.bmcl.2016.01.081 PMID:26856922

Garcés, P., Saura, P., Zornoza, E., & Andrade, C. (2011). Influence of pH on the nitrite corrosion inhibition of reinforcing steel in simulated concrete pore solution. *Corrosion Science*, *53*(12), 3991–4000. doi:10.1016/j.corsci.2011.08.002

Garcia-Ochoa, E., Guzmán-Jiménez, S. J., Hernández, J. G., Pandiyan, T., Vásquez-Pérez, J. M., & Cruz-Borbolla, J. (2016). Benzimidazole ligands in the corrosion inhibition for carbon steel in acid medium: DFT study of its interaction on Fe30 surface. *Journal of Molecular Structure*, *1119*, 314–324. doi:10.1016/j.molstruc.2016.04.057

Garimella Purna, S. K., Pollard, A., & Metzger, L. E. (2006). Effect of Formulation and Manufacturing Parameters on Process Cheese Food Functionality - I. Trisodium Citrate. *Journal of Dairy Science*, *89*(7), 2386–2396. doi:10.3168/jds.S0022-0302(06)72311-6 PMID:16772554

Gedge, G. (2008). Structural uses of stainless steel Buildings and civil engineering. *Journal of Constructional Steel Research*, *64*(11), 1194–1198. doi:10.1016/j.jcsr.2008.05.006

Geethanjali, R., & Subhashini, S. (2015). Thermodynamic Characterization of Metal Dissolution and Adsorption of Polyvinyl Alcohol-Grafted Poly(Acrylamide-Vinyl Sulfonate) on Mild Steel in Hydrochloric Acid. *Portugaliae Electrochimica Acta*, *33*(1), 35–48.

Georges, C., Rocca, E., & Steinmetz, P. (2008). Synergistic effect of tolutriazol and sodium carboxylates on zinc corrosion in atmospheric conditions. *Electrochimica Acta*, *53*(14), 4839–4845. doi:10.1016/j.electacta.2008.01.073

Gerengi, H., & Sahin, H. I. (2012). Schinopsis lorentzii Extract As a Green Corrosion Inhibitor for Low Carbon Steel in 1 M HCl Solution. *Industrial & Engineering Chemistry Research*, *51*(2), 780–787. doi:10.1021/ie201776q

Ghanbaria, A., Attara, M. M., & Mahdavianb, M. (2010). Corrosion inhibition performance of three imidazole derivatives on mild steel in 1M phosphoric acid. *Materials Chemistry and Physics*, *124*(2-3), 1205–1209. doi:10.1016/j.matchemphys.2010.08.058

Gharda, N., Galai, M., Saqalli, L., Ouakki, M., Habbadi, N., Ghailane, R., Souizi, A., Ebn Touhami, M., & Peres-lucchese, Y. (2017). Synthesis, Structural Properties and Complex Corrosion Inhibition Cu (II) With Amino Acid (DL-α-Alanine). *Oriental Journal of Chemistry*, *33*, 1665–1676.

Ghazoui, A., Bencaht, N., Al-Deyab, S. S., Zarrouk, A., Hammouti, B., Ramdani, M., & Guenbour, M. (2013). An Investigation of two novel pyridazine derivatives as corrosion inhibitor for C38 steel in 1.0 M HCl. *International Journal of Electrochemical Science*, *8*, 2272–2292.

Ghazoui, A., Zarrouk, A., Bencaht, N., Salghi, R., Assouag, M., El Hezzat, M., Guenbour, A., & Hammouti, B. (2014). New possibility of mild steel corrosion inhibition by organic heterocyclic compound. *Journal of Chemical and Pharmaceutical Research*, *6*, 704.

Ghibate,, R., Sabry,, F., & Kharbach,, Y., KandriRodi, Y., Skalli, M. K., Haoudi, A., Senhaji, O., Touzani, M., Taouil, R., Aouniti, A., & Hammouti, B. (2015). Valuation of surfactant phosphonates synthesized in the protection of metal surfaces against corrosion of mild steel in 0.5M H2SO4 media. *International Journal of Engineering Research and Applications*, *5*, 22.

Ghods, P., Isgor, O. B., Mcrae, G. A., & Gu, G. P. (2010). Electrochemical investigation of chloride-induced depassivation of black steel rebar under simulated service conditions. *Corrosion Science*, *52*(5), 1649–1659. doi:10.1016/j.corsci.2010.02.016

Ghouse, S., Reznikov, N., Boughton, O. R., Babu, S., Ng, K. G., Blunn, G., Cobb, J. P., Stevens, M. M., & Jeffers, J. R. (2019). The design and in vivo testing of a locally stiffness-matched porous scaffold. *Applied Materials Today*, *15*, 377–388. doi:10.1016/j.apmt.2019.02.017 PMID:31281871

Goel, R., Siddiqi, W. A., Ahmed, B., Khan, M. S., & Chaubey, V. M. (2010). Synthesis characterization and corrosion inhibition efficiency of N-C2 {(2E)-2-[4-(dimethylamino) benzylidene] hydrazinyl} 2-oxo ethyl benzamide on mild steel. *Desalination*, *263*(1-3), 45–57. doi:10.1016/j.desal.2010.06.033

Goldberg, A. J., & Kuhn, L. T. (2018). Biomaterials. In Regenerative Engineering. CRC Press.

González, J. A., Miranda, J. M., & Feliu, S. (2004). Considerations on reproducibility of potential and corrosion rate measurements in reinforced concrete. *Corrosion Science*, *46*(10), 2467–2485. doi:10.1016/j.corsci.2004.02.003

Gowers, K. R., & Millard, S. G. (1999). Measurement of concrete resistivity for assessment of corrosion severity of steel using Wenner technique. *ACI Materials Journal*, *96*(5), 536–541. doi:10.14359/655

Gruber, K. A., Ramlochan, T., Boddy, A., Hooton, R. D., & Thomas, M. D. A. (2001). Increasing concrete durability with high-reactivity metakaolin. *Cement and Concrete Composites*, *23*(6), 479–484. doi:10.1016/S0958-9465(00)00097-4

Gunasekaran, G., & Chauhan, L. R. (2004). Eco-Friendly Inhibitor for Corrosion Inhibition of Mild Steel in Phosphoric Acid. *Electrochimica Acta*, *49*(25), 4387–4395. doi:10.1016/j.electacta.2004.04.030

Güneyisi, E., Özturan, T., & Gesoğlu, M. (2005). A study on reinforcement corrosion and related properties of plain and blended cement concretes under different curing conditions. *Cement and Concrete Composites*, *27*(4), 449–461. doi:10.1016/j.cemconcomp.2004.05.006

Guocheng, L., Haidong, C., Chunchun, X., & Zonghu, H. (2008). Effect of Strain and Chloride Concentration on Pitting Susceptibility for Type 304 Austenitic Stainless Steel. *Chinese Journal of Chemical Engineering*, *16*(2), 314–319. doi:10.1016/S1004-9541(08)60080-4

Gupta, N. K., Quraishi, M. A., Verma, C., & Mukherjee, A. K. (2016a). Spiro [indoline-3,4¢-pyrano[2,3-c]pyrazole] Derivatives as Novel Class of Green Corrosion Inhibitors for Mild Steel in Hydrochloric Acid Medium: Theoretical and Experimental Approach. *RSC Advances*, *6*, 102076–102087. doi:10.1039/C6RA22116E

Gupta, N. K., Quraishi, M. A., Verma, C., & Mukherjee, A. K. (2016b). Schiff's bases derived from l-lysine and aromatic aldehydes as green corrosion inhibitors for mild steel: Experimental and theoretical studies. *Journal of Molecular Liquids*, *215*, 47–57. doi:10.1016/j.molliq.2015.12.027

Gurrappa, I., & Reddy, D. V. (2005). Characterisation of titanium alloy, IMI-834 for corrosion resistance under different environmental conditions. *Journal of Alloys and Compounds*, *390*(1-2), 270–274. doi:10.1016/j.jallcom.2004.08.040

Gutiérrez, E., Rodríguez, J. A., Cruz-Borbolla, J., & Alvarado-Rodríguez, J. G. (2016). Pandiyan Thangarasu, Development of a predictive model for corrosion inhibition of carbonsteel by imidazole and benzimidazole derivatives. *Corrosion Science*, *108*, 23–35. doi:10.1016/j.corsci.2016.02.036

Gu, X., Wang, F., Xie, X., Zheng, M., Li, P., Zheng, Y., Qin, L., & Fan, Y. (2018). In vitro and in vivo studies on as-extruded Mg-5.25 wt.% Zn-0.6 wt.% Ca alloy as biodegradable metal. *Science China Materials*, *61*(4), 619–628. doi:10.100740843-017-9205-x

Guzman, P., Caballero, J. L., Orozco-Hernández, G., Aperador, W., & Caicedo, J. C. (2018). Tribocorrosion Behavior of Niobium-Based Thin Films for Biomedical Applications. *Tribology in Industry, 40*(4).

Hackerman, N., Snavely, E., & Payne, J. S. (1966). Effects of Anions on Corrosion Inhibition by Organic Compounds. *Journal of the Electrochemical Society*, *113*(7), 677–681. doi:10.1149/1.2424089

Hansen, D. C. (2008). Metal corrosion in the human body: The ultimate bio-corrosion scenario. *The Electrochemical Society Interface*, *17*(2), 31.

Hansson, C. M., Mammoliti, L., & Hope, B. B. (1998). Corrosion inhibitors in concrete - Part I: The principles. *Cement and Concrete Research*, *28*(12), 1775–1781. doi:10.1016/S0008-8846(98)00142-2

Hansson, I. L. H., & Hansson, C. M. (1983). Electrical resistivity measurements of Portland cement based materials. *Cement and Concrete Research*, *13*(5), 675–683. doi:10.1016/0008-8846(83)90057-1

Han, X., Zhu, Y., Yang, X., & Li, C. (2010). Electrocatalytic activity of Pt doped TiO2 nanotubes catalysts for glucose determination. *Journal of Alloys and Compounds*, *500*(2), 247–251. doi:10.1016/j.jallcom.2010.04.019

Haschek, W. M., Rousseaux, C. G., Wallig, M. A., Bolon, B., & Ochoa, R. (Eds.). (2013). *Haschek and Rousseaux's handbook of toxicologic pathology*. Academic Press.

Ha, T. H., Muralidharan, S., Bae, J. H., Ha, Y. C., Lee, H. G., Park, K. W., & Kim, D. K. (2007). Accelerated short-term techniques to evaluate the corrosion performance of steel in fly ash blended concrete. *Building and Environment*, *42*(1), 78–85. doi:10.1016/j.buildenv.2005.08.019

HazwanHussin, M., Jain Kassim, M., Razali, N. N., Dahon, N. H., & Nasshorudin, D. (2011). The effect of tinospora crispa extracts as a natural mild steel corrosion inhibitor in 1m hcl solution. *Arabian Journal of Chemistry*, S616–S624.

He, Q.H., Leppard, G.G., Paige, C.R., & Snodgrass, W.J. (1996). Transmission electron microscopy of a phosphate effect on the colloid structure of iron hydroxide. *Water Research, 30*, 1345-1352.

Heimann, R. B. (2016). Plasma-sprayed hydroxylapatite-based coatings: Chemical, mechanical, microstructural, and biomedical properties. *Journal of Thermal Spray Technology*, *25*(5), 827–850. doi:10.100711666-016-0421-9

Henao, J., Cruz-bautista, M., Hincapie-Bedoya, J., Ortega-Bautista, B., Corona-Castuera, J., Giraldo-Betancur, A. L., Espinosa-Arbelaez, D. G., Alvarado-Orozco, J. M., Clavijo-Mejía, G. A., Trapaga-Martínez, L. G., & Poblano-Salas, C. A. (2018). HVOF HAP/Titania-Graded Coatings: Microstructural, Mechanical, and In Vitro Characterization. *Journal of Thermal Spray Technology*, *27*(8), 1302–1321. doi:10.100711666-018-0811-2

Heniegal, A. M., Amin, M., & Youssef, H. (2017). Effect of silica fume and steel slag coarse aggregate on the corrosion resistance of steel bars. *Construction & Building Materials*, *155*(September), 846–851. doi:10.1016/j.conbuildmat.2017.08.111

Hermas, A. A., & Morad, M. S. (2008). A comparative study on the corrosion behaviour of 304 austenitic stainless steel in sulfamic and sulfuric acid solutions. *Corrosion Science*, *50*(9), 2710–2717. doi:10.1016/j.corsci.2008.06.029

Hernández, M. R., & Cañas, J. C. (2006). Citotoxicidad in vitro: Sistema para la evaluación de biomateriales y equipos médicos implantables en Cuba. *Revista CENIC. Ciencias Biológicas*, *37*(3), 173–176.

Hernández-Montes, V., Betancur-Henao, C. P., & Santa-Marín, J. F. (2017). Titanium dioxide coatings on magnesium alloys for biomaterials: A review. *Dyna (Bilbao)*, *84*(200), 261–270. doi:10.15446/dyna.v84n200.59664

He, X., Jiang, Y., Li, C., Wang, W., Hou, B., & Wu, L. (2014). Inhibition properties and adsorption behavior of imidazole and 2-phenyl-2-imidazoline on AA5052 in 1.0 M HCl solution. *Corrosion Science*, *83*, 124–136. doi:10.1016/j.corsci.2014.02.004

Hirakawa, K., Jacobs, J. J., Urban, R., & Saito, T. (2004). Mechanisms of failure of total hip replacements: Lessons learned from retrieval studies. *Clinical Orthopaedics and Related Research*, *420*, 10–17. doi:10.1097/00003086-200403000-00003 PMID:15057073

Holland, R. B., Kurtis, K. E., & Kahn, L. F. (2016). Effect of different concrete materials on the corrosion of the embedded reinforcing steel. In *Corrosion of Steel in Concrete Structures* (pp. 131–147). Elsevier. doi:10.1016/B978-1-78242-381-2.00007-9

Holm, T. R., & Schock, M. R. (1991). Potential Effects of Polyphosphate Products on Lead Solubility in Plumbing Systems. *Journal - American Water Works Association*, *83*, 76–82.

Hongbo, F. (2002). *Synthesis and Application of New Type Inhibitors*. Chemical Industry Press.

Hope, B. B., & Ip, A. K. C. (1987). *The Research and Development Branch.* Report No. ME-87-09, Ontario Ministry of Transportation.

Hope, B., Ip, A. K., & Manning, D. G. (1985). Corrosion and electrical impedance in concrete. *Cement and Concrete Research*, *15*(3), 525–534. doi:10.1016/0008-8846(85)90127-9

Hou, J., Fu, X., & Chung, D. D. L. (1997). Improving both bond strength and corrosion resistance of steel rebar in concrete by water immersion or sand blasting of rebar. *Cement and Concrete Research*, *27*(5), 679–684. doi:10.1016/S0008-8846(97)00043-4

Howard, E. B., & Timothy, L. G. (Eds.). (1985). Metals Handbook. American Society for Metals.

Huang, H. H., Chiu, Y. H., Lee, T. H., Wu, S. C., Yang, H. W., Su, K. H., & Hsu, C. C. (2003). Ion release from NiTi orthodontic wires in artificial saliva with various acidities. *Biomaterials*, *24*(20), 3585–3592. doi:10.1016/S0142-9612(03)00188-1 PMID:12809787

Huang, Q. A., Hui, R., Wang, B., & Zhang, J. (2007). A review of AC impedance modeling and validation in SOFC diagnosis. *Electrochimica Acta*, *52*(28), 8144–8164. doi:10.1016/j.electacta.2007.05.071

Huda, F. S. G. A. (2018). HAP/rGO/Pd nanocomposite thin film coating on SST 304 - Synthesize, characterization, and properties investigations. *Journal of Alloys and Compounds*, *741*, 562–574. doi:10.1016/j.jallcom.2018.01.047

Hughes, B. P., Soleit, A. K. O., & Brierly, R. W. (1985). New technique for determining the electrical resistivity of concrete. *Magazine of Concrete Research*, *37*(133), 243–248. doi:10.1680/macr.1985.37.133.243

Hwang, C. J., Shin, J. S., & Cha, J. Y. (2001). Metal release from simulated fixed orthodontic appliances. *American Journal of Orthodontics and Dentofacial Orthopedics*, *120*(4), 383–391. doi:10.1067/mod.2001.117911 PMID:11606963

Ijaz, M. F., Zhukova, Y., Konopatsky, A., Dubinskiy, S., Korobkova, A., Pustov, Y., Brailovski, V., & Prokoshkin, S. (2018). Effect of Ta addition on the electrochemical behavior and functional fatigue life of metastable Ti-Zr-Nb based alloy for indwelling implant applications. *Journal of Alloys and Compounds*, *748*, 51–56. doi:10.1016/j.jallcom.2018.03.033

Indrani, S., & Emmanuel, A., N. S. K., & Eric, F. (2011). Microstructure and mechanical properties of annealed SUS 304H austenitic stainless steel with copper. *Materials Science and Engineering A*, *528*(13), 4491–4499.

ISO 8044:2015, Corrosion of metals and alloys — Basic terms and definitions. (2015). Retrieved from https://www.iso.org/obp/ui/#iso:std:iso:8044:ed-4:v1:en

Issaadi, S., Douadi, T., Zouaoui, A., Chafaa, S., Khan, M. A., & Bouet, G. (2011). Novel thiophene symmetrical Schiff base compounds as corrosion inhibitor for mild steel in acidic media. *Corrosion Science*, *53*(4), 1484–1488. doi:10.1016/j.corsci.2011.01.022

Jacobs, J. J., Gilbert, J. L., & Urban, R. M. (1998). Current concepts review-corrosion of metal orthopaedic implants. *Jbjs*, *80*(2), 268–282. doi:10.2106/00004623-199802000-00015

James, A. O., Oforka, N. C., & Abiola, K. (2007). Inhibition of Acid Corrosion of Mild Steel by Pyridoxal and Pyridoxol Hydrochlorides. *International Journal of Electrochemical Science*, *2*, 278–284.

Jiang, L., Junaid, A. S., Yangzhi, G., Hongbin, L., & Xiangkang, M. (2018). Electrodeposition of $Ni(OH)_2$ reinforced polyaniline coating for corrosion protection of 304 stainless steel. *Applied Surface Science*, *440*, 1011–1021. doi:10.1016/j.apsusc.2018.01.145

Jiang, L., Qiang, Y., Lei, Z., Wang, J., Qin, Z., & Xiang, B. (2018). Excellent corrosion inhibition performance of novel quinoline derivatives on mild steel in HCl media: Experimental and computational investigations. *Journal of Molecular Liquids*, *255*, 53–63. doi:10.1016/j.molliq.2018.01.133

Jiang, L., Xue, X., Zhang, W., Yang, J., Zhang, H., Li, Y., Zhang, R., Zhang, Z., Xu, L., Qu, J., Song, J., & Qin, J. (2015). The investigation of factors affecting the water impermeability of inorganic sodium silicate-based concrete sealers. *Construction & Building Materials*, *93*, 729–736. doi:10.1016/j.conbuildmat.2015.06.001

Jiang, Z., Wang, J., Hu, Q., & Huang, S. (1995). The influence of 1-(2-pyridylazo)-2-naphthol (PAN) on the corrosion of titanium in 10 N sulfuric acid solution. *Corrosion Science*, *37*(8), 1245–1252. doi:10.1016/0010-938X(95)00029-J

Jing, L., Lida, P., Qiang, F., Yingluo, Z., & Nan, G. (2019). Wettability and corrosion behavior of a Ni coating on 304 stainless steel surface. *Surface and Coatings Technology*, *357*, 740–747. doi:10.1016/j.surfcoat.2018.10.050

John, S., & Joseph, A. (2012). Effective inhibition of mild steel corrosion in 1 M hydrochloric acid using substituted triazines: An experimental and theoretical study. *RSC Advances*, *2*(26), 9944–9951. doi:10.1039/c2ra20663c

Johnson, K., Schultz, A. E., & French, C. (2009). *Crack and concrete deck sealant performance* (Vol. 89261). Academic Press.

Jung, M. S., Kim, K. B., Lee, S. A., & Ann, K. Y. (2018). Risk of chloride-induced corrosion of steel in SF concrete exposed to a chloride-bearing environment. *Construction & Building Materials*, *166*, 413–422. doi:10.1016/j.conbuildmat.2018.01.168

Kadiri, L., Galai, M., Ouakki, M., Essaadaoui, Y., Ouass, A., Cherkaoui, M., Rifi, E. H., & Lebkiri, A. (2018). Coriandrum Sativum. L Seeds Extract as a Novel Green Corrosion Inhibitor for Mild Steel in 1.0 M Hydrochloric and 0.5 M Sulfuric Solutions. *Anal. Bioanal. Electrochem*, *10*(2), 249–268.

Kalman, E., Varhegi, B., Bako, I., Felhosi, I., Karman, F. H., & Shaban, A. (1994). Corrosion Inhibition by 1-hydroxy-ethane-1, 1-diphosphonic acid: An electrochemical impedance spectroscopy study. *Journal of the Electrochemical Society*, *141*(12), 3357–3360. doi:10.1149/1.2059339

Kamimura, T., & Stratmann, M. (2001). The influence of chromium on the atmospheric corrosion of steel. *Corrosion Science*, *43*(3), 429–447. doi:10.1016/S0010-938X(00)00098-6

Kannan, V., & Ganesan, K. (2014). Chloride and chemical resistance of self compacting concrete containing rice husk ash and metakaolin. *Construction & Building Materials*, *51*, 225–234. doi:10.1016/j.conbuildmat.2013.10.050

Karre, R., Kodli, B. K., Rajendran, A., Nivedhitha, J., Pattanayak, D. K., Ameyama, K., & Dey, S. R. (2019). Comparative study on Ti-Nb binary alloys fabricated through spark plasma sintering and conventional P/M routes for biomedical application. *Materials Science and Engineering C*, *94*, 619–627. doi:10.1016/j.msec.2018.10.006 PMID:30423747

Karthega, M., Raman, V., & Rajendran, N. (2007). Influence of potential on the electrochemical behaviour of β titanium alloys in Hank's solution. *Acta Biomaterialia*, *3*(6), 1019–1023. doi:10.1016/j.actbio.2007.02.009 PMID:17478134

Karthik, R., Muthukrishan, P., Chen, S. M., Jeyaprabha, B., & Prakash, P. (2015). Anti-corrosion inhibition of mild steel in 1M hydrochloric acid solution by using tiliacoraaccuminata leaves extract. *International Journal of Electrochemical Science*, *10*, 3707.

Keleştemur, O., & Demirel, B. (2015). Effect of metakaolin on the corrosion resistance of structural lightweight concrete. *Construction & Building Materials*, *81*, 172–178. doi:10.1016/j.conbuildmat.2015.02.049

Kertit, S., Hammouti, B., Taleb, M., & Brighli, M. (1997). Peptidic compounds as corrosion inhibitor for iron in acid chloride solution. *Bulletin of Electrochemistry*, *13*, 241.

Khaled, Al-F., & Muhammad, A. (2018). Microstructure Refinement and Mechanical Properties of 304 Stainless Steel by Repetitive Thermomechanical Processing. *Metallurgical and Materials Transactions A*, *49*(4).

Khaled, K. F. (2008). Molecular simulation, quantum chemical calculations and electrochemical studies for inhibition of mild steel by triazoles. *Electrochimica Acta*, *53*, 3484.

Kıcır, N., Tansuğ, G., Erbil, M., & Tüken, T. (2016). Investigation of ammonium (2, 4-dimethylphenyl)-dithiocarbamate as a new, effective corrosion inhibitor for mild steel. *Corrosion Science*, *105*, 88–99. doi:10.1016/j.corsci.2016.01.006

Kim, H. M., Himeno, T., Kawashita, M., Kokubo, T., & Nakamura, T. (2004). The mechanism of biomineralization of bone-like apatite on synthetic hydroxyapatite: An in vitro assessment. *Journal of the Royal Society, Interface*, *1*(1), 17–22. doi:10.1098/rsif.2004.0003 PMID:16849149

Klinghoffer, O. (1995). In situ monitoring of reinforcement corrosion by means of electrochemical methods. *Nordic Concrete Research-Publications*, *95*(1), 1–13.

Kliškic, M., Radoševic, J., & Gudic, S. (1997). Pyridine and its derivatives as inhibitors of aluminium corrosion in chloride solution. *Journal of Applied Electrochemistry*, *27*(8), 947–952. doi:10.1023/A:1018405803182

Koch, G., Varney, J., Thompson, N., Moghissi, O., Gould, M., & Payer, J. (2016). International Measures of Prevention, Application, and Economics of Corrosion Technologies Study. NACE International.

Koch, G. H., Brongers, M. P. H., Thompson, N. G., Virmani, Y. P., & Payer, J. H. (2002). Corrosion costs and preventive strategies in the United States. *NACE International, FHWA-RD, 01*, 1–12.

Koizumi, H., Takeuchi, Y., Imai, H., Kawai, T., & Yoneyama, T. (2019). Application of titanium and titanium alloys to fixed dental prostheses. *Journal of Prosthodontic Research*, *63*(3), 266–270. doi:10.1016/j.jpor.2019.04.011 PMID:31147298

Kokalj, A. (2012). On the HSAB based estimate of charge transfer between adsorbates and metal surfaces. *Chemical Physics*, *393*(1), 1–12. doi:10.1016/j.chemphys.2011.10.021

Kong, L., Gao, Y., Lu, G., Gong, Y., Zhao, N., & Zhang, X. (2006). A study on the bioactivity of chitosan/nano-hydroxyapatite composite scaffolds for bone tissue engineering. *European Polymer Journal*, *42*(12), 3171–3179. doi:10.1016/j.eurpolymj.2006.08.009

Konstantinos & Mavredaki. (2007). Inhibition and growth control of colloidal silica: Designed chemical approaches. *Journal of Science Direct.*, *213*, 38–46.

Kopasz, J. P., Benjamin, T. G., & Schenck, D. *(2017). Bipolar Plate Workshop Summary Report.* doi:10.2172/1406358

Kowsari, E., Payami, M., Amini, R., Ramezanzadeh, B., & Javanbakht, M. (2014). Task-specific ionic liquid as a new green inhibitor of mild steel corrosion. *Applied Surface Science*, *289*, 478–486. doi:10.1016/j.apsusc.2013.11.017

Krasicka-Cydzik, E., & Amin, A. N. (2012). Anodic layer formation on titanium and its alloys for biomedical applications. In *Titanium Alloys-Towards Achieving Enhanced Properties for Diversified Applications*. InTech. doi:10.5772/34395

Krishnegowda, P. M., Venkatesha, V. T., Krishnegowda, P. K. M., & Shivayogiraju, S. B. (2013). Acalypha torta Leaf Extract as Green Corrosion Inhibitor for Mild Steel in Hydrochloric Acid Solution. *Industrial & Engineering Chemistry Research*, *52*, 722–728.

Kumari, R., & Majumdar, J. D. (2017). Studies on corrosion resistance and bio-activity of plasma spray deposited hydroxylapatite (HA) based Titania and ZrO2 dispersed composite coatings on titanium alloy (Ti-6Al-4V) and the same after post spray heat treatment. *Applied Surface Science*, *420*, 935–943. doi:10.1016/j.apsusc.2017.05.208

Kumar, S., Narayanan, T. S., Raman, S. G. S., & Seshadri, S. K. (2010). Surface modification of CP-Ti to improve the fretting-corrosion resistance: Thermal oxidation vs. anodizing. *Materials Science and Engineering C*, *30*(6), 921–927. doi:10.1016/j.msec.2010.03.024

Kumar, T., Vishwanatham, S., & Kundu, S. S. (2010). A laboratory study on pteroyl-l-glutamic acid as a scale prevention inhibitor of calcium carbonate in aqueous solution of synthetic produced water. *Journal of Petroleum Science Engineering*, *71*(1-2), 1–7. doi:10.1016/j.petrol.2009.11.014

Laabaissi, T., Benhiba, F., Rouifi, Z., Rbaa, M., Oudda, H., Zarrok, H., Lakhrissi, B., Guenbour, A., Warad, I., & Zarrouk, A. (2011). Benzodiazepine Derivatives as Corrosion Inhibitors of Carbon Steel in HCl Media: Electrochemical and Theoretical Studies, T. Zhihua, Z. Shengtao, L. Weihua, H. Baorong, Adsorption and inhibitory mechanism of 1H-1,2,4-triazol-l-yl-methyl-2-(4-chlorophenoxy) acetate on corrosion of mild steel in acidic solution. *Industrial & Engineering Chemistry Research*, *50*(10), 6082–6088. doi:10.1021/ie101793b

Labriti, B., Dkhireche, N., Touir, R., Ebn Touhami, M., Sfaira, M., El Hallaoui, A., Hammouti, B., & Alami, A. (2012). Synergism in Mild Steel Corrosion and Scale Inhibition by a New Oxazoline in Synthetic Cooling Water. *Arabian Journal for Science and Engineering*, *37*(5), 1293–1303. doi:10.100713369-012-0257-7

Lafont, M., Pebere, N., Moran, F., & Bleriot, P. (1993). Inhibition de la corrosion d'un acier au carbone par des produits dérivés de phosphonates en association avec des sels de zinc. *Revue des sciences de l'eau/Journal of Water Science*, *6*(1), 97-112.

Lake, D. L. (1988). Approching environmental acceptability in cooling water corrosion inhibitor. *Corrosion Prevention & Control*, 113.

Lakhrissi, B., Lakhrissi, L., Massoui, M., Essassi, E. M., Comelles, F., Esquena, J., Solans, C., & Rodriguez-Abreu, C. (2010). Surface and self-aggregation properties of bis-benzimidazolones derivatives of D-glucose. *Journal of Surfactants and Detergents*, *13*(3), 329–338. doi:10.100711743-010-1183-x

Larsen, M. J., & Jensen, S. J. (1986). Solubility study of the initial formation of calcium orthophosphates from aqueous solutions at pH 5–10. *Archives of Oral Biology*, *31*, 565–572. PMID:3467681

Lavanya, K., Saranya, J., & Chitra, S. (2018). Recent reviews on quinolone derivative as corrosion inhibitors. *Corrosion Reviews*, *36*(4), 365–371. doi:10.1515/corrrev-2017-0129

Law, D. W., Millard, S. G., & Bungey, J. H. (1999). The use of galvanostatic pulse measurements to determine corrosion parameters. *Durability of Building Materials and Components*, *8*, 310–319.

Lazell, E. W. (1915). *Hydrated lime: history, manufacture and uses in plaster, mortar, concrete; a manual for the architect, engineer, contractor and builder*. Jackson-Remlinger Printing Co.

Lebrini, M., Bentiss, F., Chihib, N.-E., Jama, C., Hornez, J. P., & Lagrenée, M. (2008). Polyphosphate derivatives of guanidine and urea copolymer: Inhibiting corrosion effect of armco iron in acid solution and antibacterial activity. *Corrosion Science*, *50*(10), 2914–2918. doi:10.1016/j.corsci.2008.07.003

Lgaz, H., Chung, I. M., Albayati, M. R., Chaouiki, A., Salghi, R., & Mohamed, S. K. (2020). Improved corrosion resistance of mild steel in acidic solution by hydrazone derivatives: An experimental and computational study. *Arabian Journal of Chemistry*, *13*(1), 2934–2954. doi:10.1016/j.arabjc.2018.08.004

Lgaz, H., Chung, I. M., Salghi, R., Ali, I. H., Chaouiki, A., El Aoufir, Y., & Khan, M. I. (2019). On the understanding of the adsorption of Fenugreek gum on mild steel in an acidic medium: Insights from experimental and computational studies. *Applied Surface Science*, *463*, 647–658. doi:10.1016/j.apsusc.2018.09.001

Li, C., Chen, L., Steinhuebel, D., & Goodman, A. (2016). Rapid construction of imidazopyridines from ortho-haloaminopyridines. *Tetrahedron Letters*, *57*(25), 2708–2712. doi:10.1016/j.tetlet.2016.04.118

Li, M., & Christofides, P. D. (2009). Modeling and control of high-velocity oxygen-fuel (HVOF) thermal spray: A tutorial review. *Journal of Thermal Spray Technology*, *18*(5-6), 753.

Lima, R. S., & Marple, B. R. (2007). Thermal spray coatings engineered from nanostructured ceramic agglomerated powders for structural, thermal barrier and biomedical applications: A review. *Journal of Thermal Spray Technology*, *16*(1), 40–63. doi:10.100711666-006-9010-7

Liu, F. G., Du, M., Zhang, J., & Qiu, M. (2009). Electrochemical behavior of Q235 steel in saltwater saturated with carbon dioxide based on new imidazoline derivative inhibitor. *Corrosion Science*, *51*(1), 102–109. doi:10.1016/j.corsci.2008.09.036

Liu, Y. J., & Shi, X. M. (2009). Electrochemical chloride extraction and electrochemical injection of corrosion inhibitor in concrete: State of the knowledge. *Corrosion Reviews*, *27*(1), 53–81. doi:10.1515/CORRREV.2009.27.1-2.53

Li, X., Deng, S., & Fu, H. (2011). Triazolyl blue tetrazolium bromide as a novel corrosion inhibitor for steel in HCl and H_2SO_4 solutions. *Corrosion Science*, *53*(1), 302–309. doi:10.1016/j.corsci.2010.09.036

Longhitano, G. A., Arenas, M. A., Conde, A., Larosa, M. A., Jardini, A. L., de Carvalho Zavaglia, C. A., & Damborenea, J. J. (2018). Heat treatments effects on functionalization and corrosion behavior of Ti-6Al-4V ELI alloy made by additive manufacturing. *Journal of Alloys and Compounds*, *765*, 961–968. doi:10.1016/j.jallcom.2018.06.319

López, W., & González, J. A. (1993). Influence of the degree of pore saturation on the resistivity of concrete and the corrosion rate of steel reinforcement. *Cement and Concrete Research*, *23*(2), 368–376. doi:10.1016/0008-8846(93)90102-F

Mahallati, E., & Saremi, M. (2006). An assessment on the mill scale effects on the electrochemical characteristics of steel bars in concrete under DC-polarization. *Cement and Concrete Research*, *36*(7), 1324–1329. doi:10.1016/j.cemconres.2006.03.015

Mai, W., Soghrati, S., & Buchheit, R. G. (2016). A phase field model for simulating the pitting corrosion. *Corrosion Science*, *110*, 157–166. doi:10.1016/j.corsci.2016.04.001

Majid, H. A., & Rana, A. M. A. (n.d.). Tribocorrosion. *Advances in Tribology*, *5*, 89-110.

Maliekkal, R., Kakkassery, J. T., & Palayoor, V. R. (2018). Efficacies of sodium nitrite and sodium citrate–zinc acetate mixture to inhibit steel rebar corrosion in simulated concrete interstitial solution contaminated with NaCl. *International Journal of Industrial Chemistry.*, *9*(2), 105–114. doi:10.100740090-018-0142-7

Manaranche, C., & Hornberger, H. (2007). A proposal for the classification of dental alloys according to their resistance to corrosion. *Dental Materials*, *23*(11), 1428–1437. doi:10.1016/j.dental.2006.11.030 PMID:17466365

Manivasagam, G., Dhinasekaran, D., & Rajamanickam, A. (2010). Biomedical implants: Corrosion and its prevention-a review. *Recent Patents on Corrosion Science*, *2*(1), 40–54. doi:10.2174/1877610801002010040

Mansfeld, F. (1987). *Corrosion Mechanisms*. Marcel Dekker.

Marchi, J. (Ed.). (2016). *Biocompatible glasses: from bone regeneration to cancer treatment* (Vol. 53). Springer. doi:10.1007/978-3-319-44249-5

Marcotte, T. D., Hansson, C. M., & Hope, B. B. (1999). The effect of the electrochemical chloride extraction treatment on steel-reinforced mortar Part I: Electrochemical measurements. *Cement and Concrete Research*, *29*(10), 1555–1560. doi:10.1016/S0008-8846(99)00118-0

Marín-Cruz, J., Cabrera-Sierra, R., Pech-Canul, M. A., & González, I. (2006). EIS study on corrosion and scale processes and their inhibition in cooling system media. *Electrochimica Acta*, *51*(8-9), 1847–1854. doi:10.1016/j.electacta.2005.02.104

Martínez, I., & Andrade, C. (2008). Application of EIS to cathodically protected steel: Tests in sodium chloride solution and in chloride contaminated concrete. *Corrosion Science*, *50*(10), 2948–2958. doi:10.1016/j.corsci.2008.07.012

Martinez, S., & Stern, I. (2002). Thermodynamic characterization of metal dissolution and inhibitor adsorption processes in the low carbon steel/mimosa tannin/sulfuric acid system. *Applied Surface Science*, *199*(1-4), 83–89. doi:10.1016/S0169-4332(02)00546-9

Martin, T. M., & Young, D. M. (2001). Prediction of the acute toxicity (96-h LC50) of organic compounds to the fathead minnow (Pimephalespromelas) using a group contribution method. *Chemical Research in Toxicology*, *14*(10), 1378–1385. doi:10.1021/tx0155045 PMID:11599929

Matsuno, H., Yokoyama, A., Watari, F., Uo, M., & Kawasaki, T. (2001). Biocompatibility and osteogenesis of refractory metal implants, titanium, hafnium, niobium, tantalum and rhenium. *Biomaterials*, *22*(11), 1253–1262. doi:10.1016/S0142-9612(00)00275-1 PMID:11336297

Matusiewicz, H. (2014). Potential release of in vivo trace metals from metallic medical implants in the human body: From ions to nanoparticles–a systematic analytical review. *Acta Biomaterialia, 10*(6), 2379–2403. doi:10.1016/j.actbio.2014.02.027 PMID:24565531

McGettigan, E. (1995). Factors affecting the selection of water-repellent treatments. *APT Bulletin, 26*(4), 22–26. doi:10.2307/1504446

Melero, H. C., Sakai, R. T., Vignatti, C. A., Benedetti, A. V., Fernández, J., Guilemany, J. M., & Suegama, P. H. (2018). Corrosion Resistance Evaluation of HVOF Produced HAP and Titania-HAP Coatings in Hanks' Solution. *Materials Research, 21*(2). Advance online publication. doi:10.1590/1980-5373-mr-2017-0210

Meligi-El, A. A. (2010). Corrosion Preventive Strategies as a Crucial Need for Decreasing Environmental Pollution and Saving Economics. *Recent Patents on Corrosion Science, 2*(1), 22–33. doi:10.2174/1877610801002010022

Mihaela, D., Emile, S. M. M., Anca, C. P., Alina, V., Leslie, F. P., & Mariana, B. (2018). Enhancement of the Corrosion Resistance of 304 Stainless Steel by Cr–N and Cr(N,O) Coatings. *Coatings, 8*(4), 132. doi:10.3390/coatings8040132

Mikulewicz, M., Wołowiec, P., Loster, B. W., & Chojnacka, K. (2015). Do soft drinks affect metal ions release from orthodontic appliances? *Journal of Trace Elements in Medicine and Biology, 31*, 74–77. doi:10.1016/j.jtemb.2015.03.007 PMID:26004895

Millard, S. G., Law, D., Bungey, J. H., & Cairns, J. (2001). Environmental influences on linear polarisation corrosion rate measurement in reinforced concrete. *NDT & E International, 34*(6), 409–417. doi:10.1016/S0963-8695(01)00008-1

Miranda, J. R., Sánchez, M. A., García, E. G., Velazquez, D.Y. M., Á. de J., & Ramírez, M. (2016). Mechanical Properties of SiO2 Coatings for Corrosion Protection of 304 Stainless Steel. *Characterization of Metals and Alloys*, 109–116.

Moghadasi, J., Jamiala hmadi, M., Müller-Steinhagen, H., & Sharif, A. (2004). Formation damage due to scale formation in porous media resulting from water injection. Society of Petroleum Engineers.

Moradi, M., Duan, J., & Du, X. (2013). Investigation of the effect of 4,5-dichloro-2-n-octyl-4-isothiazolin-3-one inhibition on the corrosion of carbon steel in *Bacillus* sp. inoculated artificial seawater. *Corrosion Science, 69*, 338–345.

Moradi, M., Duan, J., & Du, X. (2013). Investigation of the effect of of 4, 5-dichloro-2-n-octyl-4-isothiazolin-3-one inhibition on the corrosion of carbon steel in Bacillus sp. inoculated artificial seawater. *Corrosion Science, 69*, 338–345. doi:10.1016/j.corsci.2012.12.017

Moridi, A., Hassani-Gangaraj, S. M., Guagliano, M., & Dao, M. (2014). Cold spray coating: Review of material systems and future perspectives. *Surface Engineering, 30*(6), 369–395. doi:10.1179/1743294414Y.0000000270

Morris, W., Vico, A., Vazquez, M., & DeSanchez, S. R. (2002). Corrosion of reinforcing steel evaluated by means of concrete resistivity measurements. *Corrosion Science, 44*(1), 81–99. doi:10.1016/S0010-938X(01)00033-6

Mu, G. N., Li, X., & Li, F. (2004). Synergistic inhibition between *o*-phenanthroline and chloride ion on cold rolled steel corrosion in phosphoric acid. *Materials Chemistry and Physics, 86*, 59–68.

Mu, G., Li, X., Qu, Q., & Zhou, J. (2006). Molybdate and tungstate as corrosion inhibitors for cold rolling steel in hydrochloric acid solution. *Corrosion Science, 48*(2), 445–459. doi:10.1016/j.corsci.2005.01.013

Muhmood, L., Vitta, S., & Venkateswaran, D. (2009). Cementitious and pozzolanic behavior of electric arc furnace steel slags. *Cement and Concrete Research, 39*(2), 102–109. doi:10.1016/j.cemconres.2008.11.002

Murakawa, T., & Hackerman, N. (1964). The double layer capacity at the interface between iron and acid solutions with and without organic materials. *Corrosion Science, 4*(1-4), 387–396. doi:10.1016/0010-938X(64)90040-X

Murulana, L. C., Kabanda, M. M., & Ebenso, E. E. (2015). Experimental and theoretical studies on the corrosion inhibition of mild steel by some sulphonamides in aqueous HCl. *RSC Advances*, *5*(36), 28743–28761. doi:10.1039/C4RA11414K

Murulana, L. C., Kabanda, M. M., & Ebenso, E. E. (2016). Investigation of the adsorption characteristics of some selected sulphonamide derivatives as corrosion inhibitors at mild steel/hydrochloric acid interface: Experimental, quantum chemical and QSAR studies. *Journal of Molecular Liquids*, *215*, 763–779. doi:10.1016/j.molliq.2015.12.095

NACE Standard TM0374-95. (1997). Laboratory screening tests to determine the ability of scale inhibitors to prevent the precipitation of calcium sulfate and calcium carbonate from solution. *NACE.*

Nady, H., El-Rabiei, M. M., & Samy, M. (2017). Corrosion behavior and electrochemical properties of carbon steel, commercial pure titanium, copper and copper–aluminum–nickel alloy in 3.5% sodium chloride containing sulfide ions. *Egyptian Journal of Petroleum*, *26*(1), 79–94. doi:10.1016/j.ejpe.2016.02.008

Nakajima, A., Akita, M., Uematsu, Y., & Tokaji, K. (2010). Effect of strain-induced martensitic transformation on fatigue behavior of type 304 stainless steel. *Procedia Engineering*, *2*(1), 323–330. doi:10.1016/j.proeng.2010.03.036

Nam, S., Cho, H. W., Lim, S., Kim, D., Kim, H., & Sung, B. J. (2012). Enhancement of electrical and thermomechanical properties of silver nanowire composites by the introduction of nonconductive nanoparticles: Experiment and simulation. *ACS Nano*, *7*(1), 851–856. doi:10.1021/nn305439t PMID:23237625

Nancy, B. (2008). Stainless steel in construction: A review of research, applications, Challenges and opportunities. *Journal of Constructional Steel Research*, *64*(11), 1199–1206. doi:10.1016/j.jcsr.2008.07.011

Natarajan, P., & Ramakrishnan, K. R. (2007). LPR meter for concrete corrosion studies. *Journal Instrument Society of India*, *37*(1), 1–6.

Nathan, C. C. (1973). Corrosion inhibitors. National Association of Corrosion Engineers.

Nazeer, A. A., & Madkour, M. (2018). Potential use of smart coatings for corrosion protection of metals and alloys: A review. *Journal of Molecular Liquids*, *253*, 11–22. doi:10.1016/j.molliq.2018.01.027

Negm, N. A., El Hashash, M. A., Abd-Elaal, A., Tawfik, S. M., & Gharieb, A. (2018). Amide type nonionic surfactants: Synthesis and corrosion inhibition evaluation against carbon steel corrosion in acidic medium. *Journal of Molecular Liquids*, *256*, 574–580. doi:10.1016/j.molliq.2018.02.078

Nesic, S. (2007). Key issues related to modelling of internal corrosion of oil and gas pipelines–A review. *Corrosion Science*, *49*(12), 4308–4338. doi:10.1016/j.corsci.2007.06.006

Newton, C. J., & Sykes, J. (1988). A galvanostatic pulse technique for investigation of steel corrosion in concrete. *Corrosion Science*, *28*(11), 1051–1074. doi:10.1016/0010-938X(88)90101-1

Niinomi, M., & Boehlert, C. J. (2015). Titanium alloys for biomedical applications. In *Advances in Metallic Biomaterials* (pp. 179–213). Springer. doi:10.1007/978-3-662-46836-4_8

Niinomi, M., & Nakai, M. (2019). Ti-Based Biomedical Alloys. In *Novel Structured Metallic and Inorganic Materials* (pp. 61–76). Springer. doi:10.1007/978-981-13-7611-5_3

Nikolopoulou, F. (2006). Saliva and dental implants. *Implant Dentistry*, *15*(4), 372–376. doi:10.1097/01.id.0000239320.57403.8d PMID:17172955

Noor, E. A., & Al-Moubaraki, A. H. (2008). Thermodynamic study of metal corrosion and inhibitor adsorption processes in mild steel/1-methyl-4[4¢(-X)-styryl pyridinium iodides/hydrochloric acid systems. *Materials Chemistry and Physics*, *110*(1), 145–154. doi:10.1016/j.matchemphys.2008.01.028

Noor, U. H., Bahadir, T., & Ali, M. S. (2019). A Competitive Design and Material Consideration for Fabrication of Polymer Electrolyte Membrane Fuel Cell Bipolar Plates. *Designs*, *3*(1), 13. doi:10.3390/designs3010013

O'Brien, B. (2015). Niobium biomaterials. In *Advances in Metallic Biomaterials* (pp. 245–272). Springer. doi:10.1007/978-3-662-46836-4_11

Obot, I. B., Macdonald, D. D., & Gasem, Z. M. (2015). Density functional theory (DFT) as a powerful tool for designing new organic corrosion inhibitors. part 1: An overview. *Corrosion Science*, *99*, 1–30. doi:10.1016/j.corsci.2015.01.037

Obot, I. B., & Obi-Egbedi, N. O. (2010). Indeno-1-one [2, 3-b] quinoxaline as an effective inhibitor for the corrosion of mild steel in 0.5 M H2SO4 solution. *Materials Chemistry and Physics*, *122*(2-3), 325–328. doi:10.1016/j.matchemphys.2010.03.037

Obot, I. B., & Obi-Egbedi, N. O. (2010). Theoretical study of benzimidazole and its derivatives and their potential activity as corrosion inhibitors. *Corrosion Science*, *52*, 657–660.

Obot, I. B., Obi-Egbedi, N. O., Ebenso, E. E., Afolabi, A. S., & Oguzie, E. E. (2013). Experimental, quantum chemical calculations, and molecular dynamic simulations insight into the corrosion inhibition properties of 2-(6-methylpyridin-2-yl) oxazolo [5, 4-f][1, 10] phenanthroline on mild steel. *Research on Chemical Intermediates*, *39*(5), 1927–1948. doi:10.100711164-012-0726-3

Obot, I., & Obi-Egbedi, N. O. (2010). Inhibition of aluminium corrosion in hydrochloric acid using nizoral and the effect of iodide ion addition. *Chemistry (Weinheim an der Bergstrasse, Germany)*, *7*(3), 837–843.

Ochoa, N., Baril, G., Moran, F., & Pébère, N. (2002). Study of the properties of a multi-component inhibitor used for water treatment in cooling circuits. *Journal of Applied Electrochemistry*, *32*(5), 497–504. doi:10.1023/A:1016500722497

Oguzie, E. E., Lia, Y., & Wang, F. H. (2007). Effect of 2-amino-3-mercaptopropanoic acid (cysteine) on the corrosion behaviour of low carbon steel in sulphuric acid. *Electrochimica Acta*, *53*(2), 909–914. doi:10.1016/j.electacta.2007.07.076

Oguzie, E. E., Okolue, B. N., Ebenso, E. E., Onuoha, G. N., & Onuchukwu, A. I. (2004). Evaluation of the inhibitory effect of methylene blue dye on the corrosion of aluminium in hydrochloric acid. *Materials Chemistry and Physics*, *87*(2-3), 394–401. doi:10.1016/j.matchemphys.2004.06.003

Oguzie, E. E., Unaegbu, C., Ogukwe, C. N., Okolue, B. N., & Onuchukwu, A. I. (2004). Inhibition of mild steel corrosion in sulphuric acid using indigo dye and synergistic halide additives. *Chemical Physics*, *84*, 363.

Okafor, P. C., Ikpi, M. E., Uwah, I. E., Ebenso, E. E., Ekpe, U. J., & Umoren, S. A. (2008). Inhibitory action of Phyllanthus amarus extracts on the corrosion of mild steel in acidic media. *Corrosion Science*, *50*(8), 2310–2317. doi:10.1016/j.corsci.2008.05.009

Olasunkanmi, L. O., Obot, I. B., Kabanda, M. M., & Ebenso, E. E. (2015). Some Quinoxalin-6-yl Derivatives as Corrosion Inhibitors for Mild Steel in Hydrochloric Acid: Experimental and Theoretical Studies. *Journal of Physical Chemistry*, *119*, 16004–16019.

Olasunkanmi, L. O., Obot, I. B., Kabanda, M. M., & Ebenso, E. E. (2015). Some quinoxalin-6-yl derivatives as corrosion inhibitors for mild steel in hydrochloric acid: Experimental and theoretical studies. *The Journal of Physical Chemistry C*, *119*(28), 16004–16019. doi:10.1021/acs.jpcc.5b03285

Orellan, J. C., Escadeillas, G., & Arliguie, G. (2004). Electrochemical chloride extraction: Efficiency and side effects. *Cement and Concrete Research*, *34*(2), 227–234. doi:10.1016/j.cemconres.2003.07.001

Otmacic Curkovic, H., Stupnisek-Lisac, E., & Takenouti, H. (2010). The influence of pH value on the efficiency of imidazole based corrosion inhibitors of copper. *Corrosion Science*, *52*(2), 398–405. doi:10.1016/j.corsci.2009.09.026

Otmacic, H., & Stupnisek-Lisac, E. (2010). Copper corrosion inhibitors in near neutral media. *Electrochimica Acta, 48*(2003), 985-991.

Otmacic, H., & Stupnisek-Lisac, E. (2002). Copper corrosion inhibitors in near neutral media. *Electrochimica Acta*, *47*, 4189–4194.

Ouakki, M., Galai, M., Cherkaoui, M., Rifi, E.H., & Hatim, Z. (2020). Study on the corrosion inhibition of mild steel in two acidic mediums (HCL 1m and H2SO4 0.5m) containing apatite. *J. Chem. Technol. Metall, 2*.

Ouakki, M., Rbaa, M., Galai, M., Lakhrissi, B., Rifi, E. H., & Cherkaoui, M. (2018). Experimental and Quantum Chemical Investigation of Imidazole Derivatives as Corrosion Inhibitors on Mild Steel in 1.0 M Hydrochloric Acid. *Journal of Bio- and Tribo-Corrosion, 4*, 35.

Ouakki, M., Galai, M., Cherkaoui, M., Rifi, E. H., & Hatim, Z. (2018). Inorganic compound (apatite doped by Mg and Na) as a corrosion inhibitor for mild steel in phosphoric acidic medium. *Anal. Bioanal. Electrochem.*, *10*(7), 943–960.

Ouakki, M., Galai, M., Cherkaoui, M., Rifi, E. H., & Hatim, Z. (2018b). Inorganic compound (apatite doped by Mg and Na) as a corrosion inhibitor for mild steel in phosphoric acidic medium. *Anal. Bioanal. Electrochem*, *10*(7), 943–960.

Ouakki, M., Galai, M., Rbaa, M., Abousalem, A. S., Lakhrissi, B., Rifi, E. H., & Cherkaoui, M. (2019). Quantum chemical and experimental evaluation of the inhibitory action of two imidazole derivatives on mild steel corrosion in sulphuric acid medium. *Heliyon*, *5*(11), e02759. doi:10.1016/j.heliyon.2019.e02759 PMID:31768434

Ouakki, M., Rbaa, M., Galai, M., Lakhrissi, B., Rifi, E. H., & Cherkaoui, M. (2018a). Experimental and quantum chemical investigation of imidazole derivatives as corrosion inhibitors on mild steel in 1.0 M hydrochloric acid. *J. Bio Tribo-Corrosion*, *4*(3), 35. doi:10.100740735-018-0151-2

Ozcan, M., Solmaz, R., Kardas, G., & Dehri, I. (2008). Adsorption properties of barbiturates as green corrosion inhibitors on mild steel in phosphoric acid. *Colloid Surf. A*, *57*, 325.

Page, C. L. (2000). *Repair of corroded reinforcement in concrete using sacrificial anodes.* US Patent 6, 022, 469.

Pan, T. J., Chen, Y., Zhang, B., Hu, J., & Li, C. (2016). Corrosion behavior of niobium coated 304 stainless steel in acid solution. *Applied Surface Science*, *369*, 320–325. doi:10.1016/j.apsusc.2016.02.088

Pan, X., Shi, Z., Shi, C., Ling, T. C., & Li, N. (2017). A review on concrete surface treatment Part I: Types and mechanisms. *Construction & Building Materials*, *132*, 578–590. doi:10.1016/j.conbuildmat.2016.12.025

Papadakis, V. G. (2000). Effect of supplementary cementing materials on concrete resistance against carbonation and chloride ingress. *Cement and Concrete Research*, *30*(2), 291–299. doi:10.1016/S0008-8846(99)00249-5

Park, B. J., & Kim, Y. K. (2003). Metallic biomaterials. *Balance (Alexandria, Va.)*, *1*, 50.

Park, J., & Lakes, R. S. (2007). *Biomaterials: an introduction*. Springer Science & Business Media.

Pech-Canul, M. A., & Castro, P. (2002). Corrosion measurements of steel reinforcement in concrete exposed to a tropical marine atmosphere. *Cement and Concrete Research*, *32*(3), 491–498. doi:10.1016/S0008-8846(01)00713-X

Peralta-Lopez, D., Sotelo-Mazon, O., Henao, J., Porcayo-Calderon, J., Valdez, S., Salinas-Solano, G., & Martinez-Gomez, L. (2017). Electrochemical Study of Three Stainless Steel Alloys and Titanium Metal in Cola Soft Drinks. *Journal of Electrochemical Science and Technology*, *8*(4), 294–306. doi:10.33961/JECST.2017.8.4.294

Perveen, K., Bridges, G. E., Bhadra, S., & Thomson, D. J. (2014). Corrosion Potential Sensor for Remote Monitoring of Civil Structure Based on Printed Circuit Board Sensor. *IEEE Transactions on Instrumentation and Measurement*, *63*(10), 2422–2431. doi:10.1109/TIM.2014.2310092

Peters,W. (1970)*Chemotherapy and Drug Resistance in Malaria*. Academic Press.

Petersen, A., Rodrigues, S. R., Dalmoro, V., Falcade, T., & Tamborim, S. M. (2017). Anthocyanins as a corrosion inhibitor for 2024-T3 aluminum alloys: A study of electrochemical behavior. *Int J Corros Scale Inhib.*, *6*, 29.

Philip, D. H. (1982). Engineering Properties of Steels. American Society for Metals.

Poitout, D. G. (Ed.). (2004). *Biomechanics and biomaterials in orthopedics*. doi:10.1007/978-1-4471-3774-0

Popova, A., Christov, M., & Vasilev, A. (2007). Inhibitive properties of quaternary ammonium bromides of N-containing heterocycles on acid mild steel corrosion. Part II: EIS results. *Corrosion Science*, *49*(8), 3290–3302. doi:10.1016/j.corsci.2007.03.012

Popova, A., Christov, M., & Zwetanova, A. (2007). Effect of the molecular structure on the inhibitor properties of azoles on mild steel corrosion in 1 M hydrochloric acid. *Corrosion Science*, *49*(5), 2131–2143. doi:10.1016/j.corsci.2006.10.021

Prabakaran, M., Venkatesh, M., Ramesh, S., & Periasamy, V. (2013). Corrosion inhibition behavior of propyl phosphonic acid–Zn^{2+} system for carbon steel in aqueous solution. *Applied Surface Science*, *276*, 592–603. doi:10.1016/j.apsusc.2013.03.138

Pramod, M., Usha, N. K., Soobhankar, P., & Sudesna, R. (2019). Investigation of the effects of electrophoretic deposition parameters on 304SS steel coated with graphene oxide for PEMFC application. *Materials Today: Proceedings*. doi:10.1016/j.matpr.2019.12.360

Pyc, W. A., Weyers, R. E., & Sprinkel, M. M. (1998). Corrosion protection performance of corrosion inhibitors and epoxy-coated reinforcing steel in a simulated concrete pore water solution. Academic Press.

Qiao, L., Wang, Y., Wang, W., Mohedano, M., Gong, C., & Gao, J. (2014). The preparation and corrosion performance of self-assembled monolayers of stearic acid and MgO layer on pure magnesium. *Materials Transactions*, *55*(8), 1337–1343. doi:10.2320/matertrans.M2014117

Qinyi, W., Xiutong, W., Xiaobo, N., Xinran, L., Jing, S., Hong, L., Wencheng, W., Yanliang, H., ... Baorong, H. (2018). Characteristics and anticorrosion performance of WSe_2/TiO_2 nanocomposite materials for 304 stainless steel. *Surface and Coatings Technology*, *352*, 26–32. doi:10.1016/j.surfcoat.2018.08.004

Quartarone, G., Battilana, M., Bonaldo, L., & Tortato, T. (2008). Investigation of the inhibition effect of indole-3-carboxylic acid on the copper corrosion in 0.5 M H2SO4. *Corrosion Science*, *50*(12), 3467–3474. doi:10.1016/j.corsci.2008.09.032

Quraishi, M. A., Wajid Khan, M. A., Ajmal, M., Muralidharan, S., & Venkatakrishnaiyer, S. (1997). Influence of molecular structure of substituted benzothiazoles on corrosion inhibition and hydrogen permeation through mild steel in sulphuric acid. *British Corrosion Journal*, *32*, 72.

Ramesh, S., & Rajeswari, S. (2004). Corrosion inhibition of mild steel in neutral aqueous solution by new triazole derivatives. *Electrochimica Acta*, *49*(5), 811–820. doi:10.1016/j.electacta.2003.09.035

Ramesh, S., Rajeswari, S., & Maruthamuthu, S. (2004). Corrosion inhibition of copper by new triazole phosphonate derivatives. *Applied Surface Science*, *229*(1-4), 214–225. doi:10.1016/j.apsusc.2004.01.063

Ramezanzadeh, B., Arman, S. Y., Mehdipour, M., & Markhali, B. P. (2014). Analysis of electrochemical noise (ECN) data in time and frequency domain for comparison corrosion inhibition of some azole compounds on Cu in 1.0 M H2SO4 solution. *Applied Surface Science*, *289*, 129–140. doi:10.1016/j.apsusc.2013.10.119

Ramírez, G., Rodil, S. E., Arzate, H., Muhl, S., & Olaya, J. J. (2011). Niobium based coatings for dental implants. *Applied Surface Science*, *257*(7), 2555–2559. doi:10.1016/j.apsusc.2010.10.021

Ramya, K., Mohan, R., Anupama, K. K., & Joseph, A. (2015). Electrochemical and theoretical studies on the synergistic interaction and corrosion inhibition of alkyl benzimidazoles and thiosemicarbazide pair on mild steel in hydrochloric acid. *Materials Chemistry and Physics*, *149*, 632–647. doi:10.1016/j.matchemphys.2014.11.020

Rasheeduzzafar, D. F. H., Bader, M. A., & Khan, M. M. (1992). Performance of corrosion-resisting steels in chloride-bearing concrete. *ACI Materials Journal*, *89*(5), 439–448. doi:10.14359/2379

Ratner, B. D., Hoffman, A. S., Schoen, F. J., & Lemons, J. E. (2006). Biomaterials science: An introduction to materials in medicine. *MRS Bulletin*, *31*, 59.

Razak, H. A., & Choi, F. C. (2001). Damage assessment of corroded reinforced concrete beams using modal testing. In Structural Engineering, Mechanics and Computation (pp. 1203–1215). Elsevier. doi:10.1016/B978-008043948-8/50134-X

Rbaa, M., & Lakhrissi,B. (2019c). Novel oxazole and imidazole based on 8-Hydroxyquinoline as a Corrosion Inhibition of mild steel in HCl Solution: Insights from Experimental and Computational Studies. *S. Interfaces*, (15), 43-59.

Rbaa, M., Abousalem, A.S., Ebn Touhami, M., Warad, I., Bentiss, F., Lakhrissi, B., & Zarrouk A. (2019a). Novel Cu (II) and Zn (II) complexes of 8-hydroxyquinoline derivatives as effective corrosion inhibitors for mild steel in 1.0 M HCl solution: Computer modeling supported experimental studies. *J. Mol. Liq., 290.*

Rbaa, M., Dohar, P., Berish, A., Dagda, O., Lakhrissi, L., Gala, M., Lakhrissi, B., Ebn Touhami, M., Warad, I., & Zarrouk, A. (2020a). New Epoxy sugar based glucose derivatives as eco friendly corrosion inhibitors for the carbon steel in 1.0 M HCl: Experimental and theoretical investigations. *Journal of Alloys and Compounds*, *833*, 154949. doi:10.1016/j.jallcom.2020.154949

Rbaa, M., Abousalem, A. S., Ebn Touhami, M., Warade, I., Bentiss, F., Lakhrissi, B., & Zarrouk, A. (2019a). Novel Cu (II) and Zn (II) complexes of 8-hydroxyquinoline derivatives as effective corrosion inhibitors for mild steel in 1.0 M HCl solution: Computer modeling supported experimental studies. *Journal of Molecular Liquids*, *290*(15), 111243. doi:10.1016/j.molliq.2019.111243

Rbaa, M., Benhiba, F., Dohare, P., Lakhrissi, L., Touir, R., Lakhrissi, B., & Lakhrissi, Y. (2020b). Synthesis of new epoxy glucose derivatives as a non-toxic corrosion inhibitors for carbon steel in molar HCl: Experimental, DFT and MD simulation. *Chem. Data Collect.*, *27*, 100394. doi:10.1016/j.cdc.2020.100394

Rbaa, M., Benhiba, F., Obot, I. B., Oudda, H., Warad, I., Lakhrissi, B., & Zarrouk, A. (2019b). Two new 8-hydroxyquinoline derivatives as an efficient corrosion inhibitors for mild steel in hydrochloric acid: Synthesis, electrochemical, surface morphological, UV–visible and theoretical studies. *Journal of Molecular Liquids*, *276*, 120–133. doi:10.1016/j.molliq.2018.11.104

Rbaa, M., Galai, M., Abousalem, A. S., Lakhrissi, B., Touhami, M. E., Warad, I., & Zarrouk, A. (2019d). Synthetic, spectroscopic characterization, empirical and theoretical investigations on the corrosion inhibition characteristics of mild steel in molar hydrochloric acid by three novel 8-hydroxyquinoline derivatives. *Ionics*, *13*(8), 1–20. doi:10.100711581-019-03160-9

Rbaa, M., Galai, M., El Faydy, M., El Kacimi, Y., Ebn Touhami, M., Zarrouk, A., & Lakhrissi, B. (2017c). Synthesis and Characterization of New Benzimidazoles Derivatives of 8-hydroxyquinoline as a Corrosion Inhibitor for Mild Steel in 1.0 M Hydrochloric Acid Medium. *Anal. Bioanal. Electrochemistry*, *9*(7), 904–928.

Rbaa, M., Galai, M., El Kacimi, Y., Ouakki, M., Touir, R., & Lakhrissi, B. (2017a). Adsorption Properties and Inhibition of Carbon Steel Corrosion in a Hydrochloric Solution by 2-(4, 5-diphenyl-4, 5-dihydro-1h-imidazol-2-yl)-5-methoxyphenol. *Portugaliae Electrochimica Acta*, *35*(6), 323–338. doi:10.4152/pea.201706323

Rbaa, M., Galai, M., Faydy, E. L., El Kacimi, M., Ebn, Y., Touhami, M., Zarrouk, A., & Lakhrissi, B. (2017b). Synthesis, inhibition effects and thermodynamic studies of novel substituted quinolines on the corrosion of mild steel in 1 M HCl solution. *J. Mater. Environ. Sci*, *8*(10), 3529–3549.

Rbaa, M., Lakhrissi, B., & Lgaz, H. (2019b). Novel 8-Hydroxyqionoline Derivatives as Effective Corrosion Inhibitors for Mild Steel in HCl Solution: Synthesis, Characterization, Gravimetric, Electrochemical, Surface Morphological and UV-Visible Spectrometry. *Anal. Bioanal. Electrochem.*, *11*(6), 787–811.

Refay, S. A., Taha, F., & Abd El-Malak, A. M. (2004). Inhibition of stainless steel pitting corrosion in acidic medium by 2-mercaptobenzoxazole. *Applied Surface Science*, *236*(1–4), 175–185. doi:10.1016/j.apsusc.2004.04.016

Rengaswamy, N. S., Srinivasan, S., & Balasubramanian, T. M. (1988). Inhibited and sealed cement slurry coating of steel rebar-A state of art report. *Transactions of the SAEST*, *23*(2–3), 163–173.

Ren, Y. J., Chen, J., & Zeng, C. L. (2010). Corrosion protection of type 304 stainless steel bipolar plates of proton-exchange membrane fuel cells by doped polyaniline coating. *Journal of Power Sources*, *195*(7), 1914–1919. doi:10.1016/j.jpowsour.2009.10.003

Ribeiro, D. V., & Abrantes, J. C. C. (2016). Application of electrochemical impedance spectroscopy (EIS) to monitor the corrosion of reinforced concrete: A new approach. *Construction & Building Materials*, *111*, 98–104. doi:10.1016/j.conbuildmat.2016.02.047

Ribeiro, D. V., Souza, C. A. C., & Abrantes, J. C. C. (2015). Use of Electrochemical Impedance Spectroscopy (EIS) to monitoring the corrosion of reinforced concrete. *Revista IBRACON de Estruturas e Materiais*, *8*(4), 529–546. doi:10.1590/S1983-41952015000400007

Ridzwan, M. I. Z., Shuib, S., Hassan, A. Y., Shokri, A. A., & Ibrahim, M. M. (2007). Problem of stress shielding and improvement to the hip implant designs: A review. *J. Med. Sci*, *7*(3), 460–467. doi:10.3923/jms.2007.460.467

Rizzuti, S., De Napoli, L., & Ventra, S. (2019). The Influence of Build Orientation on the Flatness Error in Artifact Produced by Direct Metal Laser Sintering (DMLS) Process. In *Advances on Mechanics, Design Engineering and Manufacturing II* (pp. 463–472). Springer. doi:10.1007/978-3-030-12346-8_45

Roberge, P. (1995). Corrosion Inhibitors. In Handbook of Corrosion Engineering (pp. 833–862). Academic Press.

Rocha, L. A., Oliveira, F., Cruz, H. V., Sukotjo, C., & Mathew, M. T. (2013). Bio-tribocorrosion in dental applications. In *Bio-tribocorrosion in biomaterials and medical implants* (pp. 223–249). Woodhead Publishing. doi:10.1533/9780857098603.3.223

Rochdi, A., Dkhireche, N., Kassou, O., Elbakri, M., Ebn Touhami, M., Touir, R., Forsal, I., & Mernari, B. (2011). Protection d'un acier doux et du laiton dans les eaux des circuits de refroidissement par une formulation contenant un oxadiazole et un biocide non oxydant. *Proceeding of 23ème Forum sur les Impédances Electrochimiques.*

Rochdi, A., Kassou, O., Dkhireche, N., Touir, R., El Bakri, M., Touhami, M. E., ... Hammouti, B. (2014). Inhibitive properties of 2, 5-bis (n-methylphenyl)-1, 3, 4-oxadiazole and biocide on corrosion, biocorrosion and scaling controls of brass in simulated cooling water. *Corrosion Science*, *80*, 442–452. doi:10.1016/j.corsci.2013.11.067

Rochdi, A., Touir, R., El Bakri, M., Touhami, M. E., Bakkali, S., & Mernari, B. (2015). Protection of low carbon steel by oxadiazole derivatives and biocide against corrosion in simulated cooling water system. *Journal of Environmental Chemical Engineering*, *3*(1), 233–242. doi:10.1016/j.jece.2014.11.020

Roche, V., Koga, G. Y., Matias, T. B., Kiminami, C. S., Bolfarini, C., Botta, W. J., ... Junior, A. J. (2019). Degradation of biodegradable implants: The influence of microstructure and composition of Mg-Zn-Ca alloys. *Journal of Alloys and Compounds*, *774*, 168–181. doi:10.1016/j.jallcom.2018.09.346

Ronald, L. P., Clara, H., Doris, M. E., Paulo, R. R., & Angelo, F. P. (2007). A Short Review on Wrought Austenitic Stainless Steels at High Temperatures: Processing, Microstructure, Properties and Performance. *Materials Research*, *10*(4), 453–460. doi:10.1590/S1516-14392007000400021

Ruan, L., Zhang, Z., Huang, X., Lyu, Y., Wen, Y., Shang, W., & Wu, L. (2017). Evaluation of corrosion inhibition of two schiff bases selfassembled films on carbon steel in 0.5 M HCl. *International Journal of Electrochemical Science*, *12*, 103–115. doi:10.20964/2017.01.38

Rubitschek, F., Niendorf, T., Karaman, I., & Maier, H. J. (2012). Corrosion fatigue behavior of a biocompatible ultrafine-grained niobium alloy in simulated body fluid. *Journal of the Mechanical Behavior of Biomedical Materials*, *5*(1), 181–192. doi:10.1016/j.jmbbm.2011.08.023 PMID:22100093

Russo, G. M., Simon, G. P., & Incarnato, L. (2006). Correlation between rheological, mechanical, and barrier properties in new copolyamide-based nanocomposite films. *Macromolecules*, *39*(11), 3855–3864. doi:10.1021/ma052178h

Saber-Samandari, S., Saber-Samandari, S., Ghonjizade-Samani, F., Aghazadeh, J., & Sadeghi, A. (2016). Bioactivity evaluation of novel nanocomposite scaffolds for bone tissue engineering: The impact of hydroxyapatite. *Ceramics International*, *42*(9), 11055–11062. doi:10.1016/j.ceramint.2016.04.002

Saddik, R., Gaadaoui, A., Hamal, A., Zarrouk, A., Touzani, R., & Benchat, N. (2014). Synthesis, antibacterial and antifungal activity of some new imidazo[1,2-a]pyridine derivatives. *Der Pharm Lett.*, *6*, 343.

Saha, G., & Kurmaih, N. (1986). Technical Note: The Mechanism of Corrosion Inhibition by Phosphate-Based Cooling System Corrosion Inhibitors. *Corrosion*, *42*, 233–235.

Saha, S. K. R., Ghosh, P., Hens, A., Murmu, N. C., & Banerjee, P. (2015). Density functional theory and molecular dynamics simulation study on corrosion inhibition performance of mild steel by mercapto-quinoline Schiff base corrosion inhibitor. *Physica E, Low-Dimensional Systems and Nanostructures*, *66*, 332–341. doi:10.1016/j.physe.2014.10.035

Saha, S. K., Hens, A., Murmu, N. C., & Banerjee, P. (2016). A comparative density functional theory and molecular dynamics simulation studies of the corrosion inhibitory action of two novel N-heterocyclic organic compounds along with a few others over steel surface. *Journal of Molecular Liquids*, *215*, 486–495. doi:10.1016/j.molliq.2016.01.024

Salarvand, Z., Amirnasr, M., Talebian, M., Raeissi, K., & Meghdadi, S. (2017). Enhanced corrosion resistance of mild steel in 1M HCl solution by trace amount of 2-phenyl-benzothiazole derivatives: Experimental, quantum chemical calculations and molecular dynamics (MD) simulation studies. *Corrosion Science*, *114*, 133–145. doi:10.1016/j.corsci.2016.11.002

Salgado-Salgado, R. J., Porcayo-Calderon, J., Sotelo-Mazon, O., Rodriguez-Diaz, R. A., Salinas-Solano, G., Salinas-Bravo, V. M., & Martinez-Gomez, L. (2016). Effect of Ag addition on the electrochemical performance of Cu10Al in artificial saliva. *Bioinorganic Chemistry and Applications*, *2016*, 2016. doi:10.1155/2016/4792583 PMID:27660601

Salghi, R., Ben Hmamou, D., Benali, O., Jodeh, S., Warad, I., Hamed, O., Ebenso, E. E., Oukacha, A., Tahrouch, S., & Hammouti, B. (2015). Study of the corrosion inhibition effect of pistachio essential oils in 0.5 M H2SO4. *International Journal of Electrochemical Science*, *10*, 8403.

Salnikow, K., & Zhitkovich, A. (2007). Genetic and epigenetic mechanisms in metal carcinogenesis and cocarcinogenesis: Nickel, arsenic, and chromium. *Chemical Research in Toxicology*, *21*(1), 28–44. doi:10.1021/tx700198a PMID:17970581

Sánchez, M. J., Garcés, P., & Climent, M. A. (2006). Electrochemical extraction of chlorides from reinforced concrete: Variables affecting treatment efficiency. *Materiales de Construcción*, *56*(284), 17–26. doi:10.3989/mc.2006.v56.i284.15

Sandrolini, F., Franzoni, E., & Pigino, B. (2012). Ethyl silicate for surface treatment of concrete – Part I: Pozzolanic effect of ethyl silicate. *Cement and Concrete Composites*, *34*(3), 306–312. doi:10.1016/j.cemconcomp.2011.12.003

Saraswathy, V., Lee, H. S., Karthick, S., & Kwon, S. J. (2018). Extraction of chloride from chloride contaminated concrete through electrochemical method using different anodes. *Construction & Building Materials*, *158*, 549–562. doi:10.1016/j.conbuildmat.2017.10.052

Saremi, M., Dehghanian, C., & Sabet, M. M. (2006). The effect of molybdate concentration and hydrodynamic effect on mild steel corrosion inhibition in simulated cooling water. *Corrosion Science*, *48*(6), 1404–1412. doi:10.1016/j.corsci.2005.06.009

Saremi, M., & Mahallati, E. (2002). A study on chloride-induced depassivation of mild steel in simulated concrete pore solution. *Cement and Concrete Research*, *32*(12), 1915–1921. doi:10.1016/S0008-8846(02)00895-5

Sasikumar, Y., Adekunle, A. S., Olasunkanmi, L. O., Bahadur, I., Baskar, R., Kabanda, M. M., Obot, I. B., & Ebenso, E. E. (2015). Experimental, quantum chemical and Monte Carlo simulation studies on the corrosion inhibition of some alkyl imidazolium ionic liquids containing tetrafluoroborate anion on mild steel in acidic medium. *Journal of Molecular Liquids*, *211*, 105–118. doi:10.1016/j.molliq.2015.06.052

Savas, K., Tüzün, B., Kaya, C., & Obot, I. B. (2016). Determination of corrosion inhibition effects of amino acids: Quantum chemical and molecular dynamic simulation study. *Journal of the Taiwan Institute of Chemical Engineers*, *58*, 528–535. doi:10.1016/j.jtice.2015.06.009

Savio, G., Rosso, S., Meneghello, R., & Concheri, G. (2018). Geometric modeling of cellular materials for additive manufacturing in biomedical field: A review. *Applied Bionics and Biomechanics*, *2018*, 2018. doi:10.1155/2018/1654782 PMID:29487626

Schock, M.R., & Wagner, I. (1985). Internal corrosion of water distribution systems. *DVGW-Forschungstelle*, 700.

Schroeder, A., van der Zypen, E., Stich, H., & Sutter, F. (1981). The reactions of bone, connective tissue, and epithelium to endosteal implants with titanium-sprayed surfaces. *Journal of Maxillofacial Surgery*, *9*, 15–25. doi:10.1016/S0301-0503(81)80007-0 PMID:6939769

Schüürmann, G., Ebert, R. U., & Kühne, R. (2011). R. Ku?hne, Quantitative read-across for predicting the acute fish toxicity of organic compounds. *Environmental Science & Technology*, *45*(10), 461. doi:10.1021/es200361r PMID:21491860

Schweinsberg, D. P., George, G. A., Nanayakkara, A. K., & Steinert, D. A. (1988). The protective action of epoxy resins and curing agents—Inhibitive effects on the aqueous acid corrosion of iron and steel. *Corrosion Science*, *28*(1), 33–42. doi:10.1016/0010-938X(88)90004-2

Searle, A. B. (1935). *Limestone and Its Products: Their Nature, Production, and Uses*. Ernest and Benn Ltd.

Serrar, H., Galai, M., Benhiba, F., Ouakki, M., Benzekri, Z., Boukhris, S., Hassikou, A., Souizi, A., Oudda, H., & Ebn Touhami, M. (2018). Two derivatives of 7-amino-thiazolo[3,2-a] pyrimidine as inhibitors of mild steel corrosion in 1.0 m hcl solution: part i: synthesis of inhibitors and electrochemical study. *J. Chem. Technol. Metall*, *53*(2), 324–335.

Serrar, H., Marmouzi, I., Benzekri, Z., Boukhris, S., Hassikou, A., Faouzi, M. E. A., & Souizi, A. (2017). Synthesis and Evaluation of Novel Pyrido[1,2-b][1,2,4]triazine-2,6-dione and Pyrido[1,2-b][1,2,4]triazepine-2,7-dione Derivatives as Antioxidant Agents. *Letters in Organic Chemistry*, *14*(4), 267–277. doi:10.2174/1570178614666170227124523

Shaban, A., Felhosi, I., & Telegdi, J. (2017). Laboratory assessment of inhibition efficiency and mechanism of inhibitor blend (P22SU) on mild steel corrosion in high chloride containing water. *Int J Corros Scale Inhib.*, *6*, 262.

Shaban, S. M., Abd-Elaal, A. A., & Tawfik, S. M. (2016). Gravimetric and electrochemical evaluation of three nonionic dithiol surfactants as corrosion inhibitors for mild steel in 1 M HCl solution. *Journal of Molecular Liquids*, *216*, 392–400. doi:10.1016/j.molliq.2016.01.048

Shahin, M., Bilgic, S., & Yilmaz, H. (2002). The inhibition effects of some cyclic nitrogen compounds on the corrosion of the steel in NaCl mediums. *Applied Surface Science*, *195*(1-4), 1–7. doi:10.1016/S0169-4332(01)00783-8

Shainy, K. M., Ammal, P. R., Unni, K. N., Benjamin, S., & Joseph, A. (2016). Surface interaction and corrosion inhibition of mild steel in hydrochloric acid using pyoverdine, an eco-friendly biomolecule. *J Bio-Tribo Corros.*, *2*(3), 20. doi:10.100740735-016-0050-3

Shan, Z., Xinyang, S., Xiaoyang, G., Jianrong, W., Naiqin, Z., & Junwei, S. (2019). Equivalent circuit model recognition of electrochemical impedance spectroscopy via machine learning. *Journal of Electroanalytical Chemistry (Lausanne, Switzerland)*, *855*, 113627. doi:10.1016/j.jelechem.2019.113627

Shetty, S. K., & Shetty, A. N. (2017). Eco-friendly benzimidazolium based ionic liquid as a corrosion inhibitor for aluminum alloy composite in acidic media. *Journal of Molecular Liquids*, *225*, 426–438. doi:10.1016/j.molliq.2016.11.037

Shih, H., & Mansfeld, F. (1989). A fitting procedure for impedance data of systems with very low corrosion rates. *Corrosion Science*, *29*(10), 1235–1240. doi:10.1016/0010-938X(89)90070-X

Siddique, R., & Klaus, J. (2009). Influence of metakaolin on the properties of mortar and concrete: A review. *Applied Clay Science*, *43*(3–4), 392–400. doi:10.1016/j.clay.2008.11.007

Siegwart, M., Lyness, J. F., & McFarland, B. J. (2003). Change of pore size in concrete due to electrochemical chloride extraction and possible implications for the migration of ions. *Cement and Concrete Research*, *33*(8), 1211–1221. doi:10.1016/S0008-8846(03)00047-4

Simchi, A. (2006). Direct laser sintering of metal powders: Mechanism, kinetics and microstructural features. *Materials Science and Engineering A*, *428*(1-2), 148–158. doi:10.1016/j.msea.2006.04.117

Singh, T. P., Singh, H., & Singh, H. (2012). Characterization, corrosion resistance, and cell response of high-velocity flame-sprayed HA and HA/TiO 2 coatings on 316L SS. *Journal of Thermal Spray Technology, 21*(5), 917-927

Singh, A. K., & Quraishi, M. A. (2010). The effect of some bis-thiadiazole derivatives on the corrosion of mild steel in hydrochloric acid. *Corrosion Science*, *52*(4), 1373–1385. doi:10.1016/j.corsci.2010.01.007

Singh, A., Ansari, K. R., Haque, J., Dohare, P., Lgaz, H., Salghi, R., & Quraishi, M. A. (2018). Effect of electron donating functional groups on corrosion inhibition of mild steel in hydrochloric acid: Experimental and quantum chemical study. *Journal of the Taiwan Institute of Chemical Engineers*, *82*, 233–251. doi:10.1016/j.jtice.2017.09.021

Singh, D. D. N., & Venugopalan, T. (2013). Understanding the corrosion and remedial measures to control the deterioration of reinforcement steel bars by modification in their chemistry and application of surface coatings. *Transactions of the Indian Institute of Metals*, *66*(5–6), 677–687. doi:10.100712666-013-0335-x

Singh, D., Chaudhary, R. S., Prakash, B., & Agrawal, C. V. (1979). Inhibitive Efficiency of some Substituted Thioureas for the Corrosion of Aluminium in Nitric Acid. *British Corrosion Journal*, *14*(4), 235–239. doi:10.1179/000705979798358571

Singh, G., Singh, H., & Sidhu, B. S. (2013). Corrosion behavior of plasma sprayed HAP and HAP-silicon oxide coatings on AISI 304 for biomedical application. *Applied Surface Science*, *284*, 811–818. doi:10.1016/j.apsusc.2013.08.013

Singh, S. K., Tambe, S. P., Gunasekaran, G., Raja, V. S., & Kumar, D. (2009). Electrochemical impedance study of thermally sprayable polyethylene coatings. *Corrosion Science*, *51*(3), 595–601. doi:10.1016/j.corsci.2008.11.025

Sivakumar, B., Pathak, L. C., & Singh, R. (2017). Role of surface roughness on corrosion and fretting corrosion behaviour of commercially pure titanium in Ringer's solution for bio-implant application. *Applied Surface Science*, *401*, 385–398. doi:10.1016/j.apsusc.2017.01.033

Siva, P. K., Srinivasa, C. R., & Damera, N. R. (2012). A Review on Welding of AISI 304L Austenitic Stainless Steel. *Journal for Manufacturing Science and Production*, *14*(1), 1–11.

Smith, K., Schokker, A., & Tikalsky, P. (2004). Performance of supplementary cementitious materials in concrete resistivity and corrosion monitoring evaluations. *ACI Materials Journal*, *101*(5), 385–389.

Solmaz, R. (2014). Investigation of adsorption and corrosion inhibition of mild steel in hydrochloric acid solution by 5-(4-Dimethylaminobenzylidene)rhodanine. *Corrosion Science*, *79*, 169–176. doi:10.1016/j.corsci.2013.11.001

Song, H., & Saraswathy, V. (2007). Corrosion monitoring of reinforced concrete structures - A review. *International Journal of Electrochemical Science*, *2*, 1–28.

Song, H.-W., Lee, C.-H., & Ann, K. Y. (2008). Factors influencing chloride transport in concrete structures exposed to marine environments. *Cement and Concrete Composites*, *30*(2), 113–121. doi:10.1016/j.cemconcomp.2007.09.005

Song, W. H., Jun, Y. K., Han, Y., & Hong, S. H. (2004). Biomimetic apatite coatings on micro-arc oxidized titania. *Biomaterials*, *25*(17), 3341–3349. doi:10.1016/j.biomaterials.2003.09.103 PMID:15020106

Song, X. Q., Wang, Y. W., Zheng, J. R., Liu, W. S., & Tan, M. Y. (2007). Synthesis and spectroscopic properties of lanthanide nitrate complexes with a new amide-based quinoxaline-2,3-dione ligand. *Mol Biomol Spectrosc*, *68*(3), 701–704. doi:10.1016/j.saa.2006.12.048 PMID:17419093

Song, Y. W., Shan, D. Y., & Han, E. H. (2007). Corrosion behaviors of electroless plating Ni–P coatings deposited on magnesium alloys in artificial sweat solution. *Electrochimica Acta*, *53*(4), 2009–2015. doi:10.1016/j.electacta.2007.08.062

Söylev, T. A., & Richardson, M. G. (2008). Corrosion inhibitors for steel in concrete: State-of-the-art report. *Construction & Building Materials*, *22*(4), 609–622. doi:10.1016/j.conbuildmat.2006.10.013

Stern, M. (1957a). Electrochemical polarization-II. Ferrous-Ferric electrode kinetics on stainless steel. *Journal of the Electrochemical Society*, *104*(9), 559–563. doi:10.1149/1.2428653

Stern, M. (1957b). Electrochemical polarization-III. Further aspects of the shape of polarization curves. *Journal of the Electrochemical Society*, *104*(9), 645–650. doi:10.1149/1.2428438

Stern, M., & Geary, A. L. (1957). Electrochemical polarization I. A theoretical analysis of the shape of polarization curves. *Journal of the Electrochemical Society*, *104*(1), 56–63. doi:10.1149/1.2428496

Stern, M., & Geary, A. L. (1957). Electrochemical Polarization: I. A Theoretical Analysis of the Shape of Polarization Curves. *Journal of the Electrochemical Society*, *104*, 56–63.

Stratfull, R. F. (1957). The corrosion of steel in a reinforced concrete bridge. *Corrosion*, *13*(3), 43–48. doi:10.5006/0010-9312-13.3.43

Stratfull, R. F. (1973). *Preliminary investigation of cathodic protection of a bridge deck*. California Department of Transportation.

Subathira, A., & Meyyappan, R. M. (2012). Corrosion Protection of Stainless Steel by Polyaniline/Polypyrrole Composite Coating. *International Journal of Engineering Science and Technology*, *2*, 7105-7111.

Sultan, A., & Asif, M. (2018). Enhancement of Anticorrosion Property of 304 Stainless Steel Using Silane Coatings. *Applied Surface Science*, *440*, 1286–1297. doi:10.1016/j.apsusc.2018.01.203

Sun, H., Ren, P., & Fried, J. R. (1998). The COMPASS force field: Parameterization and validation for phosphazenes. *Computational and Theoretical Polymer Science*, *8*(1-2), 229–246. doi:10.1016/S1089-3156(98)00042-7

Sun, L., Berndt, C. C., Gross, K. A., & Kucuk, A. (2001). Material fundamentals and clinical performance of plasma-sprayed HAP coatings: A review. *Journal of Biomedical Materials Research: An Official Journal of The Society for Biomaterials, The Japanese Society for Biomaterials, and The Australian Society for Biomaterials and the Korean Society for Biomaterials*, *58*(5), 570–592. doi:10.1002/jbm.1056 PMID:11505433

Suzuki, T., Nishihara, H., & Aramaki, K. (1996). The synergistic inhibition effect of octylmercaptopropionate and 8-quinolinol on the corrosion of iron in an aerated 0.5 M Na2SO4 solution. *Corrosion Science*, *38*(8), 1223–1234. doi:10.1016/0010-938X(95)00172-G

Szauer, T., & Brand, A. (1981). Adsorption of oleates of various amines on iron in acidic solution. *Electrochimica Acta*, *26*(9), 1253–1256. doi:10.1016/0013-4686(81)85107-9

Szauer, T., & Brandt, A. (1981). On the role of fatty acid in adsorption and corrosion inhibition of iron by amine—Fatty acid salts in acidic solution. *Electrochimica Acta*, *26*, 1257–1260.

Tafel, J. (1905). Über die Polarisation bei kathodischer Wasserstoffentwicklung. *Zeitschrift für Physikalische Chemie*, *50*, 641–712. doi:10.1515/zpch-1905-5043

Takeuchi, M., Abe, Y., Yoshida, Y., Nakayama, Y., Okazaki, M., & Akagawa, Y. (2003). Acid pretreatment of titanium implants. *Biomaterials*, *24*(10), 1821–1827.

Talha, M., Behera, C. K., & Sinha, O. P. (2013). A review on nickel-free nitrogen containing austenitic stainless steels for biomedical applications. *Materials Science and Engineering C*, *33*(7), 3563–3575. doi:10.1016/j.msec.2013.06.002 PMID:23910251

Talha, M., Behera, C. K., & Sinha, O. P. (2015). Effect of nitrogen and cold working on structural and mechanical behavior of Ni-free nitrogen containing austenitic stainless steels for biomedical applications. *Materials Science and Engineering C*, *47*, 196–203. doi:10.1016/j.msec.2014.10.078 PMID:25492189

Tang, L., Lia, X., Lia, L., Qua, Q., Mua, G., & Liua, G. (2005). The effect of 1-(2-pyridylazo)-2-naphthol on the corrosion of cold rolled steel in acid media: Part 1: Inhibitive action in 1.0M hydrochloric acid. *Materials Chemistry and Physics*, *94*(2-3), 353–359. doi:10.1016/j.matchemphys.2005.05.015

Tang, Y., Yang, X., Yang, W., Wan, R., Chen, Y., & Yin, X. (2010). A preliminary investigation of corrosion inhibition of mild steel in 0.5 M H2SO4 by 2-amino-5-(n-pyridyl)-1, 3, 4-thiadiazole: Polarization, EIS and molecular dynamics simulations. *Corrosion Science*, *52*(5), 1801–1808. doi:10.1016/j.corsci.2010.01.028

Tang, Y., Zhang, F., Huc, S., Cao, Z., Wu, Z., & Jing, W. (2013). Novel benzimidazole derivatives as corrosion inhibitors of mild steel in the acidic media. Part I: Gravimetric, electrochemical, SEM and XPS studies. *Corrosion Science*, *74*, 271–282. doi:10.1016/j.corsci.2013.04.053

Tarfaoui, K., Ouakki, M., Galai, M., Nehiri, M., Ebn Touhami, M., Barhada, N., & Ouhssine, M. (2019). Valorization of the essential oil of Zingiber officinale by its Use as inhibitor against the corrosion of the carbon steel in acid medium hydrochloric acid 1M. *Mediterr. J. Chem*, *8*(1), 17–24.

Tawfik, S. M. (2015). Corrosion inhibition efficiency and adsorption behavior of N,N-dimethyl-4-(((1-methyl-2-phenyl-2,3-dihydro-1H-pyrazol-4-yl) imino) methyl)-N alkylbenzenaminium bromide surfactant at carbon steel/hydrochloric acid interface. *Journal of Molecular Liquids*, *207*, 185–194. doi:10.1016/j.molliq.2015.03.036

Taylor, R., Richardson, I. G., & Brydson, R. M. D. (2010). Composition and microstructure of 20-year-old ordinary Portland cement-ground granulated blast-furnace slag blends containing 0 to 100% slag. *Cement and Concrete Research*, *40*(7), 971–983. doi:10.1016/j.cemconres.2010.02.012

Tazouti, A., Galai, M., Touir, R., Touhami, M. E., Zarrouk, A., Ramli, Y., Saraçoğlu, M., Kaya, S., Kandemirli, F., & Kaya, C. (2016). Experimental and theoretical studies for mild steel corrosion inhibition in 1.0 M HCl by three new quinoxalinone derivatives. *Journal of Molecular Liquids*, *221*, 815–832. doi:10.1016/j.molliq.2016.03.083

Tecuapa-Flores, E. D., Turcio-Ortega, D., Hernandez, J. G., Huerta-Aguilar, C. A., & Thangarasu, P. (2019). The role of keto group in cyclic ligand 1,4,8,11-tetraazacyclotetradecane-5,7-dione as strong corrosion inhibitor for carbon steel surface: Experimental and theoretical studies. *Journal of Molecular Structure*, *1189*, 131–145. doi:10.1016/j.molstruc.2019.04.006

Teguh, D. W., & Kazuhiko, N. (2013). *Corrosion Resistance Enhancement of 304 Stainless steel Under droplet of Chloride Solution by Mechanical Surface Treatment*. The Minerals, Metals & Materials Society.

Teissedre, P. L., & Waterhouse, A. L. (2000). Inhibition of oxidation of human low-density lipoproteins by phenolic substances in different essential oils varieties. *Journal of Agricultural and Food Chemistry*, *4*(9), 3801–3805. doi:10.1021/jf990921x PMID:10995274

Teng, S., Lim, T. Y. D., & Sabet Divsholi, B. (2013). Durability and mechanical properties of high strength concrete incorporating ultra fine ground granulated blast-furnace slag. *Construction & Building Materials*, *40*, 875–881. doi:10.1016/j.conbuildmat.2012.11.052

Thangavel, K., Rengaswamy, N. S., & Balakrishnan, K. (1995). Influence of protective coatings on steel-concrete bond. *Indian Concrete Journal*, *69*(4), 191–196.

The Effects and Economic Impact of Corrosion. (2000). In Corrosion: Understanding the Basics (pp. 1–20). ASM International.

Thomas, M. D. A., & Bamforth, P. B. (1999). Modeling chloride diffusion in concrete effect of fly ash and slag. *Cement and Concrete Research*, *29*, 487–495. doi:10.1016/S0008-8846(98)00192-6

Tiancheng, Z. (2001). Improvement in the corrosion-erosion resistance of 304 stainless steel with alloyed yttrium. *Journal of Materials Science*, *36*(14), 3479–3486. doi:10.1023/A:1017920314397

Tijing, L. D., Kim, H. Y., Lee, D. H., Kim, C. S., & Cho, Y. I. (2010). Physical water treatment using RF electric fields for the mitigation of CaCO3 fouling in cooling water. *International Journal of Heat and Mass Transfer*, *53*(7-8), 1426–1437. doi:10.1016/j.ijheatmasstransfer.2009.12.009

Topal, E., & Gece, G. (2017). Untangling the inhibition effects of aliphatic amines on silver corrosion: A computational study. *Chem J Moldova.*, *12*(2), 64–70. doi:10.19261/cjm.2017.411

Topcu, I. B., & Boga, A. R. (2010). Effect of ground granulate blast-furnace slag on corrosion performance of steel embedded in concrete. *Materials & Design*, *31*(7), 3358–3365. doi:10.1016/j.matdes.2010.01.057

Touir, R., Dkhireche, N., Ebn Touhami, M., Sfaira, M., Senhaji, O., Robin, J.J., Boutevin, B., & Cherkaoui, M. (2010). *Study of phosphonate addition and hydrodynamic conditions on ordinary steel*. Academic Press.

Touir, R. (2016). *Nouvelles formulations pour la protection des circuits de refroidissement*. Éditions Universitaires Européennes.

Touir, R., Cenoui, M., El Bakri, M., & Touhami, M. E. (2008). Sodium gluconate as corrosion and scale inhibitor of ordinary steel in simulated cooling water. *Corrosion Science*, *50*(6), 1530–1537. doi:10.1016/j.corsci.2008.02.011

Touir, R., Dkhireche, N., Ebn Touhami, M., El Bakri, M., Rochdi, A., & Belakhmima, R. A. (2014). Study of the mechanism action of sodium gluconate used for the protection of scale and corrosion in cooling water system. *Journal of Saudi Chemical Society*, *18*(6), 873–881. doi:10.1016/j.jscs.2011.10.020

Touir, R., Dkhireche, N., Ebn Touhami, M., Lakhrissi, M., Lakhrissi, B., & Sfaira, M. (2009). Corrosion and scale processes and their inhibition in simulated cooling water systems by monosaccharides derivatives, Part I: EIS study. *Desalination*, *249*(3), 922–928. doi:10.1016/j.desal.2009.06.068

Touir, R., Dkhireche, N., Ebn Touhami, M., Sfaira, M., Senhaji, O., Robin, J. J., Boutevin, B., & Cherkaoui, M. (2010, July). corrosion inhibition in simulated cooling water. *Materials Chemistry and Physics*, *122*(1), 1–9. doi:10.1016/j.matchemphys.2010.02.063

Touir, R., El Bakri, M., Dkhireche, N., Ebn Touhami, M., & Rochdi, A. (2010). Operational parameters effect of sodium gluconate inhibition used in the protection of cooling water system. *Journal of Materials Environmental Science*, *1*(S1), 317–328.

Tsuchiya, H., Fujimoto, S., Chihara, O., & Shibata, T. (2002). Semiconductive behavior of passive films formed on pure Cr and Fe–Cr alloys in sulfuric acid solution. *Electrochimica Acta*, *47*(27), 4357–4366. doi:10.1016/S0013-4686(02)00508-X

Umoren, S. A., & Solomon, M. M. (2015). Effect of halide ions on the corrosion inhibition efficiency of different 4 organic species—A review. *Journal of Industrial and Engineering Chemistry*, *21*, 81–100. doi:10.1016/j.jiec.2014.09.033

Upadhyay, D., Panchal, M. A., Dubey, R. S., & Srivastava, V. K. (2006). Corrosion of alloys used in dentistry: A review. *Materials Science and Engineering A*, *432*(1-2), 1–11. doi:10.1016/j.msea.2006.05.003

Vallet-Regí, M. (2001). Ceramics for medical applications. *Journal of the Chemical Society, Dalton Transactions: Inorganic Chemistry*, (2), 97–108. doi:10.1039/b007852m

Vayre, B., Vignat, F., & Villeneuve, F. (2012). Metallic additive manufacturing: State-of-the-art review and prospects. *Mechanics & Industry*, *13*(2), 89–96. doi:10.1051/meca/2012003

Vedalakshmi, R., Kumar, K., Raju, V., & Rengaswamy, N. S. (2000). Effect of prior damage on the performance of cement based coatings on rebar: Macrocell corrosion studies. *Cement and Concrete Composites*, *22*(6), 417–421. doi:10.1016/S0958-9465(00)00041-X

Verma, C., Ebenso, E. E., Bahadur, I., Obot, I. B., & Quraishi, M. A. (2015b). 5-(Phenylthio)-3H-pyrrole-4-carbonitriles as effective corrosion inhibitors for mild steel in 1 M HCl: Experimental and theoretical investigation. *Journal of Molecular Liquids*, *212*, 209–218. doi:10.1016/j.molliq.2015.09.009

Verma, C., Ebenso, E. E., Bahadur, I., & Quraishi, M. A. (2018). An overview on plant extracts as environmental sustainable and green corrosion inhibitors for metals and alloys in aggressive corrosive media. *Journal of Molecular Liquids*, *266*, 577–590. doi:10.1016/j.molliq.2018.06.110

Verma, C., Ebenso, E. E., & Quraishi, M. A. (2017). Ionic liquids as green and sustainable corrosion inhibitors for metals and alloys: An overview. *Journal of Molecular Liquids*, *233*, 403–414. doi:10.1016/j.molliq.2017.02.111

Verma, C., Olasunkanmi, L. O., Ebenso, E. E., & Quraishi, M. A. (2018). Adsorption characteristics of green 5-arylaminomethylene pyrimidine2,4,6-triones on mild steel surface in acidic medium: Experimental and computational approach. *Results Phys.*, *8*, 657–670. doi:10.1016/j.rinp.2018.01.008

Verma, C., Olasunkanmi, L. O., Ebenso, E. E., Quraishi, M. A., & Obot, I. B. (2016). Adsorption behavior of glucosamine-based, pyrimidine-fused heterocycles as green corrosion inhibitors for mild steel: Experimental and theoretical studies. *The Journal of Physical Chemistry C*, *120*(21), 11598–11611. doi:10.1021/acs.jpcc.6b04429

Verma, C., Quraishi, M. A., & Singh, A. (2015a). 2-Amino-5-nitro-4,6-diarylcyclohex-1-ene-1,3,3-tricarbonitriles as new and effective corrosion inhibitors for mild steel in 1 M HCl: Experimental and theoretical studies. *Journal of Molecular Liquids*, *212*, 804–812. doi:10.1016/j.molliq.2015.10.026

Víctor, E. S., Maria, T. D. M., Isabel, L., & Gustavo, G. M. (2019). Electrochemical impedance spectroscopy (EIS): An efficient technique for monitoring corrosion processes in molten salt environments in CSP applications. *Solar Energy Materials and Solar Cells*, *191*, 157–163. doi:10.1016/j.solmat.2018.11.007

Vieira, A. C., Ribeiro, A. R., Rocha, L. A., & Celis, J. P. (2006). Influence of pH and corrosion inhibitors on the tribo-corrosion of titanium in artificial saliva. *Wear*, *261*(9), 994–1001. doi:10.1016/j.wear.2006.03.031

Vračar, Lj. M., & Dražić, D. M. (2002). Adsorption and corrosion inhibitive properties of some organic molecules on iron electrode in sulfuric acid. *Corrosion Science*, *44*(8), 1669–1680. doi:10.1016/S0010-938X(01)00166-4

Walkowiak-Przybyło, M., Klimek, L., Okrój, W., Jakubowski, W., Chwiłka, M., Czajka, A., & Walkowiak, B. (2012). *Adhesion, activation, and aggregation of blood platelets and biofilm formation on the surfaces of titanium alloys Ti6Al4V and Ti6Al7Nb*. Academic Press.

Wang, H. P., Wu, Q., Li, C. M., & Gu, N. (2013). Copper corrosion inhibition by polyaspartic acid and imidazole. *Materials and Corrosion*, *64*, 347-352.

Wang, C., Lai, C., Xie, B., Guo, X., Fu, D., Li, B., & Zhu, S. (2018). Corrosion inhibition of mild steel in HCl medium by S-benzyl-O, O¢-bis (2-naphthyl) dithiophosphate with ultra-long lifespan. *Results Phys*, *10*, 558–567. doi:10.1016/j.rinp.2018.07.002

Wang, H. L., Fan, H. B., & Zheng, J. S. (2003). Corrosion inhibition of mild steel in hydrochloric acid solution by a mercapto-triazole compound. *Chemical Physics*, *77*, 655.

Wang, L., Yin, G.-J., & Yin, J.-G. (2001). 2-Mercaptothiazoline and Cetyl Pyridinium Chloride as Inhibitors for the Corrosion of a Low Carbon Steel in Phosphoric Acid. *Corrosion Science*, *43*(6), 1197–1202. doi:10.1016/S0010-938X(00)00138-4

Wang, W., & Alfantazi, A. (2016). Electrochemical Corrosion Behavior of Niobium Alloys as Metallic Bio-implants. In *NACE International Corrosion Conference Proceedings* (p. 1). NACE International.

Watson, A. A., Fleet, G. W. J., Asano, N., Molyneux, R. J., & Nugh, R. J. (2001). Polyhydroxylated alkaloids— Natural occurrence and therapeutic applications. *Phytochemistry*, *56*(3), 265–295. doi:10.1016/S0031-9422(00)00451-9 PMID:11243453

Wei, K., Wang, Y., Lai, C., Ning, C., Wu, D., Wu, G., & Ye, J. (2005). Synthesis and characterization of hydroxyapatite nanobelts and nanoparticles. *Materials Letters*, *59*(2-3), 220–225. doi:10.1016/j.matlet.2004.08.034

Weining, S., Shufeng, Y., & Jingshe, L. (2018). Correlation between evolution of inclusions and pitting corrosion in 304 stainless steel with yttrium addition. *Scientific Reports*, *8*(1), 4830. doi:10.103841598-018-23273-x PMID:29556014

Wenner, F. (1916). A method of measuring earth resistivity. *Bulletin of the Bureau of Standards*, 469–478.

Wilk, A., Szypulska-Koziarska, D., & Wiszniewska, B. (2017). The toxicity of vanadium on gastrointestinal, urinary and reproductive system, and its influence on fertility and fetuses malformations. Advances in Hygiene & Experimental Medicine. *Postepy Higieny i Medycyny Doswiadczalnej*, 71.

Woo, R. S. C., Zhu, H., Chow, M. M. K., Leung, C. K. Y., & Kim, J.-K. (2008). Barrier performance of silane–clay nanocomposite coatings on concrete structure. *Composites Science and Technology*, *68*(14), 2828–2836. doi:10.1016/j.compscitech.2007.10.028

Xianqiang, X., Liya, F., Yiming, X., Qiqiang, Z., Jiangshan, L., Yong, W., Guihua, C., Chenglin, W., Tianding, L., Shuai, F., & Guolin, Z. (2019). Online corrosion monitoring of photoelectrochemical cathodic protection of carbon steel using particle video microscope. *Optik (Stuttgart)*, 164091.

Xiao, L., Kim, Y. S., Abbaschian, R., & Hecht, R. J. (1991). Processing and mechanical properties of niobium-reinforced MoSi2 composites. *Materials Science and Engineering A*, *144*(1-2), 277–285. doi:10.1016/0921-5093(91)90234-E

Xia, Y., Yang, Z. Y., Xia, P., Bastow, K. F., Tachibani, Y., Kuo, S. C., Hamel, E., Hackl, T., & Lee, K. H. (1998). Antitumor Agents. 181. Synthesis and Biological Evaluation of 6,7,2′,3′,4′-substituted-1,2,3,4-tetrahydro-2-phenyl-4-quinolones as a New Class of Antimitotic Antitumor Agents. *Journal of Medicinal Chemistry*, *41*, 1155. doi:10.1021/jm9707479 PMID:9544215

Xing, X., Han, Z., Wang, H., & Lu, P. (2015). Electrochemical corrosion resistance of CeO2-Cr/Ti coatings on 304 stainless steel via pack cementation. *Journal of Rare Earths*, *33*(10), 1122–1128. doi:10.1016/S1002-0721(14)60535-4

Xu, B., Liu, Y., Yin, X., Yang, W., & Chen, Y. (2013). Experimental and theoretical study of corrosion inhibition of 3-pyridinecarbozalde thiosemicarbazone for mild steel in hydrochloric acid. *Corrosion Science*, *74*, 206–213. doi:10.1016/j.corsci.2013.04.044

Xu, W., Lu, X., Wang, L. N., Shi, Z. M., Lv, S. M., Qian, M., & Qu, X. H. (2018). Mechanical properties, in vitro corrosion resistance and biocompatibility of metal injection molded Ti-12Mo alloy for dental applications. *Journal of the Mechanical Behavior of Biomedical Materials*, *88*, 534–547. doi:10.1016/j.jmbbm.2018.08.038 PMID:30223215

Xu, Y., Lu, Y., Liang, J., & Sisson, R. D. (2019). Microstructure and corrosion behaviour of additively manufactured Ti–6Al–4V with various post-heat treatments. *Materials Science and Technology*, *35*(1), 89–97. doi:10.1080/02670836.2018.1542052

Yadav, M., Kumar, S., Purkait, T., Olasunkanmi, L. O., Bahadur, I., & Ebenso, E. E. (2016). Electrochemical, thermodynamic and quantum chemical studies of synthesized benzimidazole derivatives as corrosion inhibitors for N80 steel in hydrochloric acid. *Journal of Molecular Liquids*, *213*, 122–138. doi:10.1016/j.molliq.2015.11.018

Yadav, M., Kumar, S., Sinha, R. R., & Behera, D. (2013). Experimental and Quantum Chemical Studies on the Corrosion Inhibition Performance of Benzimidazole Derivatives for Mild Steel in HCl. *Eng. Chem. Res*, *52*, 6318.

Yang, H., Yang, K., & Zhang, B. (2007). Pitting corrosion resistance of La added 316L stainless steel in simulated body fluids. *Materials Letters*, *61*(4-5), 1154–1157. doi:10.1016/j.matlet.2006.06.071

Yang, K., Zhou, C., Fan, H., Fan, Y., Jiang, Q., Song, P., & Zhang, X. (2017). Bio-functional design, application and trends in metallic biomaterials. *International Journal of Molecular Sciences*, *19*(1), 24. doi:10.3390/ijms19010024 PMID:29271916

Yangzhi, G., Junaid, A. S., Hongbin, L., & Xiangkang, M. (2016). Anti-corrosive performance of electropolymerized phosphomolybdic acid doped PANI coating on 304SS. *Applied Surface Science*, *360*, 389–397. doi:10.1016/j.apsusc.2015.11.029

Yaya, K., Khelfaoui, Y., Malki, B., & Kerkar, M. (2011). Numerical simulations study of the localized corrosion resistance of AISI 316L stainless steel and pure titanium in a simulated body fluid environment. *Corrosion Science*, *53*(10), 3309–3314. doi:10.1016/j.corsci.2011.06.006

Yean, S., Al Saiari, H., Kan, A. T., & Tomson, M. B. (2008). Ferrous carbonate nucleation and inhibition. In *The SPE International Oilfield Scale Conference.* Society of Petroleum Engineers. 10.2118/114124-MS

Yegambaram, M., Manivannan, B., Beach, G., & Halden, R. (2015). Role of environmental contaminants in the etiology of Alzheimer's disease: A review. *Current Alzheimer Research, 12*(2), 116–146. doi:10.2174/1567205012666150204121719 PMID:25654508

Yeih, W., Chang, J. J., & Hung, C. C. (2006). Selecting an adequate procedure for the electrochemical chloride removal. *Cement and Concrete Research, 36*(3), 562–570. doi:10.1016/j.cemconres.2005.12.008

Yeih, W., & Huang, R. (1998). Detection of the corrosion damage in reinforced concrete members by ultrasonic testing. *Cement and Concrete Research, 28*(7), 1071–1083. doi:10.1016/S0008-8846(98)00060-X

Yılmaz, E., Gökçe, A., Findik, F., Gulsoy, H. O., & İyibilgin, O. (2018). Mechanical properties and electrochemical behavior of porous Ti-Nb biomaterials. *Journal of the Mechanical Behavior of Biomedical Materials, 87*, 59–67. doi:10.1016/j.jmbbm.2018.07.018 PMID:30041140

Yüce, A. O., Telli, E., Mert, B. D., Kardaş, G., & Yazıcı, B. (2016). Experimental and quantum chemical studies on corrosion inhibition effect of 5,5 diphenyl 2-thiohydantoin on mild steel in HCl solution. *Journal of Molecular Liquids, 218*, 384–392. doi:10.1016/j.molliq.2016.02.087

Yusuke, T., Atsushi, N., & Tooru, T. (2006). Monitoring of Rusting of Stainless Steels in Marine Atmospheres Using Electrochemical Impedance Technique. *Journal of the Electrochemical Society, 153*(7), B278–B282. doi:10.1149/1.2202110

Yusuke, T., Atsushi, N., & Tooru, T. (2007). Pitting corrosion mechanism of Type 304 stainless steel under a droplet of chloride solutions. *Corrosion Science, 49*(3), 1394–1407. doi:10.1016/j.corsci.2006.08.016

Zeng, Q., Pan, C., Li, C., Fei, T., Ding, X., Du, X., & Wang, Q. (2018). Online monitoring of corrosion behavior in molten metal using laser-induced breakdown spectroscopy. *Spectrochimica Acta. Part B, Atomic Spectroscopy, 142*, 68–73. doi:10.1016/j.sab.2018.01.011

Zhang, D., Gao, L., & Zhou, G. (2004). Inhibition of copper corrosion in aerated hydrochloric acid solution by heterocyclic compounds containing a mercapto group. *Corrosion Science, 46*(12), 3031–3040. doi:10.1016/j.corsci.2004.04.012

Zhang, D., Gao, L., Zhou, G., & Lee, K. Y. (2008). Undecyl substitution in imidazole and its action on corrosion inhibition of copper in aerated acidic chloride media. *Journal of Applied Electrochemistry, 38*(1), 71–76. doi:10.100710800-007-9401-6

Zhang, L., He, Y., Zhou, Y., Yang, R., Yang, Q., Qing, D., & Niu, Q. (2015). A novel imidazoline derivative as corrosion inhibitor for P110 carbon steel in hydrochloric acid environment. *Petroleum, 1*(3), 237–243. doi:10.1016/j.petlm.2015.10.007

Zhang, X. L., Jiang, Z. H., Yao, Z. P., Song, Y., & Wu, Z. D. (2009). Effects of scan rate on the potentiodynamic polarization curve obtained to determine the Tafel slopes and corrosion current density. *Corrosion Science, 51*(3), 581–587. doi:10.1016/j.corsci.2008.12.005

Zhang, X., Xu, Y., & Ye, B. (2018). An efficient electrochemical glucose sensor based on porous nickel-based metal organic framework/carbon nanotubes composite (Ni-MOF/CNTs). *Journal of Alloys and Compounds, 767*, 651–656. doi:10.1016/j.jallcom.2018.07.175

Zhang, Z., Chen, S., Li, Y., Li, S., & Wang, L. (2009). A study of the inhibition of iron corrosion by imidazole and its derivativesself-assembled films. *Corrosion Science, 51*(2), 291–300. doi:10.1016/j.corsci.2008.10.040

Zhao, Y., & Jin, W. (2016). Chapter-1. In Steel Corrosion-Induced Concrete Cracking (pp. 1–17). China Science Publishing & Media Ltd. doi:10.1016/B978-0-12-809197-5.00001-3

Zheludkevich, M. L., Yasakau, K. A., Poznyak, S. K., & Ferreira, M. G. S. (2005). Triazole and thiazole derivatives as corrosion inhibitors for AA2024 aluminium alloy. *Corrosion Science*, *47*(12), 3368–3383. doi:10.1016/j.corsci.2005.05.040

Zheng, J., & Li, Y. Q. (2011). One-pot synthesis of tetrahydrobenzo[b]pyran and dihydropyrano [c] chromene derivatives in aqueous media by using trisodium citrate as a green catalyst. *Archives of Applied Science Research*, *3*(2), 381–388.

Zhou, L., Yuan, T., Tang, J., He, J., & Li, R. (2019). Mechanical and corrosion behavior of titanium alloys additively manufactured by selective laser melting–A comparison between nearly β titanium, α titanium and α+ β titanium. *Optics & Laser Technology*, *119*, 105625. doi:10.1016/j.optlastec.2019.105625

Zindani, D., Kumar, K., & Davim, J. P. (2019). *Metallic biomaterials - A review*. Mechanical Behavior of Biomaterials. doi:10.1016/B978-0-08-102174-3.00004-8

Zunita, M., Wahyuningrum, D., & Bundjali, B. (2012). Investigation of Corrosion Inhibition Activity of 3-butyl-2,4,5-triphenylimidazole and 3-butyl-2-(2-butoxyphenyl)-4,5-diphenylimidazole toward Carbon Steel in 1% NaCl Solution. *International Journal of Electrochemical Science*, *7*, 3274–3288.

Zuo, Y., Junlei, T., Chongzhi, F., Yuming, T., & Jinping, X. (2008). An electroless plating film of palladium on 304 stainless steel and its excellent corrosion resistance. *Thin Solid Films*, *516*(21), 75. doi:10.1016/j.tsf.2008.04.049

About the Contributors

Younes El Kacimi was born in Azrou City, Morocco. He received his B.Sc. in Polymer Sciences (2006) & (D.E.S.A) M.Sc. degree in Biomaterials Sciences (2008), from Ibn Tofail University. He was awarded his M.Sc. degree in Quality & Project Management (2009) from François Rabelais University. Then he started to PhD education. His research activities are focused on the analytical chamistry, electrochemical behavior of metals and alloys, corrosion and inhibition, nanomaterials and development of new surface modification processes. Younes has authored 1 Patent, 8 Books and Books chapters and many others publications in corrosion sciences, nanomaterials and analytical chemistry. He is also an editorial board member and reviewer in several international journals.

Rachid Touir was born in 1977 in Kenitra City, Morocco. He received his Ph.D. in Physics Chemistry, from Ibn Tofail University, Morocco, 2009. He had published more than 100 papers in international journals. His work has been cited 1726 total citations by 976 documents with h-index = 23 (since 2006 in SCOPUS) and presented several communications in symposia and national/international meetings. He was honored with the award for the best thesis researcher in Ibn Tofail University since the year 2009. He has developed 3 patents and he has participated in 6 national research projects. He has supervised 35 Doctoral Theses and 60 Master theses in science and didactics. He is a reviewer at several national and international journals. He is the author of a 3 books. His main research interests are in the field of Physical Chemistry, Especially electrochemistry (Corrosion and its inhibition in several media for metals and alloys using various techniques). He is also interested for electroless plating of Ni-P, Ni-Cu-P, Cu-P, Ni-Pd-Cu… and their components in several corrosive medium. He is currently interested in using some quantum chemical methods like density functional theory (DFT) and other semi-empirical methods to study corrosion inhibition efficiencies of organic compounds. He is also interested in the Didactics of physics and chemistry.

* * *

Ashraf Abousalem currently works as a QC lab manager for JOTUN Egypt. He obtained his Ph.D. in physical chemistry, electrochemistry in January 2018 for doing a comprehensive study on the corrosion inhibition action of bichalcophene compounds for carbon steel in aqueous solutions. He is currently taking part in research by designing theoretical chemical studies to explore the relationship between the molecular structure and mechanism of action at the atomic level.

Khaoula Alaoui received his B.Sc. in Chemistry (2013) and She was awarded his M.Sc. degree in Organic chemistry and Environnemental Sciences (2015) from Ibn Tofail University. Then he started to PhD education. Her research interest covers corrosion inhibition by synthesized compounds. She has published more than 14 papers, 4 books and book chapters in international journals and she presented about 30 communications and some conferences. She's also a reviewer at several international journals.

Jorge Corona-Castuera is a research at the CIATEQ, A.C. with 22 years of experience in different areas: Research, Control and Automation, Special Machines and Manufacturing. He obtained his Ph.D. in automation and control PICYT Mexico. Currently, he is working on additive manufacturing research related projects, in particular, for biomedical applications.

Md Daniyal is presently working as Assistant Professor in the Department of Civil Engineering, Lingaya's Vidyapeeth, Faridabad, India. He has worked as a Junior and Senior Research Fellow in the Department of Civil Engineering, Aligarh Muslim University (AMU), Aligarh, India and obtained Ph.D. degree in 2018. His Ph.D. thesis topic was " Performance of Chemical Corrosion Inhibitors and Nano-Admixtures in Controlling Corrosion of Cementitious Composites". He did his B.Tech (Civil Engineering) and M.Tech (structural engineering) from AMU, Aligarh in 2012 and 2014 respectively. He has qualified the Graduate Aptitude Test in Engineering (GATE-2012) and CSIR-UGC National Eligibility Test (NET-2013).

Mohamed Ebn Touhami is a research professor in Ibn Tofail University, specialist in materials sciences engineering and environment. He was born in 1970 in Kenitra City, Morocco. He received his Ph.D. in Physics Chemistry, from Ibn Tofail University, Morocco, 1996. He had published more than 160 papers in international journals. His work has been cited 2744 total citations by 1494 documents with h-index = 27 (since 1996 in SCOPUS) and presented several communications in symposia and national/international meetings. He has developed 7 patents and he has participated in 6 national research projects. He has supervised 100 Doctoral Theses and 300 Master theses in science. He is a reviewer at several national and international journals and has three books. His main research interests are in the field of chemistry especially for electroless plating of many alloys and corrosion inhibition in several industrial media. He was responsible for several Master formations. He was director of University Center for Analysis, Expertise, Technology Transfer and Incubator. Currently, he is a Dean of Faculty of Science, Ibn Tofail University, Morocco.

Astrid Giraldo-Betancur is a Physical Engineer and Master in Science-Physics graduated from the Universidad Nacional de Colombia, campus Manizales. Doctor in Science- Materials science graduated from Cinvestav Campus Querétaro. Currently, she works as CONACYT commissioned to the Cinvestav unit Querétaro. Co-founder of the Mexican Laboratory of Thermal Spray - CENAPROT. Her research area is focused on multifunctional materials; especially in bioceramics: synthesis of raw material such hydroxyapatite from natural sources and bioglasses, hydroxyapatite-based composite materials to coating manufacture using different thermal spray technologies and their characterization. Additionally, she is working on the microencapsulation processes development for controlled drug release.

John Henao is currently professor at the National Council of Science and Technology (CONACYT) unit-CIATEQ A.C. in Querétaro, Mexico, and is member of the Mexican system of researchers (SNI) level 1. He is materials engineer from the University of Antioquia, Colombia, and obtained his master degree in high performance materials engineering from the University of Limoges, France. He also obtained his PhD. Degree in Engineering and Advanced Technologies from the University of Barcelona, Spain. During the last years, He has been working on the field of thermal spray, biomaterials, and materials for energy and corrosion applications.

Brahim Lakhrissi has his expertise in Organic Synthesis, Surfactants, Complexation, corrosion inhibition, and Synthesis of bioactive molecules that can be used against bacteria, viruses, parasites and cancer. Former head of of Chemistry department, and Director of the Agroresources and Process Engineering Laboratory Expert evaluator affiliated to the CNRST - Morocco, and expert evaluator for several international scientific journals. He was warded his PhD in 1988 at the University of Nancy-1 - France, and his state doctorate in 2003 at Ibn Tofail University. He has published more than 130 articles in international journals.

Hayat Larhzil is a research professor in Regional Center for Education and Training Professions (CRMEF), specialist in materials sciences and chemical didactics. She was born in 1976 in Morocco. She received her Ph.D. in Physics Chemistry, from Ibn Tofail University, Morocco, 2007. She had published more than 20 papers in international journals. His work has been cited 76 total citations by 67 documents with h-index = 6 (since 2006 in SCOPUS) and presented several communications in symposia and national/international meetings. She has supervised 10 Doctoral Theses and 30 Master theses in science and didactics. She is a reviewer at several national and international journals. She is the author of a 2 books. Her main research interests are in the field of electroless plating of secondary and ternary alloys and corrosion inhibition in several media. She is currently interested in the Didactics of physics and chemistry.

Carlos Poblano-Salas is a Metallurgical Engineer graduated from Universidad Nacional Autonoma de Mexico (UNAM) where he also obtained a MSc degree in metallurgy. He received a PhD in Materials Science from the University of Sheffield, UK. Currently, he is director of the Advanced Materials Department at CIATEQ A.C. He has been working on the development and characterization of functional coatings by thermal spray for different applications durign the last thirteen years. His current research interests include additive manufacturing of metallic alloys and the development of novel biomedical devices.

Oscar Sotelo is currently a posdoctoral associate at the CIATEQ in Queretaro, Mexico. He obtained his M.SC. and Ph.D. in Materials Science from the Universidad Autonoma del Estado de Morelos in 2011 and 2015, respectively. He carried out his first postdoctoral studies at the Universidad Nacional Autonoma de Mexico unit Instituto de Ciencias Fisicas in 2018. He has been working on the field of corrosion at high temperature, corrosion inhibitors, thermal spray and biomaterials.

Index

N

O

S

T

U

www.ingramcontent.com/pod-product-compliance
Lightning Source LLC
LaVergne TN
LVHW060423300926
849162LV00006B/1512